ROBERT BOYLE
BY HIMSELF AND HIS FRIENDS

Brass cast by Carl Reinhold Berch of ivory medallion
by Jean Cavalier inscribed 'Robertus Boylæus 1690'.
British Museum.

ROBERT BOYLE

BY

HIMSELF AND HIS FRIENDS

with a fragment of William Wotton's lost

Life of Boyle

EDITED
WITH AN INTRODUCTION
BY

MICHAEL HUNTER

Routledge
Taylor & Francis Group

LONDON AND NEW YORK

First published 1994 by Pickering & Chatto (Publishers) Limited

2 Park Square, Milton Park, Abingdon, Oxfordshire OX14 4RN
52 Vanderbilt Avenue, New York, NY 10017

Routledge is an imprint of the Taylor & Francis Group, an informa business

First issued in paperback 2019

Copyright © Taylor & Francis 1994
Introduction and notes © Michael Hunter (1994)

BRITISH LIBRARY CATALOGUING IN PUBLICATION DATA
Boyle, Robert
    Robert Boyle: Key Biographical texts
    I. Title II. Hunter, Michael Cyril William
    192

    ISBN 13: 978-1-8519-6085-9 (hbk)
    ISBN 13: 978-0-367-87612-8 (pbk)

LIBRARY OF CONGRESS CATALOGING-IN-PUBLICATION DATA
Boyle, Robert, 1627–1691.
    Robert Boyle / by himself and his friends : with a fragment of William
Wotton's lost Life of Boyle : edited with an introduction by Michael Hunter.
        p. cm.
    ISBN 1–85196–085–6 : $75.00 (est.)
    1. Boyle, Robert, 1627–1691. 2. Scientist—Great Britain-
—Biography. 3. Scientists—Correspondence, reminiscences, etc.
I. Hunter, Michael Cyril William. II. Wotton, William, 1666–1727.
Life of Boyle. 1994. III. Title.
Q143.B77A3 1994
530'.092—dc20
    [B]   94–24414
                                                                       CIP

Typeset by Waveney Typesetters
Norwich

CONTENTS

Contents

ACKNOWLEDGMENTS

The following have kindly assisted in the preparation of this volume. A generous grant from the Royal Society paid for the transcription of the texts and for ancillary research expenses, and Kate Fleet and Martha Morris were exemplary in carrying out this work for me. I have from the outset received help and encouragement from two veteran Boyle scholars, Dr Marie Boas Hall and Dr R. E. W. Maddison. The latter sadly died just as this book was going to press. In addition, the book has benefitted from the collaborative work on a new edition of Boyle's writings and correspondence for the Pickering Masters series in which I am engaged with Edward B. Davis and Antonio Clericuzio; both of these scholars also gave me advice on specific points in the texts published below. Jan Wojcik, Nicholas Tyacke, Fred Trott, Lawrence Principe, Joe Levine, and Robert Day read sections of the Introduction and provided helpful comments on them. Teresa Bridgeman, Douglas Chambers, Janelle Evans, Richard Evans, Roger Gaskell, Vanessa Harding, Emma Mason, David Money, Malcolm Rogers, Simon Schaffer, Nigel Smith, Luke Syson, G. J. Toomer, Maarten Ultee and Christopher Wright helped me with specific queries. The Hartlib Papers Project at the University of Sheffield kindly provided print-outs from Hartlib's *Ephemerides* and related texts. The library staff of the Royal Society have been unfailingly helpful, as has John Wing, Assistant Librarian of Christ Church, Oxford. Material from the Evelyn Collection at Christ Church is cited by permission of the Trustees of the Will of the late Major Peter George Evelyn. Permission to print the texts has kindly been granted by the President and Council of the Royal Society and the Board of the British Library. The frontispiece is reproduced by courtesy of the Trustees of the British Museum; the facsimile title page on p. 36 by kind permission of Roger Gaskell Rare Books; and the plate on p. 115 by permission of the President and Council of the Royal Society.

NOTE ON CITATIONS AND ABBREVIATIONS

All books and articles referred to in the notes have been cited by short-title. Full bibliographical details will be found in the Bibliography. In addition, the following abbreviations have been used:

Add. MS	British Library Additional Manuscript
BL	Royal Society Boyle Letters
BP	Royal Society Boyle Papers
Corr.	Christ Church, Oxford, Evelyn Collection: bound volumes of correspondence
CSP Dom	*Calendar of State Papers (Domestic)*
DNB	*Dictionary of National Biography*
DSB	*Dictionary of Scientific Biography*, ed. C. C. Gillispie, 16 vols (Scribners: New York, 1970–80)
Evelyn MS	Christ Church, Oxford, Evelyn Collection, MS
GEC	G. E. Cokayne, *Complete Peerage*, new edition, 14 vols (London, 1910–59)
Letterbook	John Evelyn's Letterbook: Christ Church, Oxford, Evelyn Collection MS 39
Life	Birch, Thomas, *The Life of the Honourable Robert Boyle*, in *The Works of the Honourable Robert Boyle*, 2nd edition (London, 1772), vol. i, pp. vi–cl
OED	*Oxford English Dictionary*
Oldenburg	Hall, A. R. and M. B., eds, *The Correspondence of Henry Oldenburg*, 13 vols (University of Wisconsin Press, Mansell and Taylor & Francis: Madison, Milwaukee and London, 1965–86)
RS	Royal Society
Works	*The Works of the Honourable Robert Boyle*, ed. Thomas Birch, 2nd edition, 6 vols (London, 1772)

NOTE ON FRONTISPIECE

Though a sculpted figure representing Boyle as a child was included on his father's tomb, the earliest fully authenticated portrait of him is the drawing from the life by William Faithorne, now in the Ashmolean Museum, Oxford. From this, Faithorne engraved his well-known plate of Boyle with his air-pump in the background, the making of which is documented in letters from Robert Hooke to Boyle in 1664. Various subsequent depictions of Boyle were based on this engraving.

Then, there are various versions of a portrait of Boyle by Johann Kerseboom, one of which has been at the Royal Society since 1692; another, now in the National Portrait Gallery, was owned by Boyle's executor, Sir Henry Ashurst, while a copy in the Royal Collection appears to be that formerly owned by Boyle's physician, Sir Edmund King.[a] There are also numerous engravings derived from this exemplar, the earliest a mezzotint by John Smith dated 1689. Though the Kerseboom portrait and related items are authenticated, it is clear that the image of Boyle that they present is somewhat idealised – a common feature of the portraiture of the period.[b] Indeed, the suave figure that they show looks hardly older than the man depicted by Faithorne twenty-five years earlier, despite the fact that Boyle had been a semi-invalid for years, having suffered a serious stroke in 1670. Since one of the Kerseboom-derived engravings formed the frontispiece to Boyle's posthumously-published *Free Discourse against Customary Swearing*, issued with the co-operation of his executors in 1695, it seems likely that this rather serene image of the great man is to be seen as a pictorial equivalent of the approved view of him presented in Burnet's funeral sermon, discussed in section 4 of the Introduction, below.

The only image of Boyle which departs from this iconographic tradition is the one reproduced as the frontispiece to this book. This is a medal made in 1729 by Carl Reinhold Berch from an ivory medallion carved by Jean Cavalier in 1690.[c] Cavalier, a Huguenot artist who worked in various parts of Europe, is known to have been in London at this time, and Boyle

[a] The Riley portrait, also at the Royal Society, is very close to the Kerseboom in its imagery, and is to be seen as a further variant of the same type.

[b] See Piper, *The English Face*, chs. 5–6.

[c] For a reproduction of the reverse of the medal, and a transcription of the inscription on it, see Maddison, 'Portraiture', pp. 206–7 and plate 45.

may have agreed to sit for him (like Samuel Pepys) as a favour to a needy émigré.[a] The medal depicts Boyle in profile, and it gives much more of an idea of the dignified but slightly ravaged complexion that one would expect Boyle to have had in his later years than do the Kerseboom portraits. Arguably, it is the most convincing image of the older Boyle that we are likely to get.

NOTE

For a definitive survey of Boyle's portraiture, from which all information in this note that is not otherwise documented has been derived, see Maddison, 'Portraiture'. In my opinion, the evidence that the Lely, Soest or Mary Beale portraits there discussed are of Boyle is tenuous in the extreme. It should be added that the interpretation put forward here differs from that of Maddison.

---

[a] For Cavalier's medallion of Pepys, see Kneller to Pepys, 16 January 1690, in Pepys, *Letters and Second Diary*, pp. 213–14: one hopes that Boyle did not share Pepys's sense of grievance over Cavalier's fees.

INTRODUCTION[a]

1. THE VOLUME AND ITS CONTENTS

Robert Boyle (1627–91) was the doyen of the new, experimental science in late seventeenth-century England, author of over forty books on science, religion and their mutual relationship. His writings were read throughout Europe, forming one of the cornerstones to the establishment of experimental science in Britain and Europe during and after his own lifetime. Not surprisingly, Boyle has been the subject of a good deal of scholarly attention.[1] But, considering his significance for the history of science and of seventeenth-century English culture, biographical materials concerning him are surprisingly inaccessible and disappointingly incomplete. There are, of course, Boyle's voluminous published works, which will take up twelve large volumes in the Pickering Masters edition (forthcoming 1997). It was by his publications that Boyle intended that he should be immortalised, and in many ways the most appropriate kind of biography of him would be an intellectual one, exploring his aims, sources and achievement in these published writings, which also contain incidental information about other aspects of his life.

In addition, a good deal of Boyle's correspondence survives, though much has also been lost.[2] Many letters to and from Boyle were published as part of the first collected edition of his works brought out by the cleric and antiquary, Thomas Birch, in 1744, and many more survive, mostly in the Boyle Letters now at the Royal Society; these will be included in the complete edition of the correspondence which will also form part of the Pickering Masters edition of Boyle. This is obviously a key source, and it was appropriate that correspondence formed one of the main components of the first proper Life of Boyle to be published, namely that which Birch prefixed to his 1744 edition of Boyle. In addition, there is much to be gleaned from such sources as the Lismore Papers, now at Chatsworth, which include the correspondence of Boyle's father, the Great Earl of Cork. Any definitive biography would also, of course, utilise evidence illuminating the contemporary background.[3]

But, essential as such materials are for reconstructing the details of Boyle's career, when it comes to understanding him a special value

[a] For notes to the Introduction, see pp. lxxx–c.

attaches to memoirs consciously written to elucidate or evaluate his life, either by Boyle himself or by others who had known him well. The importance of material of this kind was recognised in Boyle's time, particularly by his contemporary, Roger North: in what was virtually the first attempt at a theoretical account of biographical writing in England, North emphasised the value both of what he called 'idiography' (his term for 'autobiography', a word then not yet invented) and of the recollections of a man's intimate friends.[4] It is the extant material of this kind relating to Boyle that this volume presents. Some has been published or utilised before, and our image of Boyle is to this extent already dependent on it. But what is remarkable is that some of the most significant of these texts here see the light of day for the first time: our understanding of Boyle will be modified and enriched accordingly.

Of the texts included here that have long been familiar, one is the sermon delivered by Boyle's friend, Gilbert Burnet, Bishop of Salisbury, at Boyle's funeral at St Martin's in the Fields on 7 January 1692, a *tour de force* which arguably did more than any other single work to form Boyle's posthumous reputation. The other is Boyle's own autobiographical *Account of Philaretus during his Minority*, an incomplete account of his early life written when he was in his early twenties, which was first published by Birch in his 1744 *Life*, and of which an improved text was included in R. E. W. Maddison's *Life of the Hon. Robert Boyle* of 1969; it is here freshly printed from the original manuscript. The *Account of Philaretus* provides an extraordinary insight into Boyle's mind at a formative stage in his intellectual development. Its significance becomes all the clearer if it is placed in the context of Boyle's intellectual evolution during the period preceding its composition, a topic on which new findings are presented in the second section of this Introduction.

In addition, in his *Life* Birch refers to and quotes briefly from a series of other documents which give a biographical assessment of Boyle, and these fall into two categories. First, there are texts which were compiled at Boyle's behest in the later years of his life. Secondly, there are papers prepared in connection with two projects for a life of the great man which preceded Birch's, but neither of which materialised: one, that of Bishop Burnet, who promised in his funeral sermon to supplement it with 'a farther and fuller account' of Boyle, the other, that of the scholar, William Wotton, author of the celebrated *Reflections upon Ancient and Modern Learning* (1694), who took over the project after Burnet abandoned it.

What is astonishing is that, although these sources have survived ever since Birch's time among his manuscripts, which he bequeathed to the British Museum, they have hitherto remained almost entirely unpublished and very little exploited.[5] Indeed, even those documents which *have* already seen the light of day – namely, the letters that John Evelyn wrote to Wotton to assist him with his Life, which were published as part of Evelyn's correspondence in the early nineteenth century – were published not from these texts but from the (slightly different) copies kept by Evelyn.[6]

Hence perhaps the most important purpose of this volume is to make these crucial sources generally available for the first time. Certain items are of minor, though not negligible significance. Thus we have some notes written at Boyle's behest by his amanuensis, Robert Bacon, which overlap with and supplement Boyle's earlier *Account of Philaretus*. There is also a letter reminiscing about Boyle written to Wotton in 1699 by Thomas Dent, a clerical protegé of the Boyle family, who, from 1690 onwards, was Rector of Stalbridge in Dorset, where the English seat of the Boyle family was located. (A comparable text which has also been included is a letter to Wotton from the Scottish divine, James Kirkwood, in which Kirkwood gave an account of Boyle's support for the distribution of bibles in the Scottish Highlands in the years around 1690. This letter, now surviving among the Boyle Letters and published by Birch, could hardly be omitted here because of its relation to Wotton's biographical project.)

Two items, however, are of crucial significance. One is a series of notes written by Boyle's friend, Sir Peter Pett, for the use of Burnet in his putative Life which contains a number of crucial insights into Boyle. Previous researchers may have been discouraged by the rather uninviting appearance of the latter part of Pett's notes, which he failed to have transcribed in the neat scribal hand of the opening section, and which are therefore in his own undisciplined handwriting, sloping up the page and with numerous deletions and marginal insertions. Most important of all is a paper comprising notes made by Burnet himself on an interview or interviews with Boyle in his later years, which is effectively an autobiographical statement by the scientist. Its significance for understanding Boyle's intellectual personality can scarcely be overestimated, as will be indicated more fully in section 3 of this Introduction, where new evidence is adduced for its authority as a statement about some of Boyle's most pressing concerns.

By the time that Birch came to write his Life, hardly anyone who had known Boyle was still alive.[7] Indeed, he himself complained how his *Life* was based on 'the best materials that could be procured at this distance of time, and without most of those advantages which Dr *Burnet* bishop of *Salisbury*, and Dr *William Wotton*, who had the same design near fifty years ago, might have obtained for the execution of it'.[8] Hence a crucial value accrues to these documents, as presenting testimony about Boyle by himself and his contemporaries which would otherwise have perished.

The last item that is presented for the first time here is somewhat different. It is not a testimony like the other items: indeed, there is no evidence that its author ever met Boyle. On the other hand, it represents part of a biography which, had it been completed and published, would have been a highly significant document in its own right, and this is the *Life of Boyle* by William Wotton. For, though only a single, incomplete section of this survives, this may be argued to represent virtually the earliest attempt in England to write an intellectual biography, a genre then in its infancy. Its publication provides an appropriate opportunity to set out what is known about the history of this abortive life, and about the earliest attempts to write Boyle's biography as a whole. It is to this, as well as to a commentary on the texts that follow, that the bulk of this Introduction is devoted.[9]

The final section of the Introduction, however, is rather different. In it, an attempt is made to assess the significance for our understanding of Boyle of the new information contained in the documents published here. The Burnet Memorandum and the notes by Pett, especially, provide evidence of complexities in Boyle's personality which have been glossed over in accounts of him hitherto. By integrating these facets of Boyle's attitudes and ideas into our overall view of him, we gain a richer and more truthful picture of this crucial figure, which helps to make sense both of his intellectual evolution and of his contemporary role.

2. BOYLE'S *ACCOUNT OF PHILARETUS* IN CONTEXT

The first text in the volume dates from several decades earlier than any of the others, and it is of a rather different kind from them. It dates from 1648 or 1649, when Boyle was twenty-one or twenty-two, and it forms an autobiographical account, written in the third person, of the first sixteen years of Boyle's life.[10] It provides an account of his birth on 25 January 1627 as the youngest son of the Great Earl of Cork; of certain episodes during his childhood; and of his education at Eton College from 1635 to 1638 and under private tuition thereafter. Then, it gives a detailed account of his travels on the continent, including a prolonged stay at Geneva, where he continued his education. It also includes his impressions of his stay in Italy, but it breaks off incomplete in the middle of Boyle's travels. Whether it is incomplete because Boyle never finished it or because its conclusion is lost is unclear: but the fact that the extant manuscript ends only part of the way down a page suggests that the former is likelier to be the case.[11]

It has not seemed appropriate to provide a lengthy commentary on the factual data about Boyle's early life in *Philaretus* here. Those who desire such an account will find one in R. E. W. Maddison's *Life of Boyle* (1969), which includes a text of the work extensively annotated with corroborative information from correspondence and other sources.[12] Here, the text is presented with only essential annotation (in this case, as elsewhere in this volume, an excessive number of repetitive notes has been avoided by giving biographical details about those who appear frequently in a Biographical Guide, pp. civ–vi). This has the effect of drawing greater attention to the form of the work and its intellectual significance, which have hitherto beeen little analysed; it is these which will be commented on here.

The date of composition of this work, 1648–9, is important: it places *Philaretus* at the climax of Boyle's earliest phase of authorship, from the mid to late 1640s, the distinctiveness and significance of which is only now being fully appreciated.[13] After returning from his travels, Boyle had settled at the family seat at Stalbridge in Dorset in 1645, where – interspersed by trips to London and elsewhere – he devoted himself to intellectual pursuits. His principal preoccupation at this time was with writing on ethical issues. From 1646 onwards he had contact with Benjamin Worsley, while his correspondence with the intelligencer, Samuel Hartlib survives from March 1647 onwards, and both Worsley and Hartlib evidently stimulated a degree of attention to more practical topics on Boyle's part, the former possibly in connection with the much misunderstood 'Invisible

College'. But Boyle's concern with such matters is easily got out of proportion by reading this period of his life through the filter of his later interests:
it is clear that, at this point, he considered his real mission in life as the
composition of treatises in which the pursuit of morality and piety would
be urged on his readers. Indeed, it is apparent from comments in letters of
members of the Hartlib Circle that they recognised and appreciated this
enterprise.[14]

A number of the writings which Boyle composed in these years have
recently been made available in print in a useful edition by John Harwood.[15]
They reveal how, at this point, Boyle's thought exemplified the humanist
values which had been purveyed in Europe since the Renaissance and
which had been merged with Protestant ideals.[16] The aspiration was to the
philosophical ideals of antiquity – particularly those of the Stoics – in
which stress was laid on the pursuit of moral balance, self control and
piety. The most important of such writings – evidently composed in 1645–6
– was Boyle's *Aretology*, a treatise on 'Ethicall Elements' which was
intended to lay down the rudiments of morality as a basis for the pursuit
of virtue: the object, in Boyle's words, was 'to call them [the ethics] from
the brain down into the breast, and from the school to the house'.[17]

Related to this were various other writings, of which a selection is also
included by Harwood. Some of these dealt with aspects of the self-
knowledge to which the *Aretology* was devoted, including essays on 'time
and idleness' and another advocating the discipline of keeping daily
memoranda.[18] Boyle also reflected on the attributes and aspirations proper
to members of his own class in an incomplete essay on *The Gentleman*,
which exemplifies the ambivalence as to how the pursuit of virtue was
supposed to relate to nobility of birth which is typical of the thought of the
day.[19] Indeed, though Boyle is not usually very explicit about his intended
audience, it seems likely that he had in mind people like himself. Other
writings were more overtly religious in tone, including essays on 'sin',
'piety' and 'the Study & Exposition of the Scriptures'.[20] It was also at this
time that Boyle first experimented with the genre represented by his *Occasional Reflections* (1665), in which he sought to 'make the little accidents
of his life, and the very flowers of his garden, read him lectures of ethicks
or divinity'.[21] From this period, we have manuscript texts, entitled *Scripture Reflections* or *Observations*, which comprise pious meditations on
biblical texts, phenomena and events, deploying a devotional form which
had been naturalised in England earlier in the century, particularly by
Bishop Joseph Hall.[22] Though it seems almost inconceivable that Boyle
was *not* influenced by Hall, he himself states that he forebore reading
Hall's meditations so as 'not to prepossess or biass my fancy', and it is
possible that his contribution to the genre emerged independently.[23]
Certainly it is striking that the extant manuscript of Boyle's *Scripture*

Reflections suggests an evolution from reflections on scriptural passages to *Occasionall Meditations* on events, the latter appearing in the middle of a text initially devoted to the former. Since some of the latter were subsequently to appear in the collection published a decade and a half later, we apparently here see the origins of that project.[24]

In addition, in an evident attempt to bring his didactic message home to his intended audience, Boyle experimented with new literary forms. One example of this is the text, 'A Mere fine Gentleman', which belongs to the tradition of 'character books' of the day, in which the failings of contemporary social groups were satirised by being caricatured.[25] Boyle also wrote verse, although this is all now lost except for one fragment recorded by Sir Peter Pett in his notes on Boyle, published below.[26] Most significant of all were Boyle's variations on the form of the romance. This evidently reflected the influence on him of his reading of the French romances which were fashionable in the circles in which he moved, and to his own appetite for which he refers in *Philaretus*.[27] Indeed, his brother, Roger Boyle, was just at this time writing the most influential English version of such a romance, *Parthenissa*, published in parts between 1651 and 1669; the first section of this was known to Boyle at least as early as 1649, when he makes specific reference to it.[28] In Boyle's case, he was quite ambivalent about romances, sometimes deprecating his enthusiasm for them as an improper diversion from more worthwhile goals.[29] On the other hand, he seems also to have seen potential benefits in exploiting the form, recognising its attraction for members of his class, and perhaps responding to the potential for including moral messages in such works of which their authors were aware.[30]

Hence such reading seems to have encouraged him to experiment with the presentation of his pious message in the form of imaginary lives and speeches. A case in point is the text comprising the address to Joseph of his seducing mistress, beginning 'Dearest Joseph': interestingly, this is to be found among Boyle's *Scripture Observations*, thus suggesting a process of evolution similar to that concerning *Occasional Reflections* which has already been noted.[31] There is also Boyle's semi-fictional 'Life' of the Old Testament King Joash, while a further example of this genre is *The Martyrdom of Theodora and of Didymus*, partially published in 1687, which Dr Johnson considered the first work 'to employ the ornaments of romance in the decoration of religion'.[32] A further, related literary development was Boyle's use of the letter form to present moralistic treatises to fictional addressees. One – dealing with the ethics of man's treatment of animals – has been published by Malcolm Oster, while another was to appear in much modified form as Boyle's first published work, *Some Motives and Incentives to the Love of God* (1659), generally known as *Seraphic Love*.[33]

This is the context of the *Account of Philaretus*. Its most immediately obvious link with this phase of Boyle's career as a writer is its literary self-consciousness, which reached its peak at the time when he was writing this work. It is thus replete with rather conceited metaphors and *obiter dicta*, such, for instance, as Boyle's phrase 'with as much successlesnesse as Diligence', or his comparison of spoilt children with confections of butter or sugar.[34] More striking still is the way in which alternative words or phrases are included in the text, pending a final decision as to which was most appropriate, a technique which can be paralleled in various of Boyle's other writings from this early phase.[35] The work is also quite self-consciously structured, with its division into sections separated by the crossing of the sea or the Alps.

Equally significant is the extent to which *Philaretus* reflects the concern for self-knowledge and character formation to be found in Boyle's moral essays of the previous few years. As in those, the whole work is organised around the pursuit of 'felicity', reflections on which recur at intervals through the text, as where Boyle comments on the advantages of a median status near the start, or when he later comments on the strengths and weaknesses of figures with whom he came into contact such as Robert Carew or Isaac Marcombes.[36] Underlying this is the theme of his earlier ethical essays, of a kind of Christianised Stoicism, an ideal of self-control, of the triumph of internalised virtue against external forces. Examples of the challenges and afflictions which he experienced include the stutter he picked up from his childhood contacts, the homosexual advances to which he was subjected in Italy, or the habit of 'raving' to which he was prone, by which he evidently meant fanciful and uncontrolled speculation, lacking in the discipline which was his proper aim.[37] There are also moralistic reflections at intervals, which again echo the sententiousness of Boyle's earlier writings, as where he contemplated the mutability of human fortunes in connection with the fate of the Boyles' mansion at Lismore; occasionally, these become quite profound, as where he reflected how men's characters were most fully revealed in their childhood or on their deathbed.[38]

This tradition of writing thus helps to explain how *Philaretus* came to take the form it did. That Boyle saw a link between it and his earlier treatises is illustrated by the fact that on one occasion he actually omitted observations on a particular point, cross-referencing his readers instead to his discussion of related themes in various chapters of a treatise entitled *The Christian Gentleman*, possibly the same as *The Gentleman*, the surviving fragment of which has already been discussed.[39] On the other hand, to write a narrative of his life was a new development, linking Boyle to the tradition of autobiographical writing of a more or less reflective kind which was beginning to develop in England at this time. It is therefore worth assessing what Boyle's inspiration might have been, concentrating on works that there is reason to believe he was aware of, rather than analogous compositions which were not published until later and by which he therefore cannot have been influenced. Indeed, the impulses to autobiographical writing at this formative stage have been obscured by conflation with the more developed tradition of the later seventeenth century, and by canvassing Boyle's potential sources, it is possible to contribute to greater precision in this matter.[40]

One potential inspiration was the literature of Greek and Roman antiquity, including the Stoic texts emphasising the need for self-knowledge which formed the basis for Boyle's Christian humanism. Certainly other autobiographers – such as Sir Robert Sibbald and Sir John Bramston – cited such texts in the course of writings of this kind, while Bramston actually used a Senecan maxim to justify his autobiographical enterprise.[41] In addition, Boyle is known to have read Diogenes Laertius' *Lives of the Philosophers*, and his desire to write an account of his own life could have owed something to this.[42] Indeed, his apology for not starting his life with a pedigree may possibly reflect his perception of the proper form of biography deriving from antiquity, paralleling John Aubrey's comment at the start of his 'Life of Mr Thomas Hobbes' that 'The writers of the lives of the ancient philosophers used to, in the first place, to speake of their lineage'.[43] A further text Boyle's fondness for which he refers to both in *Philaretus* and in the notes that he later dictated to Robert Bacon is Quintus Curtius' spirited account of the life of Alexander the Great, while another lively narrative to the influence of which he attributed his adolescent enthusiasm for history was Sir Walter Ralegh's *History of the World* (1614).[44]

None of these were autobiographies, but for writing such an account of himself Boyle had an exemplar much closer to home: the 'True Remembrances' about himself that his father had compiled in 1623, revising it in 1632.[45] An entry in Samuel Hartlib's *Ephemerides* under 1648 reveals both that Boyle was aware of the existence of this text and that he had developed biographical interests in connection with it. Hartlib records that 'The old Earl of Corke writ his owne Life, which is somwhere in Ireland which as soone as Mr Boyle hath gotten he promised to polish it and make a complete History of it'.[46] On the other hand, we do not know whether or not Boyle found the life at this point; even if he did, it is hard to postulate that it influenced the actual format of Boyle's own autobiography, as against the basic idea of writing such a text, since Cork's life is a more straightforward *res gestae*, aimed at justifying his life against his detractors.

One repeated motif in Cork's autobiography is the influence of providence. A sense of providential design was crucial to his image of himself, as Nicholas Canny has persuasively shown.[47] Indeed, there is an interesting amalgam of the role of providence and that of the *virtu* of the Italian humanists, since Cork is known to have owned and valued a copy of Guicciardini. On the other hand, Cork's 'True Remembrances' fail to provide an exemplar for the form of providentialism which is a notable feature of the *Account of Philaretus*, namely its retailing of detailed instances of providential intervention on Boyle's behalf. This must result from the influence on Boyle of a characteristic genre of autobiographical writing in seventeenth-century England, the spiritual autobiography, aimed at chronicling God's purpose for the individual in question by recounting providential escapes, spiritual trials and conversion experiences. Certainly, *Philaretus*'s account of the accidents from which Boyle escaped as a child reflects the commonplaces of such writings, in which such events were often recorded 'with more enthusiasm than judgement', in the words of one commentator.[48] Similarly, Boyle's vivid and memorable account of his conversion experience and of the 'distracting Doubts of some of the Fundamentals of Christianity' that followed it bears some comparability to the classic of such narratives, John Bunyan's *Grace Abounding to the Chief of Sinners*.[49]

In fact, Bunyan's book was not written until more than a decade after Boyle's, and most published examples of the genre of spiritual autobiography are of a later date. But the roots of such writings are to be found in the accounts of the careers and sufferings of spiritual figures which go back through the Elizabethan period to Foxe's *Book of Martyrs*.[50] What we know about the books that Cork owned makes it quite plausible that such material might have circulated in his household (though, of course, by 1648, there were other routes by which it might have reached the young Boyle).[51] Indeed, it is worth noting that Boyle's sister, Mary Rich, was later to write an autobiography of a godly kind, *Some Specialties in the Life of M. Warwicke*, compiled in the 1670s.[52] Undoubtedly there is a link between Boyle and this tradition, though its exact nature remains slightly obscure. In any case, in *Philaretus* such influence was modified by being cross-fertilised with the influence of Stoicism which has already been referred to, reflecting the extent to which Boyle's sermonising always reflected a intellectualised blend between this evangelical impulse and the traditions of antiquity. It is thus revealing that, in describing his conversion experience, Boyle abandoned his original choice of wording, 'Thus was Christ borne in [him]', instead substituting an elaborate analogy between the effects of the storm on him and of rain on the earth.[53] Similarly, whereas in one of his conceited metaphors, Boyle compared his religious doubts with tooth-ache, elsewhere he invoked Stoicism as the source of his equanimity in handling pain of this kind.[54]

Boyle's narrative also contrasts with extant autobiographies of the spiritual tradition in its form. Here, we revert to the influence of Boyle's romance reading, since *Philaretus* clearly owes a good deal to this (his liking of Quintus Curtius is perhaps also to be seen in this context, since Curtius' lively and detailed account of Alexander's life had something in common with romances). For one thing, the name of the protagonist itself seems likeliest to derive from the romance tradition: many of the characters in Roger Boyle's *Parthenissa* have fictional names of a similar kind, and the same is true of its French exemplars.[55] More significant, the work is linked to the romance tradition by its third person narrative structure, including the use of reported speech and the attempt to depict emotions, and by its style, with its long sentences with balanced clauses. There are hints of similar influence in the *Life* of Lord Herbert of Cherbury, while a comparable impulse, in this case deriving from Sir Philip Sidney's *Arcadia*, seems – quite independently – to have formed a route to autobiographical writing for Boyle's later contact, Sir Kenelm Digby.[56]

Indeed, it seems possible that romances enabled Boyle to achieve a vividness of writing that eluded him when he wrote more theoretically about ethical matters in treatises like the *Aretology*. That Boyle associated narrative writing – including about himself – with romances is suggested by a comment in one of his earliest surviving letters, that to his sister, Lady Ranelagh, of 30 March 1646. Like other early letters, this has a directness lacking from his rather stilted literary compositions of the same period, offering a spirited account of his journey from London interspersed by quite witty asides. It is thus revealing that he introduced it by offering his sister 'a piece of a real romance in the story of my peregrination hither'.[57]

In the way in which it brought together these various traditions, *Philaretus* is a significant text. Indeed, it is arguably the most successful of Boyle's writings up to this point in his life, balancing moralism with narrative in a genuinely effective whole. It forms a fitting climax to his career as a moralist, while the attitudes to which it gives expression are also revealing. We gain a real insight into the young Boyle, perceptive and sententious, yet also priggish and snobbish: thus his comment that Lyons was 'fitter for the Residence of Marchands then of Gentlemen' echoes his treatise, *The Gentleman*, in reminding us of the extent to which his moralism was juxtaposed with elitist attitudes.[58] Also of interest are such sections of the text as Boyle's account of Italy, and particularly of the relationship of the Catholic church with Galileo. Here, he displays a robust and slightly arch anti-Catholicism typical of English attitudes to the Galileo affair at the time, paralleling Jeremy Taylor's view that posterity would 'make themselves very merry with the wise sentences made lately at Rome' on this subject.[59] But to single out episodes in this way is to risk starting to recapitulate a text which readers can digest for themselves. Undoubtedly, *An Account of Philaretus during His Minority* deserves fresh scrutiny.

3. BOYLE'S LATER REMINISCENCES AND
THE 'BURNET MEMORANDUM'

After the writing of *Philaretus*, there is a long gap in Boyle's concern to record information about his own life. The climax to his early literary activities, of which *Philaretus* forms a part, preceded a sharp change in direction in Boyle's intellectual career around 1650. It was now that he developed the preoccupations which dominated the whole of his subsequent life but which had been little in evidence in his adolescent writing, centred on profuse experimentation and the use of the findings of natural philosophy to bolster Christian orthodoxy against the intellectual challenges which it faced. In addition, it was from this time that Boyle abandoned the pursuit of literary elegance which had characterised his writings in the 1640s, instead developing the more erudite, digressive style which was to remain with him until his death.[60]

Hence from 1659 onwards Boyle produced a stream of publications, including such classic works as *Certain Physiological Essays* (1661) and his experimental studies of the air, cold and colours, a programme which continued with his *Experiments, Notes, &c. about the Mechanical Origin or Production of Divers Particular Qualities* (1675) and other writings. In addition, there is the philosophical programme exemplified by such works as his *Discourse of Things above Reason* (1681) and his *Free Enquiry into the Vulgarly Receiv'd Notion of Nature* (1686). Such books may occasionally be gratuitously autobiographical in their prefatory matter, but none makes any sustained attempt to tell the story of his life.[61] Perhaps Boyle felt that what was most important about him was encapsulated in these writings: indeed, it is significant that in the various portraits of him that were painted towards the end of his life by Johann Kerseboom and John Riley, he gestures away from himself towards the book open on the table in front of him, as if to minimise the significance of the author's personality and maximise that of his published output.[62] Boyle also received coverage in such promotional writings on behalf of the new science as Henry Oldenburg's *Philosophical Transactions* and Joseph Glanvill's *Plus Ultra* (1668). The latter devoted a whole chapter to 'An Account of what hath been done by the Illustrious Mr *Boyle* for the promotion of *Useful* Knowledge'; another included an account of his forthcoming publications, and Boyle was certainly a party to this, since he provided notes on his unpublished works – including the second section of the second part of *The Usefulness of Natural Philosophy* – for Glanvill's use.[63]

We also know of a piece of biographical writing concerning Boyle by a great admirer of his, the enthusiastic and slightly eccentric virtuoso, John Beale. In a letter to Samuel Hartlib dated 18 April 1657, Beale explained how he had recently sent

> A full answere to Mr Boyle relating & offering to his Memory some Monuments Conc[erning] himselfe his right honourable father, his Ances-tors, Their Name, Seates Possessions ever since Edward the Confessors dayes, not by uncertaine Heraldry, but by undeceiving records. A motion How they may reinvest themselves of their ancient desmesnes at reasonable rates, which was much desired by his right honourable father, whilst it was the lands of my cosen Germane. Then not obtainable now easy.[64]

The document to which Beale here refers evidently drew on his own background in the Herefordshire milieu from which the Boyle family had originated, and the core piece of information that he divulged in it was included by Birch in his *Life*. He there attributed it to a 'Letter of Doctor *John Beale* to Mr *Samuel Hartlib*', no doubt referring to an item which he and his collaborator, Henry Miles, acquired from William Wotton via his son in law, William Clarke, but which no longer survives.[65] The claim was that the Boyles were descended from 'Humphrey de Biuvile', who appeared in Domesday Book as the preConquest owner of Pixely Court, near Ledbury, and it would have been of particular interest to the Great Earl of Cork, had he still been alive, who had long sought in vain to prove the ancient nobility of his line.[66] As for Boyle himself, he appears to have been sufficiently interested in the matter to keep Beale's paper – and perhaps to have an 'Extract' of it made – thus explaining how it reached Wotton and Birch, but that is as much as is known about it.

Boyle's interest in his own life seems to have revived in his later years, and two surviving documents which are printed below bear witness to this. One comprises the notes dictated by Boyle to his amanuensis, Robert Bacon, included here as Document 2.[67] It is unfortunately impossible to date this exactly. Bacon evidently started to work for Boyle in the 1670s, and this document could therefore date from any time between then and Boyle's death.[68] What it comprises are two sections of an autobiography, written, like *Philaretus*, in the third person, though here the name of the protagonist has been left blank. The first section is clearly the opening one, giving an account of Boyle's birth, ancestry and early life overlapping with that in *Philaretus*. Then, on a separate sheet, a discrete section begins, describing Boyle's return to London after his travels abroad and various events which elapsed at that point.

It has been suggested that these notes are 'obviously a continuation of the Account of Philaretus', but the fact that the first part overlaps with the earlier work suggests on the contrary that they represent fragments of an attempt to write Boyle's life afresh.[69] It is therefore instructive to note the differences between the two. Here, as earlier, the role of providence is invoked, in this case in placing Boyle in a godly, rather than a courtly, milieu on his return from his travels, an interesting reflection of Boyle's awareness of tension between his pietistic outlook and the fashionable world associated with his social class.[70] In addition, Boyle's account of his reunion with his sister is quite vivid and touching. On the other hand, there is here none of the sententious moralism of *Philaretus*, nor its elaborate style, including the duplication of words; and, though reflections are included, they are more matter of fact. Indeed, the rewritten version of Boyle's early years is so matter of fact that the revealing memories of this phase in Boyle's life included in *Philaretus* – concerning his relations with his parents, his acquisition of his stutter and the near-fatal accident which befell him – are dismissed with the formula that there was 'nothing peculiar' about his childhood. Even the overlapping account of the role of John Harrison in giving Boyle a liking for reading and of his enthusiasm for Quintus Curtius, though differently told and including information lacking from the earlier version, is dismissed as 'petty' as if it, too, might have been dispensed with. Perhaps Boyle now found his earlier persona slightly embarrassing.

Fortunately, however, in another document Boyle was much more revealing about himself. This is Document 3, referred to here as the 'Burnet Memorandum', which consists of a series of notes made by Bishop Gilbert Burnet on an interview, or interviews, with Boyle. This document is peculiarly significant in that the notes were vouchsafed by Boyle to one of his most intimate acquaintances, with whom he had 'a close and entire friendship' over three decades.[71] Burnet had first met Boyle when he came to England in 1663. Later in the reign of Charles II, Boyle gave him financial support while he compiled his well-known *History of the Reformation of the Church of England*, while, thereafter, when political circumstances forced Burnet into exile, it was to Boyle that he wrote letters describing his experiences.[72] After Burnet's return to England at the Glorious Revolution to become Bishop of Salisbury, the two men clearly saw a good deal of one another. Burnet was one of the churchmen to whom Boyle went for advice on matters on his conscience at this time, and Boyle's notes on one of the interviews that took place survives.[73] Their intimacy is further captured by a letter from Burnet to Boyle of 20 January 1691, in which he wrote: 'It is my inclination, as well as my duty, to delight in being often with the person in the world, that I esteem the most, and to whom I stand so highly engaged'.[74] It is hardly surprising that Burnet was the only churchman who received a bequest in Boyle's will, his 'great Hebrew Bible with silver Claspes'.[75]

In addition – though this has hitherto been unknown – Burnet may have played a significant role in formulating the idea of a sermon series in defence of Christianity for which Boyle provided in his will, the so-called Boyle Lectureship. At least, this is the implication of the fact that the earliest draft of the relevant codicil to Boyle's will – headed 'For a Lecture for the Christian Religion' – is in Burnet's hand, which suggests either that Burnet gave Boyle the idea, or that he helped him to formulate a project which he already had in mind.[76] What is striking is that the wording concerning the objectives of the benefaction is virtually identical in this draft with that of the widely-publicised final version, though various additions and modifications were made concerning the Lectureship's administrative details.[77] The text in Burnet's hand thus clearly states the requirement 'to preach 8 Sermons in the year for proving the Christian Religion against Atheists and Theists descending no lower to any Controversies that are among Christians'; it also replicates the final version in seeing the 'satisfying' of 'scruples' as one of the tasks of the lecturer, which could reflect the casuistical relationship between Boyle and Burnet (though there is no evidence that it came to much in practice).[78] Indeed, it is conceivable that the form that Boyle's endowment took may have owed more to Burnet's perception of the apologetic needs of the church than to Boyle's own: Boyle himself may have had alternative ideas as to how Christianity was best defended, such as by endowing a 'Collection of Bookes tending to the Truth of the Christian Religion', another idea which appears in a memorandum from his later years relating to his will and associated matters.[79]

The date of the Burnet Memorandum is unclear. It may or may not be earlier than the autobiographical notes in Bacon's hand just described. Burnet's exile from 1685 to 1688 means that it must date either from before that or from the last three years of Boyle's life. That it should date from the latter period has a certain plausibility. Thus Burnet might have compiled it because he was trawling for a biographical account of Boyle such as he was to include in his funeral sermon. There may also have been an overlap between this interview or interviews and the sessions devoted to matters on Boyle's conscience which we know occurred between the two men in those years. It is revealing that, in the course of the notes, Burnet refers to Boyle's consultation of him on a moral issue when it arose. On the other hand, there is no reason why such interviews should not have taken place earlier, and these notes could equally easily date from before Burnet's exile.[80] One clue that this is the case is the repeated reference in the document to Charles II as 'the King', as if he were still on the throne; John Fell, Bishop of Oxford (who died in 1686), is also referred to as if he was still alive. If Boyle was indeed grooming Burnet to write a posthumous life of him, this was as likely to have occurred before 1685 as later. Boyle had long been apprehensive about his health, having suffered a stroke in 1670 which brought him to 'deathes doore', in his sister, Mary's, words.[81] Moreover at this earlier date Burnet would have seemed all the more appropriate as a putative biographer, since it was in the 1670s and early 1680s that he published various *Lives* of eminent contemporaries, for which he has been acclaimed one of the principal exponents of biographical writing in his day.[82]

The Memorandum is in Burnet's hand, and its authenticity is underlined by its verbal alterations and its occasionally garbled character, presumably due to being written down during the interview, or recollected at speed while the matters were fresh in Burnet's mind. It also alludes to the fact that Burnet was himself present at the discussion between Boyle and Fell that took place in 1677 (the most recent event referred to in the document), while it refers to Burnet in the first person in connection with Boyle's consultation of him on the morality of intercourse with spirits. The fact that it was to his close friend, Burnet, that Boyle dictated these notes helps to explain the frankness of the revelations that he made in them. At the very least, he must have been a willing party to the transaction. But he may well have been its initiator, as is suggested by a clue from later documentation concerning the Memorandum which it is appropriate to deal with here.

One of the matters referred to in the Memorandum is Boyle's evident belief that certain days were unlucky, since he notes of May Day, 'observed often Inauspicious'. Such beliefs were linked to divinatory systems of thought which were widely frowned on by churchmen at the time, such as the early Stuart divine, William Ames, who saw this as symptomatic 'Of Consulting with the Devil'.[83] We know from a letter to Thomas Birch from his collaborator Henry Miles of 21 October 1741 that, in a passage of his lost life of Boyle, Dr William Wotton

> mentions the opinion Mr B. had conceivd of inauspicious days, that he accounted the 1st of May one to him in particular – the Dr dissaproves of the notion & hints of the evils which may attend such a perwasion, but sais, he is *not at liberty to omit* this, because *in the minutes he* ‹[Mr B.]› *left behind him of his Life, he desird it shoud be taken notice of.*

Miles was dismissive:

> Now I much question if he particularly desird it, nor do I know of any minutes he left behind him of his Life, but those I have, except he means those hints which Bishop Burnet took from his mouth, where this is mentiond in brief with other Circumstances, and perhaps the Dr intends no more by it than that he dictated this with other matters to the Bishop.[84]

On the other hand, since Wotton was a protegé of Burnet's, and received help from him in preparing his life, it is equally possible that, in taking the line he did, he was repeating information about Boyle's wishes that he had had from Burnet.

Hence in many ways this is the single most significant biographical source relating to Boyle. What it comprises is, first, a page of brief, numbered sentences concerning events and themes in Boyle's life, including the ones referring to May Day. These are in approximately chronological order, and in general it may be presumed that items which are not clearly dateable are intended to belong temporally at the point in the sequence at which they appear.[85] Over the page are some slightly more continuous narrative passages, including a general account of the evolution of Boyle's scientific programme from the 1650s through to the 1660s. There then follows a lengthy passage – comprising nearly half the document – which retails various stories about intercourse with spirits and the pursuit of the philosopher's stone, which Boyle evidently saw as interconnected. Then, at the end, there is further narrative concerning events in the 1660s and 1670s: Charles II's high view of Boyle; the suggestion that he might take Holy Orders after the Restoration; his role in relation to the Council for Foreign Plantations and the Company for Propagation of the Gospel in New England; and his role in subsidising publications for use in missionary work in the Near East.

So rich is this text that a full commentary on it would almost amount to an intellectual biography of Boyle.[86] The early section overlaps with *Philaretus* and retells some of the same stories, including two of Boyle's providential escapes from harm, as Henry Miles noted in his annotations to the manuscript.[87] But even this section contains some information that is not in *Philaretus*, for instance concerning Boyle's meeting with the famous Protestant divine, Jean Diodati, in Geneva, or his reading of such authors as Sir Walter Ralegh, Seneca and 'the lives of the old Philosophers', presumably those by Diogenes Laertius. In addition, because it goes further into Boyle's life, it provides crucial hints on his later development. Thus the two sentences dealing with his study of scriptural languages under Archbishop Ussher's encouragement accentuate the significance for Boyle of this episode, echoing the account of it that he included in a section of his *Essay on the Scripture* published by Birch which is otherwise lost.[88] In addition, the account of the illness that he contracted in Ireland in 1654 and its effect particularly on his eyes complements the fuller description of the episode that Boyle gave at the end of his life in the preface to the second volume of his *Medicinal Experiments*, again emphasising its significance for him.[89]

As with *Philaretus*, there is an element of reflectiveness, though the rather sanctimonious quality of the earlier text does not recur here. The comment about 'Loving trueth of a child' echoes the comparable remark in *Philaretus* which is illustrated by Boyle's ingenuous story about Lady

Ranelagh and the plum tree; but unique to this text are his observations about his restraint from play and from 'appelizing' as a child, and his slightly enigmatic comment about marriage, familiar due to being quoted by Birch.[90] Equally important is the way in which the text dwells on temptation and self-control later in his life, whether it be the bad temper which he 'Governed', or his restraint from dabbling in magic. Indeed, in one passage he went so far as to describe his curbing of his curiosity about the latter as 'the greatest Victory he had ever over himselfe'. Clearly, although Boyle ceased to write treatises on the pursuit of a Protestantised Stoic moral mean after the 1640s, he had not ceased to internalise the message that he had been putting across in them, which remained with him for the rest of his life.

Perhaps the most striking section of the text comprises its vividly told stories about magical encounters, which are at least partly autobiographical. Though most are hard to date precisely, the one concerning the pregnant clairvoyant is dateable to the Second Dutch War of 1665–7, since it was only then that Dutch ships came near enough to London for their guns to be audible from there. Hardly less significant is Boyle's account of his programme as a natural philosopher from the 1650s onwards, in which he attempts to give a kind of rationale of his progression through the various major scientific works that he published between 1660 and 1665. He also refers to his debate with Linus and Hobbes, giving, in connection with the latter, an intriguing piece of information – that he had initially suspected one view expounded by Hobbes to be the result of a misprint – which is otherwise wholly unknown.[91] The text also illuminates Boyle's relations with the Restoration establishment. Though there is reason to think that some of Boyle's statements may not be strictly true – for instance, what he says about his correspondence – even these have to be taken seriously as evidence about his perception of himself. In general, we have here a view of what seemed to Boyle to be significant in his life, giving a conspectus of his preoccupations which is all the more revealing in view of his intimacy with Burnet and the fact that he was almost certainly aware of Burnet's likely role as his biographer. It is undoubtedly the most significant new document in this volume.

4. BURNET'S 'OFFICIAL' VIEW OF BOYLE AND THE NOTES OF SIR PETER PETT

Boyle died on 31 December 1691; his funeral just over a week later, on 7 January, provided the occasion for the earliest and most influential published account of him, the sermon that Burnet delivered on that day at St Martin's in the Fields. It was doubly appropriate that Burnet should have delivered this accolade to Boyle. For one thing, as we saw in the last section, he was a close friend of Boyle's who was privy to intimate information about him on which he drew selectively in the sermon. In addition, he was an experienced panegyrist; his sermon on Boyle is merely the best known of a number of funeral orations which he delivered in the 1680s and 1690s, though an eighteenth-century commentator noted that 'there are many, who think his performance on that occasion the best he ever published'.[92] He delivered similar orations for the philanthropist, James Houblon, in 1682, for such pious ladies as Ann Seile and the dowager Lady Brooke in 1678 and 1691, and for the Archbishop of Canterbury, John Tillotson, in 1694.[93] All of these bear a family resemblance to one another, though there is some variation in format. In Boyle's case, Burnet began with a fairly lengthy, general disquisition on his text from *Ecclesiastes*, and the 'wisdom, knowledge, and joy' which God vouchsafed 'to a man that is good in his sight', before moving on to a more *ad hominem* account of Boyle. Indeed, John Evelyn seems to have seen the two sections almost as discrete parts, writing to William Wotton in 1696 how 'the Sermon you know is Printed, with the Panegyric, so justly due to his Memory'.[94] This format is paralleled in certain of Burnet's other sermons, for instance that for Houblon; on the other hand, his sermon for Tillotson had rather less by way of general reflection on the biblical text, while that for Ann Seile had virtually no biographical matter.

The funeral sermon was a well-established form of panegyric, which overlapped with biographical writing of the day, and had similarities to the formalised 'éloges' of scientists which were to be associated with the Parisian Académie des Sciences in the next century.[95] Like the latter, it owed something to the tradition of panegyric poetry with its roots in antiquity, which had been revived in the Renaissance.[96] Indeed, in the aftermath of his death, Boyle was the subject not only of Burnet's eulogy, but also of a number of such poems in Latin and English, 'in the way of what we call Lapidary Verse', in the words of Sir Peter Pett.[97] These represent a tribute to the esteem in which the great scientist was held, though the formality of the genre means that almost by definition such poems contain little of biographical value.[98]

There is an element of such formality in the sermon. Writings of this kind were intended more for edification than the provision of biographical detail, which was in any case rendered redundant by the extent to which the deceased was known to those present when the sermon was delivered. As Burnet put it in his funeral sermon for Archbishop Tillotson, the object was 'to get together some parts of his Character; and to set him out to you such as you all knew him to be'.[99] In Boyle's case, Burnet's objective was to present a memorable image of Boyle as a great and good man, his example showing 'in the simplest and most convincing of all Arguments, what the Humane Nature is capable of, and what the Christian Religion can add to it, how far it can both exalt and reward it'.[100] In the course of this, Burnet gave some telling pieces of specific information – about Boyle's manner, for instance, which Burnet found 'exactly civil, rather to Ceremony', or about the sheer scale of his charitable donations and his patronage of evangelical activity abroad.[101] But all this was subordinated to an eloquent and genuinely moving account of Boyle, and, ancillary to this, of Lady Ranelagh, who had died only eight days before him. Its message was the way in which Boyle's religious commitment enabled him to rise above the frailty of human nature, triumphing over such impediments as illness to achieve a wisdom, serenity and insight into nature that both illustrated God's gifts to him and enabled him further to serve Him. There can be no doubt of the brilliance of Burnet's depiction, nor of its effectiveness in setting out Boyle's exemplary role; indeed, it formed the principal basis for all the evaluations of Boyle to be written in the eighteenth century, and still has its legacy today.

Collation of the sermon with the Burnet Memorandum shows some passages which definitely echo it – for instance, concerning the influence on Boyle of Archbishop Ussher – and others which may do.[102] On the other hand, Burnet's statements tend to be of a more general nature, extrapolating from the specifics of the Memorandum to broader points, due either to the effect that the genre had on him, or to his being able to refer to conversations with Boyle other than the one(s) on which the Memorandum was based. A case in point is Burnet's elaboration of one topic which must at least partly be based on the Memorandum, Boyle's refusal to take Holy Orders when it was suggested to him that he might after the Restoration. The only reason that is recorded in the Memorandum is Boyle's awareness that he had 'never felt the Inward Vocation'. In his sermon, Burnet elaborated on this at length, but he also added a further reason which does not appear in the earlier text, that as a layman Boyle felt that his views would be taken more seriously than those of clergymen with a professional interest in promoting Christianity.[103] This may have been based on further conversations with Boyle, or Burnet might have drawn on Boyle's statements to similar effect in the prefaces to various of his published works.[104] Equally, however, it is possible that Burnet himself, a consummate ecclesiastical politician, embroidered Boyle's opinion with a cliché of the anticlericalism of the time which is known to have caused him concern, 'That it was their Trade, and that they were paid for it'.[105] This should induce slight caution in reading this passage entirely literally: yet, even if Burnet elaborated the point, there is no doubt that it remained in tune with Boyle's general sentiments. One here witnesses a marriage of minds similar to that seen in the institution of the Boyle Lectures, Burnet's likely part in the origination of which has already been noted, and to which he gave the first public announcement in this sermon.[106] Boyle would undoubtedly have been content with the image of himself that his friend here presented.

Burnet's sermon may have been a classic, but in it he implied that this was not to be his final word on the subject, promising 'a farther and fuller account' of Boyle.[107] That contemporaries took this literally is suggested by various announcements to this effect in the early 1690s. In January 1692 *The Gentleman's Journal* stated that 'A learned and eloquent Pen will give us his Life'; in the brief account of Boyle that the antiquary, Anthony Wood, included in his *Fasti Oxenienses* (1692), he specifically noted that a life of Boyle was 'about to be published' by Burnet; while in the English version of Lewis Moreri's *Dictionary* which came out in 1694, a lengthy account of Boyle, based principally on Burnet's sermon, was accompanied by the statement that Burnet was 'about publishing his Life, a Work worthy of such a pen'.[108] Certainly, biographical writing was as much a speciality of Burnet's as the delivery of funeral sermons, as has already been noted. By this time he had published lives of figures ranging from the Dukes of Hamilton and Castleherald to the jurist, Sir Matthew Hale, and from the Irish churchman, William Bedell, to the Earl of Rochester, his celebrated account of the latter juxtaposing an account of Rochester's career with a focus on Burnet's own attempts to retrieve his soul in his latter days.[109]

Just how committed Burnet was to producing a full-length life of Boyle is unclear, but it never materialised, and for this various reasons can be suggested. One might have been an awareness that he was ill-equipped to write an account of Boyle as a natural philosopher – a topic which he had dealt with only briefly in his sermon. He had not, however, felt disqualified from earlier writing a biography of Lord Chief Justice Hale despite admitting his 'ignorance of the Law of *England*' (he had done his best to get round the problem by including a lawyer's evaluation of Hale within his text).[110] Probably more significant was shortage of time: all of Burnet's published lives dated from earlier in his life, whereas by the 1690s he was one of the leading churchmen of the day. Indeed, his promise of such a life in the sermon was conditional upon the availability of 'more leisure and better opportunities', and he may have been aware that only a counter-revolution would have been likely to bring these about. Certainly it was John Evelyn's impression that Burnet 'declines his purpose upon the account of want of Leisure, &c.'[111]

In addition, however, Burnet may have felt that he had little to add to the evaluation of Boyle that he had already published: in his *History of My Own Time*, he wrote of his sermon how 'I gave his character so truly that I do not think it necessary now to enlarge more upon it'.[111] This is all the more plausible in view of the nature of Burnet's published biographies, which themselves tended towards the panegyric. They exemplify what has been described as the 'cult of perfection in biography', in which a didactic aim took precedence over the provision of detail about the figure in question, and in which anything which might complicate the adulatory picture presented was quietly elided.[113] Burnet's *Life and Death of Sir Matthew Hale*, for instance, is in many ways simply a book-length version of an evaluation like that of Boyle in the sermon. Burnet might well have felt that any further account of Boyle would represent an anticlimax.

Be that as it may, the comments already quoted show that there was a clear expectation in the early 1690s that Burnet would write such a life, and it was for the purpose of assisting him in this that Boyle's old friend, Sir Peter Pett, produced his biographical notes on Boyle. Indeed, in these Pett actually cited various passages in Burnet's sermon, which he sought to supplement by providing Burnet with ancillary information and advising him where he could obtain still more. These very substantial notes – one of the most extensive of all biographical accounts of Boyle – comprise Document 5. Despite the piecemeal form they take, they provide a rich and hitherto largely untapped source concerning Boyle.

Pett was a Cambridge graduate who had migrated to Oxford in the 1650s, where he became a Fellow of All Souls and an active member of the Oxford group of natural philosophers.[114] It was at this time, while Boyle was living in Oxford, that Pett met him, and it was probably from then till the early 1660s that their contact was closest. Their links arguably reached a climax in the years around 1660, when Pett was involved with Boyle in a campaign in favour of religious toleration – from which resulted Pett's *Discourse concerning Liberty of Conscience* (1661) – and when Pett was closely associated with the publication of Boyle's *Some Considerations Touching the Style of the Holy Scriptures* (1661).[115] Thereafter, the two men drifted apart, Pett going into legal practice and public office, attaining the office of Advocate General for Ireland. By the 1680s, he had become an Anglican 'in the way of Archbishop Sancroft', in his own revealing phrase.[116] On the other hand, Pett was always more of an Erastian than a high churchman, as is revealed particularly by his book, *The Happy Future State of England* (1688), in which he deployed historical and statistical evidence to try to illustrate the sheer weakness of the Catholic interest in England and hence the lack of a real threat of Catholic restoration.[117] The rather secular-minded views which Pett there put forward were not ones which Boyle would wholeheartedly have supported: there was a distinct difference in outlook between the two men.[118] Yet it is clear from his notes that Pett retained the highest possible regard for Boyle, and it was for this reason that he was so anxious to assist Burnet in giving the best possible account of his deceased friend.

The notes, described by Pett as 'paragraph[s] or memoire[s]', comprise three separate sections, two of which are incomplete. The second set comprises numbered points, dealing with topics in roughly chronological order, though sometimes one seems to have suggested another in the course of composition.[119] On the other hand, the notes show no direct evidence of having been written at different times, as suggested in an endorsement to them by Henry Miles. Indeed, it is not easy to date them exactly at all. Their references to Burnet's sermon show that they post-date it; they clearly predate Pett's death in 1699; and a date of c. 1695 may be suggested by a promise in an extant letter from Samuel Pepys to Pett to supply information which Pett records in these notes that he had requested.[120]

Pett offered Burnet all sorts of assistance. In addition to the approach to Pepys just referred to, he also offered to arrange for Thomas Hale, one of the technological improvers whom Boyle had encouraged, to wait on Burnet.[121] He also suggested how access might be gained to the records of the Royal Society or to the wills of Boyle and his father, and who should be consulted for advice on Boyle's scientific work.[122] In addition, he offered to look through the verse elogies that had been published after Boyle's death to see if there was anything of value in them, and volunteered the loan of various relevant books, including his only copy of his 1661 tract on liberty of conscience.[123] What is most important about Pett's notes, however, is the large amount of detailed information about Boyle that they give. Clearly, Pett perceived the need for the generalised tone of Burnet's funeral sermon to be supplemented by circumstantial detail, and his notes are almost entirely devoted to this. In them, he purveyed a great deal of information that is not otherwise known, including a number of anecdotes which shed crucial light on Boyle's personality and which are assessed in section 8, below. Some of this information, revealingly, he gave to Burnet 'en secret', expecting him to use his own discretion as to how much to deploy.[124] Even passing points that Pett makes are often revealing, such as his characterisation of subsidiary figures like Boyle's trustee, Sir John Rotherham, or his servant, John Warr, whom Pett describes as 'such an one as the French call tout a fait honet homme'.[125]

Pett's notes show that he had clear views about the form that a biography of Boyle should take. His sense of the need for telling detail itself reflects a key trend in biographical writing in his day, exemplified by such practitioners as John Aubrey and Roger North in contrast to the more formal tradition represented by Burnet's own lives.[126] Indeed, Pett had biographical aspirations of his own, planning a collection of the writings of Lord Falkland, prefaced by a biographical essay, though this failed to materialise. He also published the memoirs of his political ally, the Earl of Anglesey, and the miscellaneous papers of his and Boyle's mutual friend, Thomas Barlow, in which he stressed the value of preserving and publishing a man's correspondence.[127]

Like others at this time, Pett compared Burnet's task with Pierre Gassendi's in his *Life of Peiresc*, the chief existing exemplar for the biography of an intellectual as against a man of affairs. He also felt it important that Boyle's 'philosophical notions & experiments' should be summarised, together with the adulation of him by his scholarly peers, while the writings of 'pretended critics' were also to be 'consulted & animadverted upon', particularly if they had 'detracted from the usefulness of any of Mr Boyles experiments'.[128] In addition, he thought it appropriate to include a portrait, citing parallels in editions of writings by Hugo Grotius, John Selden and Thomas Barlow, and suggesting as a potential artist Robert White, who had engraved the plates of Selden and Barlow to which he referred, and whose engraved portrait of Boyle was prefixed to the edition of his *Free Discourse against Customary Swearing* which came out in 1695.[129] In Pett's view, such a portrait might properly be accompanied by adulatory verses, though he had reservations about the inclusion of too much general panegyric in the work itself, 'Poetry & History not looking well together ‹and› especially in so great a Life'.[130] In addition, he wanted Burnet to do justice to Boyle's religious views, on the grounds that his 'example' was 'conducive to true and solid piety'.[131]

In the information that Pett furnished, it is possible that certain of the priorities displayed could be Pett's rather than Boyle's own; in particular, the stress on the application of science might reflect the fact that Pett was a member of one of the leading families of shipbuilders in seventeenth-century England.[132] But there is sufficient independent corroboration of such interests on Boyle's part to render Pett's account plausible: an example of this is the leading role he took in Robert Fitzgerald's scheme for making saltwater drinkable, one of the projects to which Pett refers.[133] In any case, Pett gives equal prominence to a less predictable theme, Boyle's interest in Biblical learning. Thus he cites a tribute to Boyle's linguistic expertise from a letter from Bodley's Librarian, Thomas Hyde; he gives examples of his own of Boyle's interest in the exact original meaning of scriptural passages; and he tells how, though Boyle was in favour of the readmission of the Jews under Cromwell, he was anxious that state support should be provided for scholars who could refute the interpretations of Jewish rabbis whose proliferation that event would encourage.[134] Perhaps more surprising is the tribute that Pett paid to Boyle's 'admirable wit', which he claimed was admired by the poets Cowley and Davenant; he also referred to early literary compositions of Boyle's which he had seen, and commented in his own right on the 'excellent sprightfull vivacity of his Thoughts'.[135] All in all, Pett gives a telling account of Boyle, while this hitherto unpublished text also throws important new light on the outlook of Pett himself.

5. WOTTON'S *LIFE*, I
ITS PROGRESS AND DEMISE

As the 1690s progressed, it became increasingly clear that Burnet was not going to produce a life of Boyle. Instead, in 1696 the mantle was taken up by William Wotton. Wotton was a former child prodigy and polymath, described by John Evelyn in a letter to Samuel Pepys as 'he whom I have sometimes mentioned to you for one of the miracles of this age for his early and vast comprehension'.[136] By this time, Wotton had achieved considerable renown for his *Reflections upon Ancient and Modern Learning* (1694), in which he took up the challenge of the ancients, as presented by Sir William Temple, and made a comprehensive case for the superiority of modern science and scholarship, though accepting the primacy of the ancients in literary and artistic endeavours. In the course of this, he wrote what can be claimed to be one of the first historical accounts of the growth of scientific ideas.[137]

Wotton first had the idea of writing a life of Boyle in March 1696, but the exact stimulus to this is unclear. Since he had long been a protegé of Burnet's, and was to execute various tasks at his behest over the following decade, it is possible that Burnet might have deputed to Wotton what had become an unwelcome task.[138] Against this, however, is evidence from Wotton's correspondence with John Evelyn that the idea was his own, and that Burnet's agreement had to be sought in order to clear the way for him to realise it.[139] Since it is from letters between Wotton and Evelyn that much of our knowledge of Wotton's project is derived, it is appropriate to introduce this epistolary relationship here. It began in 1694, when Wotton was acting as tutor to the Finch family and was based at Albury in Surrey, not far from the eponymous house, Wotton, where Evelyn now lived. By this time the climax to Evelyn's career in public life – as Commissioner for the Privy Seal under James II – had passed. Though he continued to hold some responsibilities, notably as Treasurer for the new Greenwich Hospital, he now had more time for literary pursuits, and in the 1690s he reverted to his earlier career as an author, producing his *Numismata* in 1697.[140] In addition, he assisted Wotton in producing a chapter on ancient and modern gardening which was added to the second edition of his *Reflections* in 1697.[141]

Evelyn relished the company of Wotton, likening the presence of the young intellectual in rural Surrey to manna in the wilderness, and they began to correspond in 1694. The earliest surviving letters between them show that they discussed what was evidently a manuscript copy of Sir

Kenelm Digby's autobiography, which, as Wotton explained, had been 'put into my Hands by one that desired to print it'. Wotton lent it to Evelyn so that his wife could read it, adding a key to the pseudonymous figures in the work, though he seems to have felt that its suggestion of Charles I's receptiveness to Catholicism at the time of the Spanish match was too sensitive to his reputation for it to be appropriate for publication.[142] The episode appears, however, to indicate an interest on Wotton's part in matters biographical and relating to seventeenth-century England. Be that as it may, it transpires that on a visit to Evelyn in March 1696, Wotton told him about Adrian Baillet's *Life of Descartes*, of which he subsequently sent Evelyn a copy.[143] This work, which had been published in 1691, was a lavish two-decker account of the great philosopher, with something in common with Gassendi's *Life of Peiresc*. As Evelyn explained to the scholar and Boyle Lecturer, Richard Bentley, in a letter of 22 March 1696: 'I found by his [Wotton's] discourse to me, that he has a mind to do something of this nature in memory of Mr Boile, which I would by all means [promote] because I think he would perform it successfully'.[144]

There is no question that the idea was Wotton's, not Evelyn's, though Wotton was to flatter Evelyn by implying the contrary later in their correspondence, and this has misled some modern scholars into attributing more of an initiatory role to Evelyn than he really had.[145] But Evelyn was from the outset enthusiastic about the project, writing at a slightly later stage: 'Mee thinks I already see my noble Friend Mr *Boyle*, Rising againe, and made Immortal by Mr *Wotton*, & Mr *Wotton* by Mr *Boyle*'.[146] Indeed, it is even possible that at one point he offered to contribute to the work, since on 27 April 1697 Wotton wrote to him: 'Your goodness in concerning yourself so far with my intentions to write an Essay upon Mr Boyles memory as to appear in the Author's defence, obliges me to look out for no other Patron'.[147] In fact, apart from furnishing Wotton with his memories of Boyle, Evelyn's contribution was largely in commenting on Wotton's project and seeking support for it. As one of the Trustees for the Boyle Lectures, Evelyn was well-connected with the ecclesiastical grandees of the day: indeed, he had canvassed Wotton as a possible Boyle Lecturer in 1694, though nothing came of this.[148] He also offered help in approaching Boyle's relatives on Wotton's behalf, including his brother and executor, the Earl of Burlington.[149]

Evelyn sent Wotton his own reminiscences of the great man in the letter of 29 March 1696 which is printed below: this gives quite a memorable portrait of Boyle, with a potent characterisation almost comparable to Burnet's. Indeed, as Wotton later told Evelyn: 'notwithstanding the notices which Mr Boyle's own papers and the Bishop of Sarum's hints have given me, I found your informations so useful, that without them my work would be very lame'.[150] In his letter, Evelyn surveyed Boyle's life and scientific achievement, outlining his experimental programme in the context of the Oxford group of natural philosophers and, subsequently, the Royal Society, summarising his published output, and seeing him as outdoing Bacon in his experimental enquiries into nature. Continuing, 'Nor will you omitt those many other Treatises relating to Religion', he went on to deal with the nature of Boyle's religiosity and his charitable work. Evelyn also provided details not otherwise available concerning Boyle's appearance, his manner (including his stutter), and his lifestyle, giving a vivid sense of Boyle's cluttered apartments. As in other evaluations, there is a slight element of partisanship, in this case a tendency to show Boyle in the image of a virtuoso like Evelyn himself, albeit a more profound one: in particular, Boyle's hostility to the Interregnum regime may be exaggerated. Only on one point was Evelyn critical, concerning Boyle's style: otherwise, in his own words 'to draw a just Character of him, one must run thro all the Vertues, as well as all the Sciences'.[151]

Over the ensuing months, Evelyn took various steps to assist Wotton. In a letter of 12 May, he told him that he had raised the matter at a meeting of the Boyle Trustees, who

> readily embrac'd the Motion, with much Sattisfaction, & desir'd that his Grace [the Archbishop of Canterbury] should in all our Names, signify so much to my Lord Bishop of Salisbury; That in case, he had quite layd ‹aside› the Thoughts of what, he seem'd to promise in his Funebral Panegyric, he would give you the best Assistance he could, which they believed would be very Considerable.

In addition, the archbishop, Thomas Tenison, himself offered his assistance, 'Especially as to his Charities, & other Circumstances of his Sickness & Death; having bin almost daily with him all the time he lay sick', though unfortunately, if such information *was* given, no record of it has survived.[152] By the middle of June, Burnet had evidently concurred in the view that Wotton was appropriate for the task in hand.[153]

Evelyn also took the initiative in sorting out a further complication that arose at an early stage in Wotton's project. In 1695, John Williams, prebendary of Canterbury and Boyle Lecturer, who was to become Bishop of Chichester in 1696, had brought out an edition of one of Boyle's moralistic writings from the 1640s, his *Free Discourse against Customary Swearing*; he had been given the manuscript of this by Boyle's executors, Sir Henry Ashurst and the Earl of Burlington.[154] As a sequel, and in view of the scarcity of some of Boyle's books, Williams seems to have been encouraged by the Oxford don and publishing entrepreneur, Arthur Charlett, to produce a collected edition of Boyle's works in folio. Evidently Williams was to provide some kind of preface, perhaps biographical, while, possibly inspired by the example of the *Free Discourse*, Williams had the idea 'withal, that it would be convenient that some inquiry should be made of what might be found among his papers, fit for the press'.[155] When Wotton heard about this rival enterprise, he was naturally upset, suspending work on his own until the matter was resolved; Evelyn (who was informed about it by the Archbishop of Canterbury) undertook to write a tactful letter to Williams to sort the matter out 'before *Mr Wotton* should be too far Ingag'd'. It is worth quoting from this at length, since it indicates how Evelyn (and, presumably, Wotton) conceived of the putative work at this stage. After singing Wotton's praises as an author – 'one of extraordinary parts for his time' – he went on:

> Tis more than 40 yeares I have had the honor of being Acquainted with *Mr Boyle*, & partly with his studies; & tho I question not, but that besides the Enumeration of his Vertues, & singular piety, you will give the World an ample Account of his Family & Relations, Person, Education, Learning, Travells, Correspondences both Abroad & at Home; in all which, he was so exemplarily Conspicuous; Yet, when I consider those innumerable heaps of Letters & papers, & other Circumstances of his Life, to be Consulted, Read-over & digested; together with what must be gather'd from his Friends & more intimate Conversations; which instead of a *Praeface* will require a worke by it selfe, & no small Labour; I begin with his Grace [the Archbishop] to doubt we are not so perfectly assur'd, that *Dr Williams* is more at Leasure than my Lord *of Salisbery*, for such a buisy Work, as I am sure he is aboundantly able to performe it, and to be prefered to any other, if he seriously undertake it.[156]

In fact, Williams replied explaining that a life had never been his intention and leaving the way open for Wotton, though with an aside implying that even Wotton was not ideal for the job due to his lack of the 'intimate acquaintance' with Boyle which would be crucial to a proper biography.[157]

A further complication arose concerning access to Boyle's manuscripts, then in the possession of his former servant and executor, John Warr, although responsibility for them was shared by the other executors, of whom the Earl of Burlington seems to have taken a particular interest in them. Their existence was well-known at this point: not only did the idea of using them form part of Williams's project, as we have seen; in addition, Pett had informed Burnet in his notes for him that it would be impossible to write Boyle's life 'to the utmost advantage' without access to these (though the information that he gave about their sheer bulk may have helped to discourage Burnet from following his advice).[158] Wotton seems from the outset to have wished to gain access to this material, mentioning the matter in a letter to Evelyn of 7 April 1696.[159] Problems ensued, however, evidently due at least in part to a prevalent misunderstanding of the provisions concerning access to the papers in Boyle's will, to which Pett alluded in the passage in his notes already referred to. Though the manuscripts were in Warr's custody, Lord Burlington seems to have been rather obstructive to Wotton concerning access to them: Wotton explained to Evelyn how, even after he had had 'a positive promise from the executors that I should have the use of Mr Boyle's papers, my Lord Burlington at last insisted upon my giving a bond that I demanded no gratification' (he added that he had already given a note to this purpose, but not a bond).[160]

Evelyn, who heard about this through Burnet, had to write with profuse apologies that the Earl 'should weare so meane & little a Soule under such a Dignity, Out-side & other Circumstances'.[161] By the beginning of 1698, however, the problem had been sorted out, evidently through Evelyn's and Burnet's good offices, and Wotton was able to report that Burlington 'is come over so far that he has delivered up my note, and has ordered all the papers to be delivered to my order, with a promise to me of all manner of assistance and encouragement'.[162] Apparently, he had been authorised to select items that he wanted to keep for use in his *Life*, and to return the rest to John Warr: this seems to have been agreed in a letter from Wotton to Warr which was in the possession of Henry Miles in the early eighteenth century, but which no longer survives.[163]

Hence at last all the complications were sorted out, and on 2 January 1698 Wotton told Evelyn 'that now I intend to dedicate all my spare hours to this business'.[164] His letters to Evelyn suggest that he went through large quantities of the Boyle Papers and Letters at this point, while by 1699 his progress had been sufficient to justify an advertisement for his intended book in the *London Gazette*, which described it as follows:

A History of the Life, and Extract of the Writings of the Hon. Robert Boyle Esq.; Collected as well from what are in Print, as from his Manuscripts and Papers of Experiments never Published, and also from a vast number of Letters to and from the most Learned Men in Europe, his Correspondents: Undertaken by W.Wotton B.D., and Fellow of the Royal Society, at the Desire of divers Gentlemen of the said Society, and other Friends of Mr Boyle, as also of his Executors, who have for that purpose entrusted the said Papers with Mr Wotton.[165]

It was this advertisement which stimulated the arrival of one of the texts printed here, Document 7, Thomas Dent's letter to Wotton of 20 April 1699. Dent was a graduate of Trinity College Dublin who had taken holy orders and who had various links with the Boyle family: as well as holding the rectory of Stalbridge from 1690 onwards, he was chaplain to Boyle's brother, Lord Shannon. Dent's ecclesiastical career had reached its climax when he became Canon of Westminster in 1694.[166] Dent had hoped to provide Lord Shannon's memoirs for Wotton's use (Evelyn had suggested that these would be valuable in a letter to Wotton of 12 May 1696[167]); this was prevented, however, by Shannon's illness and death. Instead, Dent provided some interesting, if fairly general, information about Boyle, some told him by Shannon, some derived from his own knowledge.

Thus he gave details of the events in Ireland which formed the background to Boyle's sudden impoverishment in France in 1642; he illustrated Boyle's assiduity in reading and in spiritual exercises during his travels, no doubt on the basis of Shannon's recollections; while, interestingly, he attributed a role to Boyle's mother in his father's solicitude for him. Evidently on his own account, he furnished details of Boyle's charitable activities at Stalbridge and his provision of medication. It was probably also from his own knowledge that he vouched for Boyle's intimacy with Barlow – suggesting to Wotton that Barlow's papers might usefully illustrate Boyle's casuistical concerns – and spoke about Boyle's religiosity more generally, in a passage familiar through its paraphrase by Birch.[168] His evidence of the link between Boyle and Tenison bears out Evelyn's comment to Wotton in 1696 which has already been quoted, while he nicely caught the slight tension between Boyle and the Earl of Burlington by speaking of 'a sort of decent regard' between the two.

Dent presumed that, by the time he wrote his letter, Wotton's life was well advanced, but shortly after this there was an intermission in work on the project. In part this may have been due to the appearance in 1699–1700 of a four-volume epitome of Boyle's scientific works by the miscellaneous writer, Richard Boulton, which clearly made Wotton reconsider the rationale of his own enterprise.[169] In addition, just at this time Burnet gave Wotton a further task, of writing a book on Roman history for the use of the then heir to the throne, the Duke of Gloucester. The boy died before the work was finished, but Wotton took a good deal of trouble over it, and it was published in 1701 to considerable acclaim.[170] Those who praised the work included Evelyn, who used the opportunity to ask how Wotton was progressing with his life of Boyle, 'a subject as worthy ‹in its kind› the pen of the learned Mr Wotton, as that of the several ‹Illustrious for› Virtues among the Roman Emperors'.[171] In his reply to Evelyn of 22 January 1702, Wotton expressed the intention of working on the *Life* again, explaining how 'I have now all the materials I am to expect, and intend with all convenient speed to digest them into such an order as may make them at hand when I shall use them'.[172] In fact, however, further information materialised at this stage, in the form of the letter to him from the Scottish divine, James Kirkwood, and the second letter from Evelyn published here (Documents 6B and 8).

Kirkwood's letter, dated 22 June 1702, mainly comprised a succinct narrative of Boyle's involvement in the project for supplying bibles to the Scottish Highlanders in the years around 1690, with which Kirkwood had been closely associated; it was accompanied by transcripts of relevant texts and letters.[173] Kirkwood, whose links with Boyle had evidently begun in 1687, accompanied the detail he provided by expressing the hope that Wotton would do justice to Boyle's 'Pattern, & Example'; he also considered that the 'seasonable Expressions' concerning education that Boyle had voiced in a letter to Kirkwood of 18 October 1690 in connection with the Bible project 'may be very well Improv'd in some part, or Other of his Life'. Boyle's comments had indeed been quite revealing: he had expressed his agreement with Kirkwood that 'the education of youth' was 'a thing so important, that, till it please God to awaken men to a greater sense than they yet have of the necessity and usefulness of that, I shall scarce expect any such reformation as I wish, either of men's principles or their manners'. It is interesting that Boyle here echoes the ethos of the societies for moral reform set up in London and elsewhere at this time.[174]

As for Evelyn's further information, in part this comprised the second letter that Wotton kept among his papers, dated 12 September 1703, which is included in this volume. In addition, however, Evelyn had written a further letter to Wotton, dated 27 January 1702, evidently in response to Wotton's letter of five days earlier. This, too, contains some passages appraising Boyle, but Wotton failed to keep it, probably because such evaluation as it contained overlapped with that given in the letters that he *did* keep, while much of it was devoted to discussing and commenting on the progress of Wotton's enterprise in a manner similar to the rest of the correspondence between the two from which extracts have been quoted here. For this reason, it has not seemed appropriate to include a complete transcript of this letter, now surviving only in the rough draft of it that Evelyn kept, as one of the documents in this volume. However, the bulk of its text will be quoted piecemeal in this Introduction, mostly in connection with Wotton's objectives in his *Life*. Here, it is worth including the passage in which Evelyn expatiated on Boyle's reputation and his visitors, in terms echoing those of his other letters:[175]

There was no man whose Conversation was more Universally sought after ‹Courted› & Cultivated, by persons of the highest rank & quality:[176] Princes, Ambassadors, Forrainers, Scholars, Travellers & Virtuosi than Mr B. so as one who had not seene Mr Boyle, was look'd-on as missing one of the most valuable Objects of our Nation: It was in his Philosophical Apartment, Tapissred & furnishd with Instruments for Trials & natural Experiments,[177] perfectly becoming his Genius & Recherches & Learned diversions; that he often Entertain'd those who came to visite him, ‹Ever› with something rare or new: sedate & deliberate in discourse & so averse to dogmatizing,[178] the beeing ‹positive› sowre or censorious [?] that as I never heard him nam'd without Veneration & honour, so nor did I ever know him detract from any, but with infinite sweetnesse, and good breeding treate the person who ever differed from him in Opinion.

The letter from Evelyn that Wotton *did* keep (Document 6B) responded to a letter that Wotton wrote to him on 13 August 1703 in which, explaining how 'my design has long been resumed, and every day I do something to it', he asked for various specific pieces of information. These requests need to be quoted in full to make sense of Evelyn's reply:

1. An account of Mr. Hartlib: what countryman? what his employment? in short, a short eulogy of him, and his writings and designs, with an account of the time of his death.

2. The like of the beginnings of Sir William Petty. Those two were very great with Mr. Boyle before the Restoration.

3. Do you know anything of one Clodius a chemist? Was he (or who was) Mr. Boyle's first master in that art?

4. What was the affinity between your lady's family and Mr. Boyle? What son of that family was it that lies buried in Deptford church? and particularly all you can gather of the old Earl of Cork's original. Was Sir Geoffry Fenton Secretary of State in Ireland; if not, what was his employment? Did not he translate Guicciardini into English?

5. In what year began your acquaintance with Mr. Boyle? I find letters of yours to him in 1657.[179] Have you any letters of his; and would you spare me the use of them? they should be returned to you with thousands of thanks.

In a postcript he added:

Pray was Sir Maurice Fenton (whose widow Sir W. Petty married) a descendant of Sir Geoffrey's? or what else do you know of him?

In one of your letters to Mr. Boyle you mention a Chymico-Mathematico-Mechanical School designed by Dr. Wilkins: what farther do you know about it?[180]

Evelyn's response, printed below, was less of an overall assessment of Boyle than an answer to these questions; this has to be taken into account in interpreting it, though its slight idiosyncracy also the reflects the fact that (as he himself states) he was by now 'Octogenarius'.[181] In response to Wotton's fourth question, Evelyn enclosed a letter from George Stanhope supplying details of the inscription to the Earl of Cork's eldest son, Roger, in Deptford church; this has been included below since Wotton kept it together with Evelyn's letters.[182] In his own letter, Evelyn became more of a raconteur than earlier, retailing anecdotes about Sir William Petty and the Earl of Cork (though admitting that his information about the Earl's humble origins was perhaps uncertain and enjoining Wotton to keep it 'a seacret'), as also about Lord Burlington's homage to his brother's burial place.[183] On the other hand, he evidently thought better of including his eccentric claims about the antiquity of the Evelyn family pedigree, since these appear in the copy of the letter that he himself kept but not the one he sent to Wotton. The letter continues and extends the tendency of his earlier one to distance Boyle (and himself) from links with the Interregnum regime, emphasising Boyle's links with the Oxford group of natural philosophers (and the interests that he shared with Evelyn), and arguably being particularly unfair to Clodius, with whom Boyle had enjoyed fruitful relations in the 1650s.

In the letter that stimulated this one, Wotton had expressed his hope to show Evelyn what he had accomplished the next spring, i.e. in 1704, but, if he did, no record of this survives. After 1703 no further letters between the two men survive. Thereafter, there are just two references to the intended *Life*. One occurs among the memoranda of the Oxford antiquary, Thomas Hearne, who wrote under the date 2 May 1706: 'Mr Wotton, who writ the Reflections upon Ancient & Modern Learning, has writ the Life of the famous Mr Boyle; which being offer'd to the Royal Society, 'tis order'd to be printed by them'. Puzzlingly, no mention of this matter occurs in the Society's records at this time, despite the fact that a further reference to the Society's approval of Wotton's enterprise occurs in a letter from Wotton to Sir Hans Sloane, then Secretary of the Society, dated Easter-Eve (i.e., 23 April) 1709, in which he apologised for 'The great slowness which I have shewn in publishing what I had to say concerning the Life & writings of Mr Boyle, & that too after the R[oyal] Society had done me the Honor to express their Approbation of my design'.[184]

Wotton's letter to Sloane illustrates a further burst of activity on the book at the time he wrote it. Noting that his procrastination was 'what I hardly know how to offer anything in excuse for', he continued: 'The best excuse I can make is, that I now am wholly employd in it: that in a few months I hope to bring up with me a very considerable part of it'. He went on to explain his intended approach to the work (see below, section 6), adding a postscript in which he alluded to the difficulties that he encountered in trying to do work of this kind at his Buckinghamshire living of Milton Keynes:

> I find one of the greatest difficulties to me that live at a distance in the Countrey, is to procure those materials which I think are absolutely necessary for my design. Considering my circumstances I have spared no expence. It is in your power to do me the greatest Kindness, if you dare trust me for a few months (longer I desire not), with some few printed Books that I can't buy, nor know else where to borrow. They shall be faithfully returned & safely. What I have now in view that I most want, are Guerick's Experimenta Madgeburgica; & Galileo's Italian Tracts in 4to about Pendulums, & the weight of the Air &c.: Mr Goodwin my Bookseller will safely convey them. It will be an infinite Kindness. I have already one or two Books which shall be returned of yours.[185]

It is interesting that the works specified are ones by Boyle's predecessors in the study of pneumatics who are referred to in the extant section of Wotton's work, published here as Document 9. This indicates that this text dates from this phase of Wotton's work, as is confirmed by the fact that in it Wotton made repeated reference to Francis Hauksbee the elder's *Physico-Mechanical Experiments on Various Subjects*, published in 1709. This described experiments displayed by Hauksbee before the Royal Society, to which he was operator from 1704 to 1713, most of them using an improved air-pump based on Boyle's (for whom he may conceivably earlier have worked).[186] Indeed, it is quite likely that it was the appearance of Hauksbee's book which stimulated Wotton to write on this particular topic at this point.

In fact, Wotton's *Life* was never completed. Even in his letter to Evelyn of 13 August 1703, he sounded rather depressed about the whole project, writing:

> It is now so long since I first mentioned to you my design of giving some account to the world of the life and writings of Mr. Boyle, that I question not but you have long since looked upon it as a vain brag of an impertinent fellow, who, when he had once appeared in public, thought he might be always trespassing upon their patience. The discouragements I met with since I undertook it were so many, that I have often wished that I had let it alone, or never thought of it. And I was ordered to pursue another scent by the Bishop of Salisbury, which it pleased God to make unsuccessful.[187]

Though this particular 'scent' came to nothing, in 1704 he published his *Letter to Eusebia*, an attack on John Toland, which doubtless also distracted him from the Boyle life, and this was followed by other controversial writings, including a defence of his *Reflections*.[188]

His final abandonment of the project probably occurred a few years after the burst of activity in 1709 which has just been dealt with. In 1714 he committed 'such indecent Actions' (in Thomas Hearne's words) that he was forced to leave his living at Milton Keynes and flee to Wales, where he lived under a pseudonym.[189] Evidently linked to this hasty departure (and presumably following the 1709 bout of work, though it is unclear why the section that he compiled then survives) an unfortunate fate befell his papers. This was outlined by his son-in-law William Clarke in a letter to Thomas Birch of 7 September 1741: 'He had made some progress in that Work, but when he was forced to leave his House, these papers fell into such Hands, as never returned them'.[190] Clarke had given an overlapping account of the matter in a life of Wotton which he had written at Birch's behest two years earlier, in which he stated that most of Wotton's papers about Boyle 'were unhappily either lost or destroyd, & he was so much affected by this Misfortune, to have spent so much Time to no Purpose, that he had not Resolution enough to think of turning ‹all› the same Books & papers over a second Time, & beginning again'.[191] Hence, though Wotton did not die till 1727, thus inhibiting attempts to write a definitive life during that time, he never returned to the project during the last eighteen years of his life.

6. WOTTON'S *LIFE*, II
FROM PANEGYRIC TO INTELLECTUAL BIOGRAPHY

What do we know about Wotton's book and its intended content? Clues are available from the *London Gazette* advertisement, already quoted; from the extant fragment reproduced below; and from various hints in letters and other sources. We may begin with certain evidence that survives from the eighteenth century. In the life of Wotton that he wrote at the behest of Birch, William Clarke stated that Wotton 'had made a Considerable ‹Progress› in his Collections for that Purpose, as appears from some parts of his Adversaria, which I have seen'.[192] Though this could allude to no more than the material which still survives, it is clear that both Miles and Birch saw a biographical manuscript apart from that printed below which has since disappeared. This is described both by Clarke and Miles as Wotton's 'Sketch',[193] and some notes by Miles on its content survive. From these and from Birch's reference to it at one point in his *Life*, it is apparent that, in contrast to the thematic approach of the extant fragment, this was a narrative of Boyle's life. It was evidently divided up into chapters, within which were paragraphs or sections – Birch thus cites it in the form 'c. iii. § i'[194] – while Miles's notes suggest that the first fourteen pages, on which he recorded various queries, took the story of Boyle's life up to the 1650s, dealing with his birth, parentage, travels and place of residence.[195]

Possibly this formed an introductory section to the *Life*, giving a chronological outline preparatory to more thematic chapters and/or the publication of manuscript material. One presumes that it was here that Wotton made the comment about Boyle's belief in day fatality referred to above in connection with the Burnet Memorandum, since this would have arisen naturally in connection with an account of Boyle's early life.[196] On the other hand, Miles's assessment of the document was very critical, which suggests that it might rather have been a separate, briefer survey of Boyle's life, possibly even a preliminary draft. Miles wrote:

> Dr Wotton's Sketch (as tis calld) is very inaccurate & has Severall Errors in it; his mind, I believe, must have been much distracted, either by his ill health or necessary Avocations or both; for he mentions things over again within a little compass & yet there are a great number of scratches & blotting out ‹or amendments› in almost every line.[197]

A perhaps more reliable source of information about the rationale and intended content of Wotton's *Life* is Evelyn's correspondence with Wotton, together with the *London Gazette* advertisement for the book and the letter from Evelyn to John Williams quoted in the previous section. Evelyn clearly took it for granted that the book would at the same time be a record and a kind of didactic celebration of Boyle, the intention being to 'consecrate the Memory' of the great man by 'publishing to the World so Illustrious an Example'.[198] Moreover, this almost certainly reflects Wotton's own thinking, for he asserted the exemplary value of biography in the *History of Rome* which interrupted the project around 1700, seeing lives such as those of different emperors that he there provided as 'the properest Instances to set Virtue and Vice, and the Consequences of them both in a clear and full Light'.[199] Such an aim seemed especially appropriate in the 1690s and 1700s, widely perceived by contemporaries as a degenerate era in which virtues like Boyle's were 'so rarely found among persons of his Rank', in Evelyn's words. Thus in 1702 Evelyn urged Wotton on with his biography of Boyle 'to give the degenerate Age this bright Example', while Wotton was similarly concerned about the prevalence of 'Lewdness and Irreligion'.[200] Parenthetically, it is worth noting that John Williams had specifically presented Boyle's *Free Discourse against Customary Swearing* as a tract for the times in the 1690s, although it had been written half a century earlier.[201]

Traditionally, the staple of such didactic biographical writing had been the lives of men of affairs. In Boyle's case, however, what was required was the biography of an intellectual, and it is worth stressing that intellectual biography was a genre which at this point was in its infancy.[202] One exemplar – to which Pett had referred in his notes for Burnet, as we have seen, and which Burnet had himself cited in one of his earlier biographies – was Gassendi's *Life of Peiresc*. Evelyn, too, made reference to this, both in his letter to Wotton of 29 March 1696 and in the letter of 27 January 1702 in which he urged Wotton on with his task.[203] Gassendi had specifically felt the need to justify writing the biography of a virtuoso rather than a man of action, and Evelyn took a similar line in relation to Boyle in his letter to Wotton, echoing the view of Boyle as an exemplar of the public usefulness of intellectual activity which he had earlier expressed in his *Publick Employment and an Active Life, preferr'd to Solitude* (1667):[204]

> Since tho' perhaps his sedate temper &[205] weake habite of body often confind him to Retirement might not ‹fully› answer,[206] the expectations of those who looke for extraordinary[207] passages & exploits ‹in the lives› of Heros, greate Ministers of State, & Politicians: which this worthy[208] Gent: (neglecting the trouble & noise of pomp) affected not, Yet so far was he from dyable [?] Sloth; that tho of so delicate a Contexture, we find him continualy employd, in the best of[209] Actions,[210] Labour of Love,[211] Expansive Charity, Benificent, & of signal piety, shining as in his life so in all his Writings, which renderd his Conversation, Innocent, Easy[212] Serious, without affectation or

moroseness, Infinitly Obliging, Generous & Bountiful without Ostentation & shew, yet discreetly frugal, that he might be so ‹well› distinguishing of merit, & real Worth.

'In a word', Evelyn went on, 'his whole Course was a Conjugation & series of the most Eminent Vertues & Intrinsiq Worth, with all the Accomplishments of a Consumate Christian & Philosopher', and he ended by acclaiming 'our Second *Verulam*, the Illustrious *Boyle*'.

Just what form, however, was it appropriate for the biography of an intellectual to take? In the undeveloped state of the genre, this was far from clear. Gassendi had devoted most of his account of Peiresc to a chronological narrative of his hero's life, divided into five books, followed by a sixth book of a more descriptive and analytic nature, which gave a perceptive account of his subject's personal and intellectual characteristics.[213] The same form was taken by Baillet's *Life of Descartes*, Evelyn's and Wotton's familiarity with which has already been noted: indeed, Evelyn actually compared facets of Boyle and Descartes on the basis of Baillet's life.[214] Clearly, Wotton and Evelyn took it for granted that, like such exemplars, Wotton's life would have in it a good deal of *res gestae*. Evelyn presumed that Wotton would include details of Boyle's 'Birth, Family, Education & other lesse necessary Circumstances for Introduction', together with less familiar 'passages of his Life' thereafter; Kirkwood's data was similarly intended to enable a narrative of his charitable activities to be included, with a gloss on his virtues in this respect.

But, beyond this, how were an intellectual's achievements best summarised? Gassendi had included an appendix devoted to 'The Prayse of Peireskius' by learned contemporaries,[215] and in the case of Wotton's life a possible analogy to this was an intended component advertised in the *London Gazette* – a collection of letters from 'the most Learned Men in Europe, his Correspondents'. Though this might be seen simply as reflecting the impulse to full documentation in biographical writing at the time, in this case it was probably also intended to show the high esteem in which Boyle had been held by his peers.[216] No doubt in connection with this, Wotton made a close study of Boyle's letters, the surviving texts of many of which are endorsed in his characteristic hand; also still extant is his careful catalogue of nearly 400 letters, arranged mainly in chronological order.[217] In addition, the advertisement indicated Wotton's intention to include an 'Extract' of Boyle's writings, 'Collected as well from what are in Print, as from his Manuscripts and Papers of Experiments never Published', and this ambition and its significance require comment.

The idea that use should be made of Boyle's unpublished papers had been in Wotton's mind from the start of the project, as we saw in the last section. Indeed, when his book seemed to be threatened by Boulton's,

Wotton consoled himself by noting: 'I don't apprehend that the Epitom-izer of Mr Boyle's ‹Works› will do me any harm: our designs are different & I have MSS: which he has not'.[218] Exactly how Wotton intended to deploy the Boyle Papers in his book is uncertain. As with the letters, he had undoubtedly studied these quite assiduously: the extant endorsements in his hand throughout the archive testify to the care with which he examined and tabulated Boyle's papers, and to the accurate and well-informed approach he brought to bear on them.[219] In addition, William Clarke's list of the manuscripts which Wotton retained for use in his *Life* shows that these covered a wide range of topics, though it provides no hints as to which of them he considered appropriate for publication.[220] Its only possible clue from this point of view is its bias towards material relating to Boyle's *The Christian Virtuoso*, something which Henry Miles confirmed when he gained access to Wotton's manuscripts in 1742 and discovered that 'the greatest quantity' of them related to that work.[221] Possibly, this suggests that it was Wotton who initially had the idea which Miles was to bring to fruition, of publishing the parts of that work which Boyle had left in manuscript.

The significance of Wotton's interest in unpublished material by Boyle lies in the fact that anthologising may at this juncture have seemed the most appropriate way of celebrating an intellectual. In seeking the roots of intel-lectual biography, Paul Korshin has pointed to the role of so-called '-ana' books, collections of extracts from the writings and conversation of great scholars like Isaac Casaubon or J. J. Scaliger, as showing the first stirrings of an impulse to study a man's ideas rather than simply to tabulate his life.[222] Though it is unfortunate that this side of the scheme is so poorly evidenced, Wotton's intention to publish material from the Boyle Papers is perhaps to be seen in a similar light: it is interesting that John Williams had shown similar aspirations in the abortive project discussed in his letter to Evelyn in 1696.

In addition, however, Wotton seems to have thought in terms of giving a kind of summary or exposition of Boyle's ideas, perhaps including a systematic view of his philosophy, something which had already seemed desirable during Boyle's lifetime.[223] In this, too, Wotton was encouraged from the outset by Evelyn, who referred in his letter of 29 March 1696 to his expectation that Wotton would provide 'a fresh Survey of the progresse he has made in these Discoveries'.[224] As the author of the *Reflections on Ancient and Modern Learning*, Wotton was well-equipped for this: he has been seen as a pioneer in his 'use of a continuous historical method applied to science' in that work, which had included a panegyric of Boyle's 'genuine Histories of the *Phænomena* of Natural Bodies'.[225] If this was Wotton's intention from the outset, the appearance of Boulton's epitome – which itself reflected the systematising impulse – evidently made him

1

reconsider it. On 22 January 1702 he wrote to Evelyn: 'His works having been epitomated by Mr Bolton after a sort, I am at a loss whether I shall interweave a kind of a system of his philosophy into the Life as I at first designed, or only relate matters of fact', by which he evidently meant biographical data to the exclusion of intellectual considerations.[226] He requested Evelyn's advice on this point, who responded: 'I Confess ‹a› Systeme of Mr Boyls Philosophy (tho' it had not ben Imperfectly set forth) would in my poore opinion, be more agreable, & hence another aime', going on to allude to Gassendi's *Peiresc* and his later *Lives of the Astronomers*, 'filld as they are with extraordinary Invention & Improvements'. In the light of such precedents, he considered it highly appropriate for Wotton to give an exposition of 'the Investigation of Profitable & rare Experiments, in which the penetrating & sagacious Mr Boyle, so hapily succeeded, by a felicity in a manner peculiar to him, & this adorn'd with Exemplar[y] Vertues'.[227]

That Wotton *did* persevere with this side of his enterprise is indicated by the people about whom he asked Evelyn in 1703, since his queries clearly indicate curiosity about the formative intellectual influences on Boyle. His ambition to write a kind of intellectual biography is made especially clear by what he said in his letter to Sir Hans Sloane of 23 April 1709. In it, he explained how 'I do not barely design to give an Account of what Mr Boyle had done already, (for that is what every body already know's,) but to compare his Inventions & Discovery's with the Discovery's & Inventions that have bin made since. This is a wide field, & I am extreamly fearfull that I shall miss my way in it. I will do my endeavor however to be as exact as I can, & to let my Friends who have done me the Honor to judg favorably of me, see, that I am as careful to preserve their Esteem; as I can be of any thing which the most nearly concerns myself'.[228] It is precisely this ambition to contextualise Boyle's work and outline its later corollaries which the extant chapter, Document 9, exemplifies. Wotton's aim, as he explained in it, was 'to shew how far Mr Boile opend the way, & what has bin since raised upon his Foundation'.[229]

How did he go about this? The chapter is focussed on Boyle's *New Experiments Physico-Mechanical, Touching the Spring of the Air and its Effects* (1660), interweaving into an account of this details of works by Boyle on related topics; in addition, Wotton discussed comparable experiments by such contemporaries of Boyle's as Blaise Pascal and the Accademia del Cimento in Florence, and, more recently, by Francis Hauksbee, the publication of whose *Physico-Mechanical Experiments on Various Subjects* in 1709 may partly account for Wotton's interest in this topic at this point, as we saw in section 5. The chapter opens with a general account of the significance of the air, building on but extrapolating from the comparable account in Boyle's own preface (insofar as Wotton extrapolated from it, it may be possible to detect the influence of the relevant passage in the Burnet Memorandum, which he had in his possession).[230] Wotton then gave an account of the intellectual problem that Boyle was attempting to solve, as exemplified by the work of Galileo, Torricelli and von Guericke (whose books he had sought to borrow from Sloane). This is a model of historical lucidity, in contrast to Boyle's rather cursory description of his predecessors, though obviously indebted to it.[231] Indeed, this passage could be seen as expanding the much briefer account of modern achievements in pneumatics that Wotton had included in his *Reflections upon Ancient and Modern Learning*, and we are again reminded that this is author of the *Reflections* by his reference to the ignorance of the ancients on such matters.[232] Then, after a description of Boyle's apparatus drawn directly from the *New Experiments*, we launch into a recapitulation of Boyle's experiments (more concise than Boyle's own), interspersed by more general 'historical' passages and by the assessment of related experiments of Pascal, the Cimento and Hauksbee.[233] Wotton was still in the full flood of his exposition at the point where the surviving text is truncated.

Wotton's account is undoubtedly quite accomplished. It represents a highly creditable attempt to get to grips with the technicalities of Boyle's work and of current understanding of it, with echoes of the exposition of Newton's arguments in *Principia* which his friend, Richard Bentley, had included in his well-known Boyle Lectures.[234] On the other hand, Wotton's chapter has definite drawbacks. In contrast to a popularising treatment like Bentley's, his account – after his initial introduction – was almost too detailed. Indeed, it raises the question of just how viable such a treatment would have been if applied to Boyle's writings as a whole, since Wotton managed to take up 40,000 words discussing a very small part of Boyle's output: he would have produced a massive book if he had devoted comparable attention to all Boyle's works, quite apart from the annexes of texts to which he apparently aspired.

Looked at from the viewpoint of the late twentieth century, a further problem arose from the very enthusiasm for the progress of knowledge in scientific and learned matters which had underwritten Wotton's earlier *Reflections*, and which reappears here: he more than once states his presumption that his contemporaries' achievement would later be superceded, as where he noted how 'the clearing of that Question is probably left to Posterity'.[235] For as a result, he laid stress principally on an exposition of what he saw as stages in the advancement of solid knowledge, dismissing what he perceived as obselete and irrelevant. This is illustrated by his treatment of Boyle's controversy with Linus and Hobbes which (as recent scholarship has underlined) played a significant role in the formulation of Boyle's ideas.[236] In his *Reflections*, Wotton had referred to this, writing of Boyle's findings:

> How far they may be relied upon, appears from this; That though *Hobbes*
> and *Linus* have taken a great deal of Pains to destroy Mr *Boyle*'s Theory, yet
> they have had few or no Abettors: Whereas the Doctrine *of the Weight and*
> *Spring of the Air*, first made thoroughly intelligible by Mr *Boyle*, has
> universally gained Assent from Philosophers of all Nations who have, for
> these last XXX Years, busied themselves about Natural Enquiries.[237]

By now, however, Wotton had evidently decided that Boyle's triumph was
so complete that these antagonists could simply be ignored. Nor was it
only the Hobbes controversy that Wotton omitted, but much of Boyle's
contemporary context altogether. Indeed, it is probably significant that the
one set of parallel experiments that he dealt with at some length, though
echoing Boyle's criticism of their limitations, were those of the Accademia
del Cimento, a further acknowledged milestone in the early history of
experimental science (unless it was simply because he could get hold of the
book in which they were published).[238] Otherwise, Boyle's references to
contemporaries were mainly suppressed, and even his own deductions and
speculations were curtailed, evidently on the grounds that the controversies to which they related had been closed by Boyle's and his successors'
experimental findings.[239] Thus in relation to the crucial Experiments 32
and 33, Wotton ignored several pages of 'reflexions' by Boyle, evidently in
view of the fact that the more recent research of Hauksbee 'quite puts an
End to all the Hypotheses of those who would solve these Phænomena by
other Principles than the Pressure & Gravitation of the ‹Air›'.[240]

As this indicates, Wotton clearly believed that Boyle was best championed by expounding his experimental contribution to a growth of knowledge that was still in progress. Thus he was particularly interested in the
way in which the principles discovered through the air-pump had since
been more widely deployed in the form of the barometer.[241] Above all, he
compared Boyle's findings with those of Hauksbee, the most recent in the
field, his account of which conveys a sense of genuine excitement about the
latest scientific discoveries. Not surprisingly, he repeatedly sang Hauksbee's praises, often seeing his work as superseding Boyle's. Moreover, in
one passage he cited both Hauksbee and the Newtonian, John Keill,
concerning Boyle's experiments on the cohesion of marble plates, which
Boyle had attributed wholly to the effects of a vacuum; Wotton, on the
other hand, was modishly able to invoke the force of attraction as expounded by Newton in this connection, Boyle being seen as 'very probably
mistaken' in this regard.[242] Yet the effect was one which is noticeable in
eighteenth-century accounts of Boyle more generally, namely to diminish
him by comparison with his successors, instead promoting him as a
worthy figure whose work had been superseded at a detailed level.[243]

Wotton was evidently at least partly aware of this difficulty. When he
first introduced Hauksbee's work, he added a rider to the effect that Boyle

deserved some of the 'Glory' for 'the progresses which other Men have made' using techniques that he had pioneered.[244] Indeed it is even possible that his awareness that Boyle was losing out from the mode of exposition that he had adopted was one of the factors which led him to abandon the work. Another may have been the vast amount of space that he had devoted to this particular topic, since his coverage was not complete even then, the surviving text failing to cover certain matters which he promised to deal with, such as a comparison of Pascal's doctrine with Boyle's.[245]

Whatever its flaws, however, the importance of this text is undoubted, as arguably the first attempt at intellectual biography in a modern sense in Britain. Indeed, it is perhaps unfair to criticise Wotton for failing to see that greater justice might have been done to Boyle's achievement by a painstaking historical reconstruction of the controversies in which he was engaged, rather than by simply tabulating his contribution to the progress of knowledge, when this was something which has only been achieved in our own day. It is also salutary to recall that we have to wait till the nineteenth or even the twentieth century for a remotely comparable study of Boyle. Quite apart from its significance for the study of Boyle, this hitherto unknown text adds a new dimension to our understanding of Wotton and the intellectual history of his day.

7. FROM WOTTON TO BIRCH

The first published life of Boyle appeared in 1715, but it owed nothing to Wotton. It was by the writer Richard Boulton, the appearance of whose four-volume epitome of Boyle's scientific writings in 1699–1700 has already been referred to in connection with Wotton's reaction to it. Boulton was an Oxford M.A. who had embarked on a career as a medical author with two learned works published in 1697 and 1698, but who in 1698–9 became involved as a pamphleteer in the virulent medical disputes then raging. Indeed, his Boyle epitome served a useful function for him in that it was said at the time that it was the advance on this which provided the cash needed to print a vindication of himself against his enemies – his *Letter to Dr Charles Goodall*, his erstwhile patron, who he discovered had been knifing him in the back.[246] His edition of Boyle bears witness to a recognition among booksellers of an appetite for Boyle's writings in an accessible form, and it was followed in 1715 by a sequel in the form of a three-volume epitome of Boyle's theological writings, most of the first volume of which comprised a 372–page life of Boyle. In it, Boulton apologised for the fact that it was not 'so compleat as we could wish' due to the loss of appropriate materials, but he rather pointedly justified his efforts 'since no Progress hath been made to draw up a Character that hath been so famous all over the World, and deserved so well of Mankind, and might be so Useful and Exemplary'.[247]

Evelyn had been concerned in the 1690s that a kind of Gresham's law might operate in relation to Boyle's life, through some 'Impertinent Scribler' imposing on the world.[248] Now it certainly had, since it has to be said that, biographically, Boulton's attempt is virtually worthless. Indeed, his ability to spin so many pages out of so little material is a tribute to his superlative qualities as a hack. The biographical information in the 'Life' is almost entirely taken from Burnet's funeral sermon, the message of which is amplified so that Boyle comes across as 'the Universal Example', a figure almost too good to be true: 'his whole Life was spent in doing Good to Himself and Mankind'.[249] Very occasionally, Boulton added pieces of information which he had gleaned from other printed sources, though these sometimes misled him, as in his erroneous statement that Boyle acquired a university education at Leiden, which was derived from the brief account of Boyle in Anthony Wood's *Fasti Oxonienses*, and which was to be repeated by various writers thereafter.[250] Otherwise, Boulton padded out his data from Burnet with verbiage or with information from or allusion to passages in Boyle's works, which were also recapitulated at length in their own right, including a 120–page summary of Boyle's articles in *Philosophical Transactions*.[251] Indeed, Boulton did not always even refer to Boyle's writings when he used them, and it is clear that, after doing so much epitomising, he could write out Boyle's ideas by almost instant recall. Even the specimens of Boyle's conversation were derived from his books, while Boulton also managed by such means to put together a section giving Boyle's thoughts on various moral and social topics.[252] There are just one or two exceptions to this general derivativeness: at one point Boulton wrote of Boyle having a chapter of Genesis read to him in Hebrew weekly, 'as I am informed by one, who he for some Years employed every Saturday Morning for that Purpose'; but later, even where he said 'I have been told', his account added nothing of substance to that of Burnet.[253]

Boulton was succeeded by a comparable writer, Eustace Budgell, who included an account of Boyle in his *Memoirs of the Lives and Characters of the Illustrious Family of the Boyles, Particularly of the Late Eminently Learned Charles Earl of Orrery*, first published in 1732, which reached a third edition in 1737. The third edition was provided with an appendix comprising an epitome of Burnet's funeral sermon, together with the codicil to Boyle's will establishing the Boyle Lectures, an interesting illustration of the sense that fuller information about him was needed. In addition, the text of the work from the first edition onwards included a thirty-one page generalised account of Boyle's life.[254] This owed much to Burnet and more still to Boulton. On the other hand, Budgell's account is at least mercifully succinct by comparison with Boulton's wordy and sanctimonious tome, a contrast well-illustrated by the way in which Boyle's 'thoughts' were adapted into pithy *obiter dicta* on love, marriage and government.[255] In so far as Budgell made statements not paralleled in previous accounts – for instance, that Boyle more than once refused the offer of a peerage – it is impossible to tell whether this was based on genuine information, which he could have derived orally from members of the Boyle family, or whether he was simply extrapolating from comparable comments by Burnet; the same is true of Budgell's slightly folklorish account of Boyle's sense of obligation to conceal certain findings, such as poisons, that he made in the course of his experiments.[256]

Indeed, in many ways this sketch of Boyle is the weakest section of Budgell's book: of its other components, his account of the Earl of Cork was the first to use his *True Remembrances*; he also gave quite a creditable narrative of Broghill's life, drawing on Thomas Morrice's (then unpublished) memoirs of him; while the book reached its climax with its lengthy account of Charles Boyle, 4th Earl of Orrery. On the other hand, it is rather ironic that a memoir of the great scientist should have been juxtaposed with an account of the Phalaris controversy written from the point of view of the antagonist of Richard Bentley, which included a text of some of the (spurious) letters of Phalaris for readers' benefit. It was hence implicitly critical of Wotton – Bentley's closest ally in the affair – and by extension of Boyle and the other moderns whom Wotton had championed in the Battle of the Books, though the only hint that Budgell gave of this antagonism was where he noted how Boyle's *Occasional Reflections* had been the subject of 'the *Ridicule* of a *certain Writer*, who knows how to expose the least Indecorum in the strongest Colours', namely Swift.[257]

It was just at the time of the appearance of Budgell's book that Boyle first attracted the attention of Thomas Birch, whose name has ever since been indissolubly linked with Boyle's. Birch, who had been ordained in 1730–1, evidently discovered his penchant for antiquarian compilation through his involvement in producing the English edition of Pierre Bayle's *General Dictionary, Historical and Critical* published in ten volumes between 1734 and 1741. This extended Bayle's text with a large number of extra biographies, notably of British worthies whom it was thought that he had unjustly neglected; it was prefaced by a dedication to the Royal Society and its President, Sir Hans Sloane, which claimed that 'the bare Names of NEWTON and BOYLE raise the most exalted Ideas, and image to us something more than human'. The book has been acclaimed as 'the first attempt in England to apply the inductive method to biography', and certainly much use was made in the additional lives of original letters and the like, as well as extant secondary sources. Its format followed that of Bayle, presenting 'critical Remarks' on each of its subjects in lengthy footnotes.[258]

Out of a total of 889 new lives, Birch wrote 618: though a vast amount of data went into these, there was inevitably relatively little time for analysis or reflection, and the Boyle life is a case in point.[259] Much of it was a compilation based on the extant biographical sources, Burnet, Wood, Boulton and Budgell (including their errors, as concerning Boyle's Leiden education). In addition, Birch made use of an important publication dealing with Boyle which had come out in 1725, the epitome of Boyle's *Philosophical Works* by the writer, Peter Shaw. Though the biographical content of Shaw's work is negligible, its significance for the history of Boyle's eighteenth-century reputation is undoubted, since it disseminated Boyle's writings more widely, while Shaw's attempt to 'methodize' them may have helped to enhance their accessibility.[260] Its importance for Birch was particularly great, since it was to Shaw that he owed virtually all he knew about Boyle's scientific achievement. Indeed, he chiefly relied on Shaw in a lengthy footnote in which he attempted to furnish a survey of Boyle's 'Philosophical Discoveries', an attempt illustrating the same impulse to write an intellectual biography that has already been observed in Wotton.[261]

On the other hand, it has to be said that Birch's account is not very successful. The existence of Shaw's epitome at least enabled him to give a connected exposition of Boyle's writings in the order in which Shaw had abridged them, in contrast to his lives of such authors as Bacon and Newton, where he did little more than purvey lumps of undigested source material and quote the evaluations of earlier writers.[262] On the other hand, his account of Boyle's ideas was rather idiosyncratic and unbalanced, giving disproportionate attention to certain works, such as Boyle's brief tract *Of the Systematical or Cosmical Qualities of Things* (1671), while others were neglected. Moreover, though he also quoted Shaw's defence of Boyle from charges of credulity, he somewhat neutralised this by devoting two further footnotes to a lengthy account of Boyle's alchemical interests, which he evidently found intriguing; among other things, he here published for the first time Newton's 1676 letter to Oldenburg commenting on Boyle's experiments on the incalescence of mercury.[263]

The prolific career as an author and editor which Birch began with the *Dictionary* continued from the 1730s until his death in 1766. His industry was undoubted, even though the shapelessness of his books and the limitations of his style have often been criticised. In Dr Johnson's view, Birch was 'as brisk as a bee in conversation; but no sooner does he take a pen in his hand, than it becomes a torpedo to him, and benumbs all his faculties'.[264] A number of Birch's contributions to the *Dictionary* formed the basis of lives prefaced to editions of the works of the author in question – including Bacon, Chillingworth and Milton – and it was this treatment which he resolved in the late 1730s to apply to Boyle. The proposed Boyle edition was more ambitious than any that Birch had undertaken hitherto, and it benefitted from his collaboration with the nonconformist minister, Henry Miles, who had independently become interested in Boyle and had gained custody of a substantial part of his papers.[265] The two men met on 7 August 1738 under the aegis of the scholar, John Ward, and they formed an effective partnership, Birch's role being that of getting material into print and Miles's of doing research on Boyle's manuscripts.[266]

Initially, Miles's efforts were devoted to the papers that he had acquired from Thomas Smith, the beneficiary of the estate of Boyle's executor, John Warr. However, Birch and he became aware that other papers had been in the hands of Wotton, and, after various enquiries, some of them of the biographer, Samuel Knight, they located these in the hands of Wotton's son-in-law, William Clarke.[267] Negotiations ensued – complicated by the interest in Wotton's book claimed by its putative publisher – and finally, in 1742, rather late in the preparation of the edition, Birch and Miles managed to obtain the papers relating to Boyle which Wotton had had.[268] These included the principal biographical documents published here, which can be identified in the inventory of the material that Clarke made prior to handing it over.[269] After making some notes on them, Miles handed these over to Birch, whose familiarity with this material is revealed by a number of references to it in his published *Life*.[270]

The papers from Clarke also included 'between 3 & 400' letters, which complemented a collection of nearly 1200 that Miles had already found in the Warr/Smith portion of the archive.[271] Miles passed a large number of such letters on to Birch, who quoted many of them verbatim in his *Life*; a further selection of what Miles called Boyle's 'Literary Correspondence' was included in the last volume of the edition of Boyle's *Works*, thus going some way towards realising Wotton's goal of publishing Boyle's correspondence, as expounded in the 1699 advertisement for his *Life*.[272] In addition, in 1741 Miles had provided Birch with a transcript of Boyle's *Account of Philaretus*, which he must have found in the Warr/Smith component of the papers. This item had evidently eluded Wotton in his work on the archive, perhaps because the first leaf of it was displaced: Miles's transcript originally began at the third page of the text, but he subsequently came across the opening leaf and was able to add a copy of it to the beginning of his text before supplying it to Birch.[273]

Indeed, it is to Miles that most that is of value in Birch's *Life* is due. Various papers by him survive tabulating and inventorising Boyle's extant remains, while his correspondence with Birch throws further light on his activities.[274] He explained what he saw as his role in a letter to Birch of 16 August 1742:

> I am taking all the pains I can to furnish you with Stones and Timber, you must be the Architect, & I shall think my Self very happy if I can ease you of the trouble of collecting materials, in any measure, knowing you will have work eno' on your hands to form them meet for the Edifice, and to put them together.[275]

Evidently he hoped that, using these materials and other relevant data, Birch would accompany his edition by a full and authoritative biography, so that both edition and life would supercede all those attempted hitherto: he was thus critical of the 'Carelessness' of such predecessors as Boulton.[276] His aspirations are further revealed by a letter to Birch of 11 November 1742 which evidently enclosed the chapter from Wotton's *Life* published here, which he must have come across among the papers received from Clarke. Noting how Wotton's 'design seems to have been to give the World an account of the improvements made by Mr Boyle in Experimental Philosophy, & of the Improvements made by others on his foundation', he added: 'which example I presume you will follow, tho' perhaps with less prolixity'. In addition, observing how 'the Dr has been careful (as far as I have read) to make very honourable mention of such Gentlemen as have made advances in several kinds of experiments, since Mr B[oyle]'s death, which he cou'd not do for want of instruments', he presumed that Birch would further update this where appropriate – though his further comments on fields where no such improvements had been made suggest that he might have been better equipped for this task than Birch himself.[277]

In fact, the *Life* which materialised is rather a disappointment. Birch began it on 9 June 1743 and finished it in 27 July the same year, making only minor revisions thereafter before consigning it to the press.[278] It is informative and reasonably authoritative on basic biographical matters, thanks not least to all the data which Miles had supplied. Thus Birch's use of *Philaretus* allowed him silently to suppress such errors in Boulton's and Budgell's lives as that concerning Boyle's Leiden education. Though he introduced one influential new misunderstanding himself by identifying the 'Invisible College' of which Boyle formed part in the mid 1640s with the scientific group which met in London at that time, this was a reasonable enough surmise in the circumstances.[279]

On the other hand, Boyle's *Life* is distinctly wooden and two-dimensional; indeed, he himself was slightly apologetic about it, noting his lack of the advantage of proximity to Boyle that Burnet and Wotton had had, though claiming that 'the genius and abilities of a *Gassendus* would be requisite to do justice to a character superior even to that of the celebrated *Peireskius*'.[280] From the point of view of interpreting Boyle, the *Life* was if anything retrogressive compared with Birch's earlier *Dictionary* article. Thus, whereas in that he had attempted to summarise Boyle's philosophy as a whole in the Bayle-like footnote already mentioned, here he made virtually no attempt at analysis. Instead, he simply gave a rather staid account of the content of each of Boyle's books at the point in his chronological narrative at which their publication occurred: the bulk of this material was derived from his earlier footnote, but not even all the information in that was included.[281]

Beyond this, Birch attempted to give a view of Boyle's 'character' in a 14-page peroration at the end of the work, but this, too, is a severe disappointment. His assessment of Boyle as a scientist was derived almost entirely from Shaw's 'General Preface' to his epitome of Boyle's *Philosophical Works*, preceded by brief adulatory passages from such recent natural philosophers as Boerhaave and Redi: Miles's hope that, like Wotton, Birch would collate Boyle's findings with those of more recent natural philosophers was entirely disappointed. Otherwise, Birch's principal source in this section was Burnet's funeral sermon. Though he quoted or paraphrased a few letters and included a handful of short passages from the biographical texts published here, his use of these was severely limited.[282]

In part, this was because Birch was rather unimaginative with this material. He had, it is true, drawn on it for information about a few episodes in Boyle's life which he had dealt with at the appropriate chronological point in his earlier narrative. On the whole, however, he saved it for his peroration, and, since the tone of this was almost as generalised as that of Burnet's funeral sermon, this had the effect of inhibiting his use of it for more than brief illustrative instances of general points. A more serious problem was the extent to which the data was filtered through the image of Boyle which Birch wished to present, which meant that passages which were at odds with this were either ignored or consciously suppressed. For instance, though Birch enthused in general terms about the debt of artisans and others to Boyle's findings, he ignored the evidence for Boyle's direct involvement in technical matters given by Pett, evidently because his preferred image of Boyle was of a lofty thinker, above 'the vulgar herds of chemists, naturalists, and philosophers'.[283] Boyle's correspondence on such matters was probably also treated selectively, as was certainly the case with letters concerning such topics as his relations with his family or his management of his estates. Birch and Miles evidently thought that readers would find material of this kind trivial, but their neglect of it had an unfortunate effect of encouraging an unduly otherwordly image of Boyle, made worse by the fact that many of the letters that they decided against publishing were subsequently lost.[284]

As for Boyle's religiosity, though this figured prominently, Birch and Miles deferred to what they saw as the 'taste' of the age in what they included. On 10 August 1743, when sending Birch his comments on the draft of the *Life*, Miles noted that he had in his possession the manuscripts of various cases of conscience proposed to Thomas Barlow, with Barlow's answers. He added that they were 'too long to be inserted in the life, and the resolutions being in the manner of the schoolmen, it was thought by a very judicious friend ‹to whom I shewd the MS› better to omit em as not suited to the genius of the present age', continuing: 'whether you will take notice of such things being now in MS or no is submitted to your Judgment'.[285] Birch did insert a brief reference in the *Life* to the fact that such manuscripts existed. But he left it at that, which is regrettable not only because Boyle's casuistry – a key facet of his religiosity – was dealt with so cursorily, but also because, presumably because their content seemed so unpalatable, the manuscripts in question have since disappeared.[286]

There is also the passage from the Burnet Memorandum concerning day fatality, Miles's comment on Wotton's treatment of which was quoted in section 3 above. Here, conscious censorship is undoubtedly in evidence, for Miles went on: 'Now I woud not wish to see this in his Life, it may have bad Consequences to some who have a Veneration for Mr B., & the omitting it, will do no injury to his Memory – I submit this to your Consideration & better Judgment'.[287] Indeed, this suggests that Birch's *Life* was regressive compared with Wotton's in its honesty about Boyle, since Wotton was at least intending to include this passage, even if animadverting on what he saw as its undesirable corollaries; Birch, on the other hand, followed Miles's recommendation and suppressed it.

He did the same with the prominent component of the Burnet Memorandum dealing with magic. It is clear from Birch's peroration that one of the issues that concerned him was Boyle's reputation for credulity: in addition to citing Shaw's refutation of this at length – as he had earlier in his article in the *General Dictionary* – he added further comments to the same effect of his own, going out of his way to differentiate Boyle from 'mystical and enigmatical writers' who were doomed to obscurity while Boyle's reputation lived on.[288] Hence the last thing he wanted was to emphasise this aspect of Boyle and, perhaps realising the inconsistency between such protestations and the prominence accorded to alchemy in his earlier account of Boyle, he now redistributed the material on this subject piecemeal through the *Life* so as to render it almost invisible. Not surprisingly, letters illustrating Boyle's interests in such matters were also jettisoned, many letters from alchemists and similar figures not only being witheld from the edition but disappearing altogether.[289]

Due to these combined factors, Birch's citations from the sources published here were on so limited a scale that they would have led no-one to suspect their true value. As a result, they have been lost from sight virtually ever since. Most biographical accounts in the aftermath of Birch's were content to follow his, including the most notable, that contributed to *Biographia Britannica* (1747–66) by the miscellaneous writer, John Campbell.[290] This is a learned and to some extent independent evaluation of Boyle, but it was largely a reworking of the materials published by Birch.[291] Thereafter, Birch's account invariably formed the basis of the lives of Boyle, as of other worthies, which proliferated in the burgeoning literature of biographical compendia in the late eighteenth century, and the same remained true in the nineteenth.[292] Even in the twentieth, surprisingly little use has been made of the material published here, and it has not always been properly understood even on the rare occasions when it has been cited.[293] For this, Birch's adequate but uninspired and partial *Life* must bear considerable responsibility.

8. THE BOYLE WE HAVE LOST

What is the implication of the information presented here for our understanding of Boyle, and how does it relate to prevalent views of him? Many
of our current ideas about Boyle derive from the work of Marie Boas Hall
(née Marie Boas), author of a whole series of studies of Boyle published
mainly in the 1950s and 1960s. Of particular significance was her *Robert
Boyle and Seventeenth-century Chemistry* (1958), an intellectual biography of the type to which Wotton evidently aspired in the abortive project
surveyed in section 6 above. This book sought to elucidate Boyle's aims
both in his chemical investigations and in his experimental work more
generally by setting them in their intellectual context; it remains an insightful study, although, after thirty-five years, it inevitably looks dated in the
light of new thought on scientific change in the period, and more detailed
archival investigation. In the same work, Boas Hall also gave an overall
sketch of Boyle's intellectual evolution, in which she made some piecemeal
use of the texts first published here, though she failed to deploy them in a
sustained way.[294]

On the other hand, partly because her preoccupation was with the
development of modern science, Boas Hall never gave as much attention as
she might have done to the integration of Boyle's scientific work into his
thought as a whole. Insofar as she did, her picture tended to mirror that
outlined by Birch two centuries earlier, and the same is true of the principal twentieth-century attempt at an overall interpretation of Boyle, the
biography that Louis Trenchard More brought out in 1944, a well-
intentioned and informative but essentially superficial work.[295] As for the
Life of the Honourable Robert Boyle published by R. E. W. Maddison in
1969, though of immense value for the precise detail that it purveys
concerning many episodes in Boyle's life, it unfortunately flags when it
comes to presenting a view of Boyle's intellectual personality.

More recently, two books have sought to reinterpret Boyle's science by
giving a broad reading of his ideas in their contemporary setting: J. R.
Jacob's *Robert Boyle and the English Revolution* (1977) and Steven
Shapin and Simon Schaffer's *Leviathan and the Air-Pump: Hobbes, Boyle
and the Experimental Life* (1985). Jacob's view is in a sense the converse
of that of Boas Hall, since, by his own admission, he largely takes Boyle's
science for granted.[296] His contribution, as he saw it, was to extrapolate
from it to elucidate the broader political strategy on Boyle's part into
which it fitted, aimed at overcoming subversion and aligning science with
the values of the burgeoning mercantilist state of the Restoration period.

In Jacob's reading, Boyle appears as a figure whose 'involvement in science, church, and state drew to him a large circle of friends, made him a patron of writers and thinkers whose views reflected his own, and caused him to confront and deal with some of the central issues of the day: the relations between church and state, the nature of government, the purpose of empire, and the relations between religion, science, and the state'.[297] This view thus extends to fresh spheres the image of the prudent and rational scientific strategist to be found in books like those of More and Boas Hall. On the other hand, since it is partly based on such of the information published in this book as was made available by Birch, it is important to assess it in the light of the fuller version presented here.[298] The result, as we shall see, is to suggest that Jacob's interpretation of Boyle is at best a highly partial one.

Shapin and Schaffer also see Boyle as seeking to use scientific knowledge for broader political purposes, but they link such aims more closely with his actual scientific practice than is the case with Jacob, emphasising Boyle's attempt to privilege 'matters of fact' as an area where agreement might be reached, whatever controversy might remain over interpretation, and elevating the 'experimental space' as a place withdrawn from the hubbub of the world where solutions to worldly problems might yet be found. In the course of this, Shapin and Schaffer have raised new and quite profound questions about the valuation of different types of intellectual activity in Boyle's period, and the manner in which knowledge claims were assessed; their volume has been deservedly influential. Yet, for all its conceptual sophistication, their interpretation turns out to have a surprising amount in common with traditional views as far as its image of Boyle is concerned. It differs from them mainly in attaching significance to facets of Boyle's practice which had previously tended to be taken for granted, seeing as contingent and manipulative what had hitherto been deemed as self-evident, almost natural. Thus Boyle continues to emerge as a kind of intellectual statesman, his laborious execution and prolix reporting of his experiments and his insistence on 'manners in dispute' being seen as strategic devices, aimed to vindicate his ideal of a balanced and sober scientific polity against alternative traditions of intellectual dispute.

Indeed, though there are discrepancies between these views, reflecting the contrasting preoccupations of their authors, they all owe a good deal to the image of Boyle purveyed by Burnet and repeated by Birch on the basis of selective use of the documents published here. All present Boyle as a purposeful, rational and effective figure, with clear goals of scientific empiricism and religious moderation which can be aligned with an essentially establishmentarian position in late seventeenth-century England. Yet it may be argued that, though they contain a substantial amount of truth, such views of Boyle are incomplete and they can be crucially supplemented by data available from the records published here. Of course, these documents bear out many facets of the image of Boyle which has been traditionally accepted, though sometimes with interesting modifications. But they also add an extra dimension, giving greater complexity and depth to our view of Boyle, and thereby making better sense both of his intellectual personality, and of features of his life and work which might otherwise seem puzzling.

At the outset, it is worth stressing that this does not depend on championing certain contemporary commentators on Boyle against others. The credentials of the authors whose writings appear below have been scrutinised in earlier sections of this introduction so that appropriate allowance can be made for such preoccupations as Burnet's pastoral concern for the public use of natural philosophy in general and Boyle's exemplary life in particular; Pett's stress on matters of practicality; and Evelyn's anti-Puritan gloss on Boyle's early career. Yet these complications are tangential, and there is a real coherence, both between the statements of the various authors, and between them and Boyle's own views as expressed in his early *Account of Philaretus* and, later, in the autobiographical remarks contained in the Burnet Memorandum. Indeed, perhaps unfortunately, we lack an entirely critical account of Boyle such as the biographer, Roger North, might have written: at least, this is the implication of his comment on an anecdote concerning Boyle's valetudinarianism, that 'Mr Boyle had such a party, that all he did was wise and ingenious'.[299] Yet what is revealing is that, even within these accounts written by Boyle himself and his close friends, there is evidence for a more complicated personality than has hitherto been acknowledged; indeed, in some respects what emerges is positively convoluted. Let us therefore take the different facets of Boyle that they reveal in turn, investigating how they fit together and how they correlate with existing interpretations.

* * * *

It is appropriate to begin with the view of Boyle himself presented in his youthful *Account of Philaretus*, which may be supplemented by the much later self portrait provided by the Burnet Memorandum. As we have seen, the author of *Philaretus* was preoccupied by the pursuit of virtue and the acquisition of self-knowledge, inspired by the mixture of Stoic and Christian ideas that were the stock in trade of Renaissance Protestant humanism.

The concern for self-control which this encouraged has its echoes both in Boyle's later comments about himself to Burnet, and in a number of points in the posthumous memoirs of him. Thus his 'very strict rules of diet', Boyle told Burnet, 'he observed like a Philosopher'; he comparably commented on his success in controlling his temper and in withstanding the powerful temptation to dabble in magic.[300] Moreover, this picture of balanced austerity and self control – in matters sartorial, medical, social and intellectual – recurs in accounts like Evelyn's letters to Wotton and Burnet's funeral sermon. Indeed, in Burnet's portrait of Boyle as a man who, with God's assistance, acquired a wisdom and knowledge which raised him above lesser mortals, we see an image that shares something with the idealised view of the philosopher to be found in the *éloges* of the Académie des Sciences in eighteenth-century France.[301]

This represents the lifelong legacy of the earliest phase of Boyle's intellectual career, when he saw his mission as that of a moralist. But it is equally important to stress the transition that occurred in the years around 1650, when he first acquired the role of natural philosopher. It was at this point that he discovered the excitement of experiment, and, at the same time, that he became convinced that the study of nature by such means provided a road to truth which eluded those who relied on unbridled reason.[302] This is a change on which the documents below comment only obliquely: thus the Burnet Memorandum appears to give a tantalising hint of Boyle's perception of the way in which he had changed between this phase of his career and the earlier literary one, while it also throws important light on the parallel process by which Boyle became learned in the scriptural tongues and in biblical scholarship; Pett, too, gave some graphic examples of the effect of studies of the latter kind on Boyle's outlook.[303] In addition, in his interview with Burnet, Boyle interestingly linked the addition of medicine to his scientific concerns with the illness which he contracted during his visit to Ireland in the early 1650s.[304]

What receives greater emphasis in the Burnet Memorandum is the series of books which Boyle wrote in the late 1650s and which were published in the early 1660s. These formed the basis for his entire subsequent programme as a scientist, in which he sought through profuse experimentation to lay the foundation for a new understanding of nature. In his conversation with Burnet Boyle gave an interesting narrative of his progress from one work to another within this group, while their underlying ethos was well-expressed by Pett, who strikingly recalled Boyle's gratitude to providence for sending him into the world at a time 'when real knowledge is in so triumphant a state and Experimental Philosophy crowned with so much success', an achievement to which he was able to make a significant contribution.[305] It is also interesting how in the Burnet Memorandum Boyle noted – in juxtaposition with his experimental investigations

– the attention that he received from foreign visitors. Despite the extent to which this might have distracted him from his laboratory work, Boyle clearly enjoyed such manifestations of the acclaim as a scientist that he achieved in his lifetime, a generalised account of which was given by Evelyn.[306] A later and more detailed echo of this is to be found in Wotton's exposition of Boyle's classic experiments with the air-pump in Document 9.

These texts also provide evidence concerning Boyle's aims in his experimental work. His hostility to scholasticism is in evidence in the Burnet Memorandum, though it is interesting that he there presented this as a commonplace attitude among the gentry which he was trying to harness to his purposes.[307] A presumption of the disharmony between scholasticism and Christianity appears in Pett's notes, where he noted how Boyle complained how scholastic metaphysics trivialised God's name, while a more general sense of the harmony between Boyle's experimentation and his religious goals is to be found in Burnet's sermon.[308] In his conversation with Burnet, Boyle also explained why he wrote his *Sceptical Chymist*, bearing out the recent claim that his priority in that work was to persuade practical chemists to join 'a litle more Philosophy with their Art'.[309] In addition, he referred to his quarrel with Hobbes over the interpretation of pneumatic phenomena: interestingly, he there implies that he was impelled in the dispute as much by Hobbes' arrogance as by his 'Principles', though this arguably illustrates how contemporaries commonly saw Hobbes' manner as integral to the pernicious nature of the ideas that he purveyed.[310]

Light is also thrown on other aspects of Boyle's motivation in these writings, including ones illustrating his links with his social and cultural milieu. The Burnet Memorandum makes it sound as if his primary goal in his scientific writings was to enthuse the gentry with such interests, a sense of mission which can be paralleled in scattered comments in Boyle's published writings, though it is not normally given such a prominent role in accounts of his aims.[311] It is also interesting how he there linked his scientific concerns with his membership of such bodies as the Council for Foreign Plantations, specifically mentioning in this connection his 'General Heads' for inquiries about the natural history of foreign countries, published in *Philosophical Transactions* in 1666.[312] Equally noteworthy is Boyle's reference in the same document to an essay that he had intended to include in his treatise on *The Usefulness of Natural Philosophy* – though in fact it was never published – about the potential for economic improvement of the transplantation of commodities and practices from one country to another.[313] Indeed, even if they reflect a degree of retrospective rationalisation on Boyle's part, these passages apparently bear out J. R. Jacob's reading of him as 'promoting the interrelated purposes of science, trade, empire, and religion'.[314]

In addition, Sir Peter Pett emphasised Boyle's enthusiasm for the direct technological spin-off of natural knowledge, instancing the experimental help which Boyle gave such entrepreneurs as Robert Fitzgerald in his project for making salt-water drinkable, or Thomas Hale in his efforts to prove that the use of lead for covering ships' hulls was beneficial. Indeed, Pett averred that Boyle's 'last Command' to him concerned a related matter, the improvement of shipbuilding in the national interest.[315] Pett also expatiated on the Christlike quality of Boyle's concern to dispense medication – citing the views of the early seventeenth-century Italian cleric and natural philosopher, Giovanni Ciampoli, in this connection – while Thomas Dent comparably averred that it was not least for this purpose that Boyle kept a laboratory. An interesting echo of this is to be found in the stress on the medical spin-offs of scientific findings in the Burnet Memorandum, where even Boyle's pneumatic investigations are presented as having the potential to reveal 'many usefull things for the Regiment of our health'.[316]

Undoubtedly Boyle was deeply committed to the utility of natural knowledge. But these documents also reveal that, in practice, complications arose which are elided in accounts like that of Jacob. For one thing, the Burnet Memorandum records that Boyle failed to complete the essay for *Usefulness* with the 'mercantilist' theme that has just been referred to, attributing this to his ill-health and the fact that many of his relevant papers 'were either losst or stolln'. Boyle was also surprisingly reticent in publishing on medical topics considering the value he placed on knowledge of this kind, in this case evidently largely because his fear of offending vested interests overcame his natural philanthropic impulse.[317] As for Boyle's links with the agencies of overseas enterprise, it is worth noting that what Boyle told Burnet about his links with these emphasised an altruistic motive on his part, and elsewhere he stressed that he belonged to the board of the East India Company out of 'the desire of knowledge, not profit'.[318] In other words, for all his enthusiasm for national prosperity, Boyle arguably retained more of the reserve towards commercial activity that he had earlier displayed in the *Account of Philaretus* than Jacob's reading of him might imply.[319]

A more serious problem about Jacob's view of Boyle's preoccupations is what Boyle told Burnet about his involvement with the Council for Foreign Plantations. It is true, as Jacob has documented, that for a while Boyle was one of the more active members of that body, and the Burnet Memorandum shows his genuine pride in what he was able to achieve during that time.[320] But his participation in it lasted for only three years, and the Memorandum explains why: as Burnet put it, 'some upon private reasons or humours obstructing it he gave over further medling in that Councill'. A similar attitude is revealed by Boyle's remark to Sir Peter Pett about how, for all his enthusiasm for the growth of knowledge in his age, 'he tooke notice to me how every great New Invention, necessarily crossing the private Interest of many particular persones was thereby hindred in its birth and growth, by such interested persons'.[321] Such passages are crucial, for they express the exact opposite of the view which Jacob repeatedly associates with Boyle, namely that 'private interests were the foundation of public good'.[322] Yet my own reading of Boyle's statements on such subjects bears out the opinion that Pett attributes to him, rather than Jacob, suggesting that Boyle was always predominantly aware of the tendency for private interests to detract from the public good. Regretfully, I have come to the conclusion that a fundamental misunderstanding of Boyle's priorities lies at the heart of Jacob's reading of him, apparently largely due to a mistaken extrapolation to Boyle of views expressed by his colleagues with which there is no reason to believe that he unreservedly agreed.[323]

The key text in this connection is a work by Pett, his *Discourse concerning Liberty of Conscience* (1661), which argued in favour of a policy of toleration in the aftermath of the Restoration. It is appropriate here to turn to this and the issues it raises. Jacob's account of the episode in which Boyle and his friends, Pett and Thomas Barlow, agreed to write treatises on this subject is based on Birch's summary of Pett's account of it in his *Life*, and it is therefore not surprising that the two are fairly similar.[324] What is not clear, however, is whether it can be presumed, as Jacob supposes, that we can extrapolate directly to Boyle's own views from those that Pett expressed in his tract. It is true that Pett tells us that Boyle agreed to let him read his manuscript to him before it was published and to advise him as to what should be deleted or altered. But it would be naive to take this as proving that Boyle's views on the merits of toleration were necessarily the same as Pett's. This was, after all, a treatise aimed to persuade others; in addition, it was consciously delimited to considerations of 'political interest'.[325] Moreover, though Pett may have seen such issues as the most crucial, reflecting his own, rather Erastian approach to such matters, later to be displayed in his *Happy Future State of England* (1688), it is doubtful if Boyle saw such considerations as more than contingent. His own views on this subject seem to have been more spiritual and altruistic, more concerned with the mandate of Christ, who is barely mentioned in Pett's text.[326] This is reflected in Burnet's comment in his funeral sermon on Boyle's view of the persecution which such toleration was aimed to avoid: 'I have seldom observ'd him to speak with more Heat and Indignation, than when that came in his way'. Pett cited this by way of introduction to his account of the episode, presumably because he concurred with it.[327] Boyle might have agreed that toleration was incidentally good for trade and stability (if it was the will of providence, such compatibility would not have surprised him). But his chief aim was to do what was spiritually right, and this did not necessarily coincide either with what was comfortable or what was politically expedient.

Indeed, if examined carefully, Boyle's religious position proves more complex than might appear at first sight, showing a discrepancy from the status quo not dissimilar to that in evidence in his withrawal from colonial administration due to the predominance of 'interest'. It is hardly surprising that all the commentators on Boyle whose opinions appear here were at pains to stress his deep religiosity, as evidenced by both his personal practice and his extensive patronage of missionary and charitable work. They also underlined his hostility to religious persecution, while insisting on his commitment to a visible church, a topic on which Pett has some particularly interesting remarks based on conversations on this subject that he had with Boyle during the Interregnum.[328] On the other hand, though stressing Boyle's conformity, they could hardly disguise certain complications in this, which it is worth dwelling on here. Indeed, had Boyle accepted the bishopric which we learn from the Burnet Memorandum that he was offered shortly after the Restoration, he would have made a distinctly odd member of the episcopal bench, not least due to his disinterest in many of the issues which were most passionately debated by churchmen of the day.[329]

Part of the difficulty in understanding Boyle's churchmanship results from the confusion caused by the widespread but vague use of the term 'Latitudinarian' to describe the religious outlook of Boyle and others associated with the new science: this bears an imperfect relation either to the way in which the word was used at the time, or to the most crucial divisions in English religious life in these years.[330] It is certainly true that there is a conspicuous lack of high churchmen among Boyle's clerical friends – although he came into contact with ecclesiastics like John Fell in connection with his missionary concerns – while Pett interestingly comments on Boyle's avoidance of links with devotees of the Anglican liturgy during the Interregnum.[331] It is also true that his closest contacts at the end of his life included churchmen regularly identified as 'Latitudinarians', notably Gilbert Burnet and Edward Stillingfleet, both of whom displayed the pliability of outlook which contemporaries saw as characteristic of Latitudinarianism.[332] They thus showed themselves notably flexible in the ease with which they reconciled themselves to the 1688 Revolution, 'throwing in every argument they thought would serve' in their defence of it, while Stillingfleet's rather Erastian attitude towards dissent had earlier altered to suit the political mood.[333] Possibly, Boyle was attracted to such men because their very flexibility encouraged them to tell him what he wanted to hear, though it is worth pointing out that it is Birch's quotation of a passage from Thomas Dent's letter to Wotton out of context that makes it appear to state that it was particularly for his views on dissent that Boyle respected Stillingfleet.[334]

On the other hand, it is possible to discern a distinct contrast in outlook between Boyle and these modish episcopal advisors of his later years, and it is revealing that his closest ecclesiastical contact for much of his life was a man of a rather different outlook, Thomas Barlow, a Calvinist episcopalian who typified what has been described as the 'conformist' wing of the Restoration church.[335] Throughout the period, Barlow was prominent behind the scenes defending a vision of Anglicanism as an integral part of the Reformed communion and attacking new trends in the church, particularly the 'Arminian' tendencies which are often seen as typically 'Latitudinarian'.[336] Though in 1660, at Boyle's behest, he wrote a treatise about religious toleration and the circumstances in which it might be appropriate, his treatment of the issue was hedged around with qualifications, and the non-publication of the tract is probably more significant than its original composition.[337] What above all attracted Boyle to Barlow was his brilliance as a casuist, but there is reason to believe that Barlow's rather austere churchmanship struck a sympathetic chord in Boyle more generally.[338] Moreover, it is revealing that Boyle was also connected with two other men whose churchmanship was very close to Barlow's, namely Robert Sanderson and John Gauden.[339]

But what is equally notable about Boyle's religious contacts and sympathies is the way in which they straddled the divide between the Anglican church and dissent. Thus Pett stressed how the trustees nominated by Boyle in his will included more dissenters than Anglicans, revealingly using the word 'latitude' to describe Boyle's practice in this regard.[340] Earlier, Boyle had had links with such leading nonconformists as Richard Baxter – who, according to Pett, was one of Boyle's casuistical confidants – and John Howe, who claimed that he wrote one of his books at Boyle's behest.[341] Indeed, in the spectrum of religious liberalism in Restoration England recently sketched by Richard Ashcraft, Boyle was arguably closer to the nonconformists that he was to Latitudinarian Anglicans like Joseph Glanvill.[342] Also interesting is the further section of Pett's notes in which he illustrated how, for all his stress on conformity, Boyle's church attendance was slightly flexible, with some similarity to the practices which Patrick Collinson has described as 'voluntary religion' in the context of the Elizabethan and Jacobean church, and which were anathema to high churchmen.[343] Revealingly, Pett left it to Burnet's 'judgment' 'Whether any thing of this his practice is proper to be published'.[344]

Moreover, as perhaps befitted so passionate an advocate of religious toleration, Boyle's contacts went beyond Presbyterians to men of more extreme religious positions. A case in point is Oliver Hill, elected a Fellow of the Royal Society on Boyle's recommendation in 1677. Hill, who had formerly been an associate of the Interregnum Hermeticist, John Everard, was described at the time as 'a Great Mystick', and he made himself unpopular at the Society's meetings in 1677–8 due to the arcane views about the natural realm which he propounded.[345] There is evidence for similar sympathies on Boyle's part at an earlier date, and, although it has been claimed that his philosophical concerns were predicated on a 'dialogue with the sects', this seems to fall far short of the truth.[346] In fact, Boyle seems to have a considerable amount of respect for the opinions of those outside the religious consensus, perhaps because his religious preoccupations made him more comfortable with interpretations of doctrine based on claimed personal inspiration than with an excessive stress on reason. Though one interesting confrontation with a notoriously heterodox religious thinker is recorded by Pett – Boyle's brush with Sir Henry Vane the younger over biblical interpretation – what one sees is not a distaste for religious questioning, but for what Boyle saw as the trivialisation of a key biblical text, something which greatly concerned him in its own right.[347] Moreover, by way of commentary on this, Pett stressed the *independence* which Boyle had shown in taking this line, and it is worth pointing out that his aristocratic status enabled him to be eclectic in such matters to an extent that might have been difficult for someone of less exalted status.

What comes across from Pett's notes is that he clearly admired in Boyle a depth of religiosity which went beyond the superficial to an almost disconcerting extent. One facet of this which Pett took the trouble to confirm, despite the fact that Burnet had already stressed it in his funeral sermon, was the fact that 'the very Name of God was never mentioned by him without a Pause and a visible stop in his Discourse'.[348] This is a piece of information about Boyle that has become well-known due its being picked up not only by subsequent biographers but also by *The Spectator* in the early eighteenth century.[349] Yet, though respectfully cited by such commentators – and even emulated by Barlow, as Pett explained[350] – such behaviour was arguably somewhat strange. There is a slightly obsessive element in Boyle's religiosity which is illustrated even by the rather awed tone in which his friends reported on it.

One symptom of this was Boyle's extreme reluctance to take oaths, which meant that – uniquely among those nominated to the post during the seventeenth century – he found it impossible to become President of the Royal Society in 1680. Indeed, though less extreme, his position on such matters had something in common with that of the Quakers.[351] Boyle's sensitiveness on this issue is illustrated below in his careful statement to Burnet that he 'made no vowes not knowing how his Circumstances might change but usually gave the 20th' – presumably meaning that he voluntarily devoted a twentieth of his income to charitable purposes, but avoided a formal pledge to do so. Further evidence for his abnormal solicitude about the solemnity of the obligation which oaths involving God's name established is provided by another relevant source,

Boyle's extant notes on the interviews which he had with Stillingfleet and Burnet in 1691 to deal with matters on his conscience.[352]

It is appropriate to consider these interviews here, for they illustrate Boyle's acute 'scrupulosity' in his later years on a range of ethical and spiritual dilemmas, certain of which recur in the documents below. Boyle's notes show that, as his death approached, he still struggled with religious doubts, echoing those which he had so graphically described in the aftermath of his conversion experience in his *Account of Philaretus*.[353] In addition, he clearly shared with many earnest Christians of his day an anxiety that he might have committed the Sin against the Holy Ghost, as described in chapter 12 of St Matthew's Gospel. As his friends saw, there could be no doubt whatever about the genuineness of a faith so repeatedly scrutinised, but it probably made Boyle an uncomfortable colleague. Thus Burnet in his sermon refers in a veiled way to Boyle's reservations about the residual 'abuses' in Anglicanism, a tantalising reference from which it is impossible to extrapolate, but which presumably goes beyond his hostility to 'debates' and to persecution. In addition, it seems likely that he retained the disdain for those who took their religion on trust which he expressed in *Philaretus*.[354]

Equally significant is the evidence that exists of Boyle's acute scruples over matters of conscience concerning his own affairs. Much space was devoted to these in the interviews with Stillingfleet and Burnet, and they are also referred to in Pett's notes. One matter that appeared in both places was the grant to Boyle as part of the Restoration settlement of impropriations on former church lands. This clearly lay on Boyle's conscience, partly because those in possession of such rights could be regarded as guilty of the sin of sacrilege, and partly because he had had the benefit of perquisites which might otherwise have been used for ecclesiastical ends. Pett's awareness of this is clearly indicated by his assurance to Burnet that he was 'able to wash my hands in innocence' concerning the relevant trusts in Boyle's will for which he had been responsible.[355]

Pett also has a very interesting story relating both to Boyle's casuistry and his relationship with his father. This concerns the slightly ambiguous reference to 'the charges of unfaithful Mammon' made by the divine, Robert Sanderson, in the preface to the casuistical treatise, *Ten Lectures on the Obligation of Humane Conscience*, which Boyle paid him a pension to prepare for publication in 1659.[356] As Pett noted, this was taken by 'malevolent Critics' as a reference to the suspicion that the Earl of Cork's landed estates included ecclesiastical property that he had sacrilegiously annexed. It thus revived speculation about the accusations by his enemies, Strafford and Laud, that he was guilty of 'a direct rapine upon the patrimony of the church', though Cork always protested his innocence of any abuse.[357] Pett realised that this was especially problematic for Boyle, partly because of his almost naive respect for his father and belief in his integrity, which is borne out elsewhere in Pett's notes, and partly because, as one of his heirs, Boyle was himself implicitly party to Cork's actions.[358] Pett's remarks thus provide further evidence of the dilemmas which confronted Boyle, helping to explain the preoccupation with matters of conscience and the extreme scrupulosity in dealing with them which is revealed by his casuistical interviews (and, for that matter, his codicil-laden will[359]).

Linked to such moralistic concerns is the extraordinary insight provided by the Burnet Memorandum into Boyle's ambivalent attitude to magic. This document reveals his fascination with stories of people who had had contact with supernatural beings, even recording Boyle's reaction when he was offered the opportunity to consult such spirits himself through a magical glass. On the other hand, it illustrates an equally strong conviction that there were *ethical* reasons why, even if knowledge might be acquired by such means, it might not be licit to achieve this: in other words, that such insights might be diabolical temptations which it was proper to resist.[360] This moralistic view of intercourse with the spiritual realm on Boyle's part is particularly significant because, as Burnet's notes reveal, Boyle clearly saw a link between this and the alchemical transmutation of base metals into gold, cases of which are there also mentioned. Boyle seems to have believed that spiritual beings might purvey alchemical insights. This being the case, it raised acute difficulties as to whether, even if transformations might be alchemically achieved, it was legitimate to do so. It also raised the related issue of whether it was proper to divulge information so discovered. This helps to explain Boyle's somewhat in-voluted attitude towards alchemy, which clearly fascinated him, yet about which he was more than usually reticent.[361]

In all, these documents illustrate a profound and troubled side to Boyle's personality in addition to the serene and effective one which has formed the basis of the traditional image of him. Yet both have to be taken into account properly to make sense of him. Indeed, the two are often intercon-nected. Perhaps most significant is the link which may be postulated between Boyle's intensity in salving his conscience and the assiduity in experimentation which is his chief claim to retrospective fame. The two are explicitly connected by the Burnet Memorandum, with its statement that 'He made Conscience of great exactnes in Experiments'.[362] Arguably it was the mental habits encouraged by Boyle's casuistry – his insistence on satisfying himself, if necessary reconsidering an issue over and over again – which explains the intensiveness of his experimentation. He also used the language of casuistry in describing his experimental work, repeatedly speaking of 'scruples' and 'scrupulosity' in this context.[363] More generally, a combination of wide curiosity with deep scrupulosity underlies all Boyle's mature writings on scientific and religious topics. Whatever he studied, he studied with the same assiduity with which he examined his conscience, never being satisfied until he had got to the bottom of a question. The uncomfortable intensity of his religious commitment thus finds its echo in his ability to penetrate to the very heart of philosophical issues that he confronted, which recent scholarly work on him has repeatedly revealed.[364]

Such a view of Boyle also makes sense of his mature style, which even Evelyn admitted, in his otherwise laudatory account of Boyle, was 'not answerable to the rest of his greate parts'.[365] For this is characterised by a tortuousness – an 'inability to bring a sentence to a successful conclusion', in R. S. Westfall's words[366] – which resulted from Boyle's proneness to obscure his chief point by adding digressions intended to illustrate it more fully, or qualifications designed to clarify his meaning. Though Steven Shapin has written about what he calls Boyle's 'literary technology', to those familiar with Boyle's convoluted style and his shapeless books, this strikes a somewhat false note.[367] Yet such convolution is highly significant, for it mirrors the difficulty that Boyle experienced in trying to do justice to the complexity of his thoughts on philosophical and scientific issues which were highly controversial in his day, a problem made worse by his acute scrupulosity. There was also a tension between Boyle's wish to purvey the vast quantity of experimental and experiential data that he accumulated during his career, and his awareness of the literary elegance to which he had aspired in such early writings as the *Account of Philaretus* but for which he had increasingly little time in his later years.

Such difficulties are reflected in the prefaces to his books, which retail profuse and overlapping excuses for their shortcomings, ranging from lack of time to the carelessness or dishonesty of his assistants.[368] Indeed, these verge on the obsessive, and they reached a climax with a publication devoted exclusively to this subject in 1688, a broadsheet 'Advertisement' which Boyle had specially printed 'About the loss of many of his writings, addressed to *J.W.* to be communicated to those of his friends, that are virtuosi; which may serve as a kind of preface to most of his mutilated and unfinished writings'. This is a text which has long been familiar, since it was reprinted by Birch as an appendix to his *Life*: yet, with its graphic account of manuscripts destroyed by chemicals in the laboratory crowning Boyle's earlier descriptions of his literary disasters, it must surely rank as one of the oddest publications of the seventeenth century.[369]

*   *   *   *

How did this complex personality originate, and when did it reach its mature form? Some aspects of this question will always remain obscure, due to our inability to place an early modern figure like Boyle on a psychiatrist's couch. But certain phases of Boyle's development seem fairly clear, and it is worth trying to sketch these here. First, there is Boyle's Anglo-Irish upbringing, and the influence of his father, whose complex personality has been so well sketched by Nicholas Canny.[370] Though Boyle left Ireland at the age of eight, returning only briefly thereafter, his earliest years were spent in the culturally mixed ambience of the planter class there. Indeed, it is worth recounting what must be his first recorded memory, since it was overlooked by Canny and provides better evidence than any he actually cites for the Cork family going native. This is the story included in Aubrey's *Brief Life* of Boyle, that 'He was nursed by an Irish nurse, after the Irish manner, wher they putt the child into a pendulous satchell (insted of a cradle), with a slitt for the child's head to peepe out'.[371] On the other hand, Boyle specifically noted that he never bothered to learn Irish, and, although Canny's emphasis on the extent to which facets of his outlook had precedents among the planter class is perceptive – especially as an antidote to undue emphasis on the influence on him of the Hartlib circle – the significance for Boyle of the Anglo-Irish experience can also be exaggerated, in view of the brevity of his exposure to it.[372]

More important is the influence on Boyle of the Earl of Cork, a domineering father for whom his children retained affection and respect even when their wills clashed with his, as was the case with Boyle's sister, Mary.[373] Boyle's awareness of his father's special fondness for him is borne out by various of the documents published here, and he was to reciprocate this by the later faith in his father's integrity that has already been noted.[374] Yet Boyle must have been equally strongly affected by the enforced separation which his father imposed on him, as on his other sons. This was due partly to Cork's wish to curb the growth of excessive affection between parent and child, and partly to his desire to gain for his sons the refinement of manners and the social acceptance, of his own lack of which he was so acutely aware. Hence, from an early age, Boyle was sent away from home, an experience that undoubtedly had a formative influence on the boy.[375] A further legacy of Boyle's childhood years was the stutter which he retained for the rest of his life and which Lorenzo Magalotti found rather disconcerting when he met Boyle in 1668, noting how it 'seems as if he were constrained by an internal force to swallow his words again and with the words also his breath, so that he seems so near to bursting that it excites compassion in the hearer'.[376]

Boyle's education at Eton and on the continent is fully described in his *Account of Philaretus*, which unselfconsciously manifests the studiousness and priggishness which characterised him as an adolescent. In addition, as we have seen, *Philaretus* itself exemplifies the career as a writer of moral treatises which Boyle adopted in the 1640s and which, though abandoned shortly after *Philaretus* was written, left the permanent legacy of concern for self-knowledge and self-control which we have already encountered. It was almost certainly also at this time that Boyle's interest in casuistry began, perhaps growing out of the ethical concerns displayed by his early writings. His moral essays of this date thus contain passages which closely echo the sentiments of contemporary casuistical treatises, while it is also revealing that in his unpublished essay 'Of the Study & Exposition of the Scriptures', he urged that the individual layman should be taught to use the Bible for himself, so that 'the Texts of Scripture will readily apply themselves, & much more exquisitely fit his private Cases'.[377]

By this time, Boyle had passed puberty, and the documents below provide evidence about his sexual and marital experiences, supplementing recent conclusions on such matters drawn particularly from Boyle's literary compilations of the 1640s.[378] These experiences may well have contributed to Boyle's later *persona*, though claims of a causal link between his celibacy and his later scientific career – as against the precocious moralistic writing which preoccupied him in that decade – have unfortunately proved ill-founded.[379] In *Philaretus*, Boyle reports on the homosexual advances to which he was subjected in Florence, the only actual sexual experience on his part of which we have record, which may have affected him more deeply than his comments there imply.[380] As for heterosexual relationships, it is certainly true (in John Evelyn's words) that 'among all his Experiments, he never made that of Marriage', while the Burnet Memorandum notes in passing that he was still a virgin when middle-aged. The latter also contains perhaps Boyle's most crucial extant statement on this subject, namely that he 'abstained from purposes of marriage at first out of Policy afterwards more Philosophically and upon a Generall proposition with many advantages he would not know the persons name'.[381]

Though this key passage is unfortunately slightly obscure, it is worth trying to correlate it with what is otherwise known about Boyle's marital affairs.[382] The initial abstention 'out of Policy' probably refers to Boyle's rejection of the match arranged for him by Cork, as for his other children, on a one-off basis: in Boyle's case, this was with Anne, daughter of Edward, Lord Howard of Escrick. The case in which 'he would not know the persons name', on the other hand, is otherwise undocumented.[383] But the intermediate phrase, 'afterwards more Philosophically', presumably alludes to the point at which Boyle 'made a short ‹yett› everlasting turne from all impressions by Ladies Eyes', in the words of Sir Peter Pett, instead devoting himself to a 'Love causing in him a noble heate without the trouble of desire', as expressed in *Seraphic Love*.[384] That Boyle was self-conscious about this commitment to the single life to the point of touchiness is suggested by a hitherto unknown letter to him from John Evelyn. This responded to Boyle's lost reply to an earlier letter of Evelyn's, inspired by the publication of *Seraphic Love*, in which Evelyn had complimented the work, but sang the praises of the married state in implicit criticism of Boyle's celebration of celibacy. It is apparent from Evelyn's fulsome apology that his original letter must have caused real offence.[385]

On the other hand, other aspects of Boyle's personality developed later. It has already been noted how the period when Boyle emerged as a natural philosopher rather than a sanctimonious moralist was around 1650, and it is important to recognise that negative as well as positive facets of his later make-up date from this phase of his intellectual evolution.[386] In particular, the stylistic equivocation which is so typical of Boyle's mature works – and which echoes the mental convolution seen in the extant records of his casuistical encounters – is not paralleled in early writings like *Philaretus*. This has been illustrated with particular clarity by Lawrence Principe's discovery of a manuscript of Boyle's *Seraphic Love* in the form in which it was initially circulated in 1648. Comparison of this with the text as published in 1659 reveals how sharp and positive points were weakened by qualification, and lengthy digressions added which detracted from the clarity of Boyle's message.[387] Evidently hesitation and prevarication were traits that were greatly intensified in Boyle by experiences subsequent to his early moralistic phase, including his exposure to the world of scholarship in the years around 1650, which seems likely to have made him more aware than previously of the sheer complexity of most issues, and increasingly cautious about committing himself.[388] By analogy, it seems probable that his scrupulosity in his personal affairs also became more rather than less convoluted as his life went on, as was probably also true of his attitude to magic.

In addition, Boyle's health evidently seriously deteriorated in the years around 1650. As a child, he was positively healthy-looking. Indeed, his ruddy and vigorous countenance was commented on both by himself and others, in contrast to the pale and emaciated figure that Evelyn described to Wotton.[389] Though he had an attack of ague in 1638, he seems to have been relatively free from illness for most of the 1640s, apart from 'disquiet' from the stone in 1647.[390] In 1649, however, he had a serious attack of quotidian ague, which by his own admission brought him near to death.[391] Then, in Ireland in 1654, he had the most debilitating illness in his life, the significance of which was such that he dwelt on it in his interview with Burnet; he also gave an even longer account of it in his preface to the second volume of his *Medicinal Experiments*, published posthumously in 1693. It seems to have been his perception that it was at this point that he became an almost permanent invalid, his eyesight, especially, never fully recovering.[392] Burnet specified in his funeral sermon that that Boyle had 'laboured under such a feebleness of Body' for 'almost Forty years', while, interestingly, Evelyn linked Boyle's stutter with 'those frequent attacques of Palsey' which he attributed to the intense experimentation which Boyle began at much the same time.[393]

This experience of illness, too, may have contributed to the obsessive element in Boyle's later persona. It is probably significant that it was in 1653 that William Petty is to be found criticising Boyle's valetudinarianism, and this sentiment was echoed much later by Roger North, who told mockingly of Boyle's 'chemical cordials calculated to the nature of all vapours that the several winds bring', which he took according to the direction of the prevailing wind.[394] North also recorded how Boyle 'would speak with no physician unless he declared, *in verbo medici*, that he had seen none with the smallpox in so long time', and Boyle's abnormal solicitude about the pox is borne out by the report of the German traveller, Caspar Lindenberg, on a curious preliminary ritual which tempered Boyle's keenness to welcome foreign visitors to admire his laboratory: 'when he sent his servant to announce his visit, Boyle returned the answer that he would be pleased to accept, provided that he had not recently been in a place where the smallpox or chickenpox was epidemic because he was very afraid of that illness'.[395]

As these examples show, the mature Boyle did not take shape overnight, and one advantage of material like that presented in this volume is to induce a sense of dynamism in our view of Boyle, in contrast to the rather static image which the traditional interpretation can easily encourage. Even more important is the need for account to be taken of Boyle's 'negative' as well as his 'positive' characteristics, since only thus will we understand his complex personality. Views of Boyle which echo the accepted image of him – the great experimenter who was at the same time a pious and morally upright man – are fine as far as they go, but they do not give us the whole Boyle. They elide the struggles and tensions with which Boyle had to live, and they fail to do justice to the less 'functional' aspects of his persona.[396] Only when these are reinstated will we fully understand him. Moreover, as this Introduction has sought to indicate, the texts presented here are indispensable to such a truer appreciation. Their publication is long overdue.

NOTES TO INTRODUCTION

1. The Volume and its Contents

[1] For a recent conspectus, see Hunter, *Robert Boyle Reconsidered*, especially the bibliography.

[2] Details will be provided in the Pickering Masters Boyle (1997). See also below, n. 274.

[3] For fullest attempt to do this so far, see Maddison, *Life*, which largely (but not wholly) subsumes a series of studies earlier published in article form: for a complete list, see Hunter, *Robert Boyle Reconsidered*, pp. 221–2. It should be noted, however, that Maddison made only minimal use of the documents published for the first time here: see below, n. 293.

[4] North, *General Preface*, pp. 77–9.

[5] For quotations from these documents by Birch and others, see below, nn. 270, 293.

[6] Evelyn, *Memoirs* (1818), vol. ii, pp. 302–8; *Memoirs* (1827), vol. iv, pp. 403–16. The former text almost certainly comes from Evelyn's Letterbook, no. 754 (though see below, p. 89 n.). The latter derives from a loose copy by Evelyn: by this time he had stopped keeping a letterbook. This item is now Add. MS 28104, fols 21–2, having been purchased at Young's sale in 1869: it probably reached Young from the collection of William Upcott. It is endorsed: 'Copy to Mr Wotton, in answer to one of his in order to the History of the life of Mr Boyle &c: which I first put him upon' and 'Mr Wotton' sideways along the edge of the page. Both items appear to precede the copy sent.

[7] Birch did, however, quote information about Boyle which he derived orally from Edmond Halley (d. 1743), who 'related to me [Birch] his conversation with him [Boyle] upon that subject', namely 'the possibility of the transmutation of metals into gold', Boyle's 'persuasion' of which 'was avowed by himself' to Halley (*Life*, p. cxxxi), and from Sir Hans Sloane (d. 1753), who told him how Boyle's 'constitution was so tender and delicate, that he had divers sorts of cloaks to put on when he went abroad, according to the temperature of the air; and in this he governed himself by his thermometer' (ibid., p. cxxxvi). The information from Halley (slightly differently worded) had earlier been included in Birch's account of Boyle in the English edition of Bayle's *Dictionary*, vol. iii, p. 557: see below, pp. lvii–viii.

[8] *Life*, p. iii.

[9] Much useful information is provided by Maddison, 'A Summary', though the author was unfortunately unaware of crucial material, notably the Wotton-Evelyn letters in the Evelyn Collection at Christ Church, Oxford.

2. Boyle's *Account of Philaretus* in Context

[10] The principal dating clues are Boyle's statements (1) that five sisters were living (p. 2), which dates it before 1656, when his sister Joan died; (2) that he burnt his

poems on coming of age (p. 11), which dates it to after January 1648; and (3) that he had been a stranger to ague ever since the attack documented in the text (p. 9), which means that it must date from before the severe attack he experienced in July 1649. Cf. Maddison, *Life*, p. 1. See also Oster, 'Biography, Culture and Science', p. 216 n. 7, who recapitulates this, but inexplicably claims, in connection with (3), that Boyle 'seems to forget' the very ague attack since which Boyle noted that he had been free from this illness; it is also unclear why the expression 'during his Minority' is relevant to the dating of the text, as there claimed.

[11] Its intended terminal point is unclear. The way in which it is divided into sections by the crossing of seas and/or the Alps suggests that it might have been his return to England. On the possibility that the notes in Bacon's hand might be an extension of it, see below, p. xxiii.

[12] Maddison, *Life*, pp. 2–45. A further, overlapping commentary will be found in Oster, 'Biography, Culture and Science', pp. 184f.

[13] See Principe, 'Early Boyle'; Hunter, 'How Boyle Became a Scientist'.

[14] See, e.g., John Hall to Hartlib [1647], Hartlib Papers 60/14/36A. See further Hunter, 'How Boyle Became a Scientist'.

[15] Harwood, *Essays and Ethics*.

[16] Ibid., xxiiif. See also Todd, *Christian Humanism*, esp. chs 2–3. For a discussion from a more literary viewpoint, see Greene, 'Flexibility of the Self'.

[17] Birch, *Life*, p. xxxiv.

[18] Harwood, *Essays and Ethics*, p. 185f.

[19] BP 37, fols 160–3. On this ambivalence, see Kelso, *Doctrine of the English Gentleman*, esp. pp. 22–4, 29–30; Hexter, 'Education of the Aristocracy', esp. pp. 66–7; Caspari, *Humanism and the Social Order*, pp. 18–20, 189–91, 233f., 339. The argument by Jacob, *Robert Boyle*, pp. 48–9, that this is an early essay expressing views that Boyle was to reject in the *Aretology* is circular, in that the only evidence of early date is the presumption that the ideas contained in it are incompatible with those of the *Aretology* and thus must have been superseded by that work. It is worth pointing out that the text is incomplete, and since the surviving fragment deals with the gentleman's 'Birth' and 'Fortune', the degree to which it stresses heredity is not surprising; but even here Boyle states that 'My Dessein is, to turn both the Greatnesse & the Obscurenesse of Men's Birth, into Motives unto Vertu' (fol. 161ᵛ).

[20] Harwood, *Essays and Ethics*, pp. 143ff.; BP 3, fols 91–2 (not to be confused with Boyle's later 'Essay of the holy Scriptures', on which see Hunter, 'How Boyle Became a Scientist'). For other texts from this period, see the items noted as being in the hand of 'early Boyle' in Hunter, *Letters and Papers*, *passim*.

[21] *Works*, vol. ii, pp. 336, and 323–461 *passim*.

[22] BP 7, fols 128f.; 14, fols 1f. On Hall's significance, see Fisch, 'Bishop Hall's Meditations', and *Jerusalem and Albion*, pp. 48f.

[23] *Works*, vol. ii, p. 327.

[24] BP 14, fols. 7ᵛ–8, 9, 10–11ᵛ and ibid., fols 1f. *passim*; *Works*, vol. ii, pp. 383, 385. For further *Scripture Reflections*, see BP 7, fols 128f.

[25] BP 37, fol. 169. On the genre, see Boyce, *Theophrastan Character*.

[26] Below, p. 70. Cf. also pp. 10–11, and his reference to 'Divers little essays, both in verse and prose' which he had scribbled: Birch, *Life*, p. xxxiv.

[27] See below, pp. 8, 15. Boyle's attitude to romances and the influence of the French romance on him will be more fully illustrated by Lawrence Principe in a forthcoming study.

[28] Boyle to Broghill, 20 December 1649, *Works*, vol. vi, p. 50. See also *Works*, vol. ii, p. 248, and the letters between Boyle and John Mallet of 22 January 1654 and 23 March 1655, Add. MS 32093, fol. 318, and *Works*, vol. vi, pp. 634–5. Harwood, *Essays and Ethics*, p. liin., is thus wrong to claim that there is no evidence that Boyle had read *Parthenissa*. It should be noted, however, that Harwood's account of Boyle's reading is here as elsewhere unfortunately vitiated by his erroneous belief that a catalogue of John Warr's books is in fact of Boyle's: see ibid., pp. 249–81, and Hunter, *Letters and Papers*, pp. xxi-ii.

[29] See Harwood, *Essays and Ethics*, pp. li-iii.

[30] See Bannister, *Privileged Mortals*, p. 96 and *passim*.

[31] BP 14, fol. 2–4, published in Harwood, *Essays and Ethics*, pp. 283–5, though, by publishing this text on its own, Harwood unfortunately obscures the significance of this juxtaposition.

[32] BP 7, fols 288–90; *Works*, vol. v, pp. 255–311; Boswell, *Life of Johnson*, vol. i, p. 312. I have recently rediscovered an early version of the whole of *Theodora*, which I hope will be published in due course.

[33] Oster, 'Beame of Divinity'; for details of others, mainly in BP 37, see Oster, 'Biography, Culture and Science', pp. 198f. For *Seraphic Love* see Boyle, *Works*, vol. i, pp. 243–93; for the discovery of an early manuscript of this treatise in its original form, see Principe, 'Early Boyle', which also briefly indicates its links to the genre of the romance.

[34] Below, pp. 3, 4.

[35] For instance, *The Gentleman*: see above, p. xvi. See also Harwood, *Essays and Ethics*, p. lxviii.

[36] Harwood, *Essays and Ethics*, p. 3 and *passim*; below, pp. 2–3, 6, 11–12.

[37] Below, pp. 3–4, 20, 8, 12. See Harwood, *Essays and Ethics*, pp. xlviii-liii, 190–6; Jacob, *Robert Boyle*, p. 29; see also *Works*, vol. v, pp. 175–7.

[38] Below, pp. 2, 5.

[39] Below, pp. 16–17: for the suggestion that the essays on 'piety', 'sin' and 'valour' in RS MS 196 formed part of this work, see Maddison, *Life*, p. 33n, and Harwood, *Essays and Ethics*, p. 169n. See also Boyle's reference to his *Essays* on p. 14.

[40] For general accounts see Delany, *British Autobiography*, esp. pp. 27f., 107f., and Watkins, *Puritan Experience*, esp. ch. 2. See also the works of Bottrall and Ebner included in the bibliography. For a brief discussion of *Philaretus* in this context, see Oster, 'Biography, Culture and Science', pp. 178–9, though this presumes that the genre was more clearly defined by the 1640s than was actually the case. Oster refers to Browne's *Religio Medici* as a possible exemplar, but there is no evidence that Boyle knew it; in any case, there is very little similarity between this and *Philaretus*.

[41] Sibbald, *Memoirs*, p. 61; Bramston, *Autobiography*, p. 4, and note also the ethos expressed in ibid., pp. 1–4, *passim*. See also Delany, *British Autobiography*, pp. 111, 143, 147–8.

[42] *Works*, vol. i, p. 355. See also below, p. 26, and Harwood, *Essays and Ethics*, p. xxxv.

[43] Aubrey, *Brief Lives*, vol. i, p. 322.

[44] Below, pp. 7, 24, 26.

[45] Grosart, *Lismore Papers*, 1st series, vol. ii, pp. 100–17; Birch, *Life*, vi–xi. See Canny, *Upstart Earl*, 152 n.4 and *passim*.

[46] Hartlib, *Ephemerides* (University of Sheffield), s.v. 1648 (31/22/2B). It is alternatively conceivable that this could be the book of evidence of Cork's innocence from the charges levied against him by Wentworth (see Canny, *Upstart Earl*, p. 22), but this seems less likely. For Boyle's later interest in Cork's life, see the comments of Pett, p. 77, below: it is interesting that Pett's remarks imply that he inherited his father's apologetic intent.

[47] Canny, *Upstart Earl*, pp. 28–9 and ch. 3 *passim*.

[48] Mendelson, 'Stuart Women's Diaries', p. 188. On providentialism in general see von Greyerz, *Vorsehungsglaube und Kosmologie*. On the interest of the Hartlib circle in the 'Registring of Illustrious Providences', see Harwood, *Essays and Ethics*, p. xxxvi.

[49] Below, pp. 15–17; Bunyan, *Grace Abounding*. For the genre as a whole, see Watkins, *Puritan Experience*.

[50] See Haller, *Rise of Puritanism*, esp. ch. 3; Collinson, 'Magazine of Religious Patterns'.

[51] Canny, *Upstart Earl*, pp. 27–8. The household of Lady Ranelagh is another potential source, in view of Boyle's comments on p. 25, below.

[52] See Crocker, ed., *Autobiography*. See also Walker, *Virtuous Woman Found*; Mendelson, *Mental Life of Stuart Women*, ch. 2.

[53] Below, p. 16 (and p. 154 n. 220).

[54] Below, pp. 17, 26.

[55] For the alternative – not mutually exclusive – suggestion that it derives from Aristotle, *Nicomachean Ethics*, I, viii, see More, *Life and Works*, p. 3n.; Oster, 'Biography, Culture and Science', p. 217 n. 8.

[56] Herbert, *Life*, p. 5 and *passim*; Ebner, *Autobiography*, pp. 115–16; Digby, *Private Memoirs*; Delany, *British Autobiography*, pp. 128–32. On a manuscript continuation of the *Arcadia* thought to have been written by a member of the Digby family, see Potter, *Secret Rites and Secret Writing*, pp. 93–4. The *Arcadia* was also known in Boyle's milieu: Cork had given a copy of the work to his daughter, Mary, when she was ten years old (Canny, *Upstart Earl*, p. 87), while Boyle cites it in his letter to [Mrs Dury] of 15 April 1647, along with the romance *Polaxandre* by de Gomberville (BL 1, 109).

[57] Birch, *Life*, p. xxvii. On romances and historicity, see Bannister, *Privileged Mortals*, ch. 6; McKeon, *Origins of the Novel*, pp. 52f.

[58] Below, p. 14; see also p. 19. Cf. above, p. xvi.

[59] See Feingold, 'Galileo in England', pp. 419–20.

3. Boyle's Later Reminiscences and the 'Burnet Memorandum'

[60] See Hunter, 'How Boyle Became a Scientist'.

[61] Perhaps the fullest is the account of Boyle's health in the preface to the second volume of his *Medicinal Experiments* (1693): see below, n. 89.

[62] See Maddison, 'Portraiture', pp. 159f., 183–4.

[63] Glanvill, *Plus Ultra*, pp. 92f. Ibid., pp. 104–6 is taken verbatim from BP 8, fol. 1, which is endorsed 'Receaved from Mr Boyle April 25th 1666. by me, Henry Oldenburg': the significance of this document will be assessed in my forthcoming study of the making of *The Usefulness of Natural Philosophy*.

[64] Hartlib Papers 62/15/1B. On Beale's 'cosen German', Lady Speke, see Beale to Evelyn, 28 May 1664, Corr. no. 40. Beale seems to have believed that he was related to Boyle. For fuller information see Stubbs, 'John Beale', pp. 465–6.

[65] *Life*, p. v. This information is not to be found among the Beale-Hartlib letters in the Hartlib Papers, to which Birch in any case did not have access; neither is it to be found in any of the Beale letters in BL 7. This suggests that it must have been included in the item described as 'Extract of Mr Beale's papers about Mr. B. Ancestors' in the list of biographical writings about Boyle that Miles and Birch had from William Clarke, which cannot now be located (Add. MS 4229, fol. 70). A further related item referred to by Clarke in his letter to Bowyer in Nichols, *Literary Anecdotes*, vol. iv, p. 454, was described by him as 'Memoirs of the Family'; he noted that it was missing, something which 'I have heard Dr Wotton often lament'. Its nature is unfortunately unknown.

[66] See Canny, *Upstart Earl*, ch. 4, esp. pp. 42–3. Humphrey de Biuvile does indeed appear in tit. 28 of the Herefordshire entry in Domesday Book, but whether there was any link between him and the Boyles is another question.

[67] That it was dictated is particularly clearly illustrated by the mishearing of 'Mansion' as 'mention' on p. 23 (cf. p. 157 n. 7; conceivably, this gives a clue to Boyle's accent and intonation). This therefore militates against the alternative possibility that Bacon was retranscribing a document of earlier date. It should be noted that Harwood is wrong in believing these to be *Bacon*'s recollections (*Essays and Ethics*, pp. xlix, 194n): here, as elsewhere, he was merely writing down Boyle's words.

[68] Hunter, *Letters and Papers*, p. xxxi.

[69] Boas, *Boyle and Seventeenth-century Chemistry*, p. 12n.

[70] In paraphrasing this document, Birch refers only to the army, not the court and the army (*Life*, p. xxvii). On this basis, Jacob has a passage concerning Boyle's failure to join the army: *Robert Boyle*, p. 13. This section of Document 2 was printed in full in Maddison, *Life*, pp. 53–4.

[71] Foxcroft, *Supplement to Burnet's History*, pp. 464–5.

[72] Burnet, *History of the Reformation*, vol. ii, sig. a2; Birch, *Life*, p. cxx; Burnet, *Some Letters*, passim.

[73] Hunter, 'Casuistry in Action', esp. pp. 141–2.

[74] *Works*, vol. vi, p. 627.

[75] Maddison, *Life*, pp. 259–60.

[76] BP 4, fol. 166: this is clearly the earliest draft, preceding the two copies surviving in ibid., fols. 179–80 (both lacking the title), which are in two different scribal hands. This is shown by the fact that, whereas in the version in Burnet's hand, the antagonists are specified as 'Atheists and Theists' only, in the others 'Pagans, Jews, and Mahometans' are added, as in the final version (see Maddison, *Life*, p. 274). In addition, the scribal copies add 'real' before 'scruples', and substitute 'Objections' for 'books' in section 3 (the wording of which is otherwise unchanged between Burnet's draft and the final version, as is section 2). They also add more detail concerning the logistics of the scheme: whereas Burnet's version simply ends:

'The endowment £50', this is extended in the scribal drafts to state that it would be in houses in London. They also add that the Lectureship was to be for a seven-year term, which could be extended for a further seven years. For a statement in the funeral sermon implying that Boyle had had the idea for some time, though only executing it in his will, see below, p. 48.

77 The drafts state that 'The designe is to fixe a character upon an Eminent Clergyman of London', but the arrangements are left rather vague: the lecture was to be 'on some setled day of the week and in some certain Church the moneths excepted to be the three hot moneths and December'. The further administrative details are as follows: 'The person to be first nominated by the Donor and Trustees to the number 5 with a power of substitution who with the Bishop of London and Dean of St Pauls for the time shall in all time coming choose the person by this standard that he is the man whom they in their consciences think the fittest for it and that they shall receive no solicitations from any hand but choose all Motu proprio'.

78 Maddison, *Life*, p. 274. See also previous note for the appeal to conscience in the drafts which does not recur in the final version.

79 BP 36, fol. 116. However, another memorandum refers to 'The Lecture for Christianity' (ibid., fol. 87).

80 This represents a modification of the assertion made in the first sentence of Hunter, 'Alchemy, Magic and Moralism'. A date prior to Burnet's exile is possibly also suggested by a story concerning Boyle and transmutation told by J. J. Manget in the preface to his *Bibliotheca Chemica Curiosa* (vol. i, sig. †4), which has come to my attention since that article was written. Manget states that he was given this information in 1685, describing his informant – who actually showed him some of the gold that resulted from the projection – as 'Vir Reverendiss. ac Illustriss. magno suo merito jam Episcopatum in Anglia occupans': this can only be Burnet. A few of the details in Manget's story echo those in the case of transmutation recounted on p. 30, below, but others differ, and it seems to describe a further episode of a similar kind. It is arguably unlikely that Boyle would have given Burnet general information such as is contained in the Memorandum if Burnet was already privy to data such as he gave to Manget in 1685.

81 Maddison, *Life*, pp. 145–7.

82 See Stauffer, *English Biography before 1700*, pp. 163–5, 233–5. See further below, pp. xxxi–ii.

83 Ames, *Conscience*, vol. iv, pp. 30–1.

84 Add. MS 4314, fol. 70. The underlining presumably denotes Miles' quotation of Wotton's exact words.

85 The principal exception to this is no. 23, relating to his writings of the late 1640s, which appears to backtrack from the previous ones, which clearly relate to Boyle's illness in Ireland in 1654 and its aftermath. However, the further phrase which Burnet added but then deleted may suggest that Boyle deliberately intended to contrast the stage in his development at which the works in question were written and that at which they were prepared for publication. In addition, no. 26 is forward-looking, presumably because it is a comment ancillary to no. 25.

86 For a commentary on the central component dealing with Boyle's attitude to magic, see Hunter, 'Alchemy, Magic and Moralism'.

[87] See below, p. 28.

[88] *Life*, pp. xlviii-l. See also Hunter, 'How Boyle Became a Scientist'.

[89] *Works*, vol. v, pp. 315–16. By reading these two texts together, and referring to ancillary evidence, it is possible to give an exact date for this crucial ailment. There are two key pieces of information: one (given in *Medicinal Experiments*) that the initial fall from a horse occurred in Ireland; the other, that the ensuing illness developed into a fever the following winter. We know from a letter from John Mallet to Boyle of 27 March 1655 that he had been ill in London in the winter of 1654–5 (*Works*, vol. vi, p. 634), while the London bills of mortality confirm that 1654 saw an unprecedented number of deaths from 'fever' (Appleby, 'Nutrition and Disease', p. 20; see also ibid., p. 15). Hence the original fall and illness must have occurred in the summer of 1654.

[90] Below, pp. 4–5, 27–8. Birch, *Life*, p. cxxxviii.

[91] I am grateful to Simon Schaffer for his advice on this point in a private communication. For background see Shapin and Schaffer, *Leviathan and the Air-Pump*.

4. Burnet's official view of Boyle and the notes of Sir Peter Pett

[92] *Biographia Britannica*, vol. ii, p. 930.

[93] Details are given in the bibliography. All but the earliest were published in matching format by Richard Chiswell.

[94] See below, pp. 37 (and 37–57 *passim*), 90.

[95] Stauffer, *English Biography before 1700*, pp. 273–8; Paul, *Science and Immortality*.

[96] See Hardison, *The Enduring Monument*; Paul, *Science and Immortality*; Garrison, *Dryden and the Tradition of Panegyric*.

[97] Below, p. 62.

[98] Fulton, *Bibliography of Boyle*, pp. 172–4 (nos. 304–11); some of the broadsheet elogies are there reproduced in facsimile.

[99] Burnet, *Sermon [for the] Archbishop of Canterbury*, p. 3.

[100] Below, p. 56.

[101] Below, pp. 51, 49, 52.

[102] See below, pp. 47 (Ussher and biblical study); 49 (Pococke and Turkish translation of Grotius); 49 (hostility to persecution); 50 (offer of holy orders); and possibly also 50 (attitude to separatist assemblies); 54 (withdrawal from 'Affairs and Courts').

[103] Below, pp. 33, 50.

[104] E.g., *Style of the Scriptures*, *Works*, vol. ii, p. 249, or *Christian Virtuoso*, ibid., vol. v, p. 509.

[105] Below, p. 50. For such clichés, see Hunter, 'Problem of Atheism', pp. 145, 149, 151. For Burnet's concern, see Goldie, 'Priestcraft and the Birth of Whiggism', p. 220.

[106] Below, pp. 48–9, and above, pp. xxiv–v.

[107] Below, p. 47.

[108] Maddison, *Life*, p. 191; Wood, *Fasti Oxenienses*, vol. ii, p. 287; Moreri, *Dictionary* (1694), vol. ii, s.v. 'Robert Boyle' (the other source used was the catalogue of Boyle's books appended to various of his publications in his later

years; the reference to Burnet's intended life is omitted from the otherwise identical account of Boyle in the 1701 edition of the *Dictionary*). See also Evelyn to Williams, 15 June 1696, Letterbook no. 759.

[109] For details of these works, see Bibliography. See also above, p. xxvi.

[110] Below, pp. 54–5; Burnet, *Life and Death of Hale*, sig. A8v, pp. 171–81.

[111] See below, p. 47; Bentley, *Correspondence*, vol. i, p. 118. See also Evelyn to Williams, 15 June 1696, Letterbook no. 759.

[112] Burnet, *History of My Own Time*, vol. i, p. 344.

[113] Stauffer, *English Biography before 1700*, p. 253 and *passim*; Hunter, 'Dilemma of Biography'.

[114] Wood, *Athenae Oxenienses*, vol. iv, pp. 576–80; *Life and Times*, vol. i, p. 201.

[115] On the former, see the account in Jacob, *Robert Boyle*, pp. 133f., though for a caveat see Hunter, 'Conscience of Robert Boyle', pp. 152–3. On the latter see *Works*, vol. ii, pp. 251–5 and Pett's edition of Anglesey's *Memoirs*, sigs. A6–7.

[116] Below, p. 78.

[117] See esp. Goldie, 'Sir Peter Pett'; see also Hunter, *Science and Society*, p. 125. Pett's characterisation of himself, quoted in the text, is slightly at odds with Goldie's (though Goldie rightly indicates his predominant commitment to a secular 'civil religion'). In addition, Goldie follows Jacob in conflating Pett's position with Boyle's, and in presuming undue unanimity on the part of the Royal Society in espousing such positions.

[118] See further sect. 8, below.

[119] Below, pp. 74, 69.

[120] See below, pp. 58, 62, 63, 72, 82; Pepys to Pett, 16 November 1695, Pepys, *Private Correspondence*, vol. i, p. 111. It might be this letter to which Pett refers on p. 82, below; Pett's letter to Pepys does not appear to survive. See also Pett to Pepys, 3 May 1696, ibid., p. 115, which suggests that the information had still not arrived by then. Though he was unaware that the sealed packet deposited by Boyle at the Royal Society had been opened, an event which occurred in February 1692 (Maddison, *Life*, p. 202), it is doubtful if this proves anything about the date of the text; I have not been able to trace the intended collection of texts concerning liberty of conscience to which he also refers: below, pp. 62–3, 73.

[121] Below, p. 59.

[122] Below, pp. 62–3, 76, 78.

[123] Below, pp. 62, 73, 75.

[124] Below, pp. 76, 77.

[125] Below, pp. 79, 77.

[126] See Hunter, 'Dilemma of Biography'.

[127] Barlow, *Genuine Remains*, sig. A2v, pp. 324f., and *passim*; Anglesey, *Memoirs*, Ep. Ded. and *passim*; Trevor-Roper, *Catholics, Anglicans and Puritans*, pp. 173, 312. See also below, p. 67, where Pett recounts an anecdote about Falkland told him by Boyle; Boyle's memoirs of Falkland were among those he was hoping to use in his abortive life.

[128] Below, pp. 62, 79.

[129] Below, p. 62; Maddison, 'Portraiture', p. 162. The work by Barlow was his *Cases of Conscience*: in associating the publication of this with Pett, J. R. Jacob has perhaps confused it with Barlow's *Genuine Remains* (*Robert Boyle*, p. 134).

[130] Below, p. 62.

[131] Below, p. 69.

[132] See DNB. See also Hale, *Account*, esp. pp. ix-x, xiii. That there was also a link between Pett and Hale's *Account* is suggested by the repeated echoes of the latter in Pett's notes, and by Hale's various citations of Pett's *Happy Future State of England*: ibid., pp. xxxiii, xxxviii, xxxixf., xlvi.

[133] See Maddison, 'Salt Water Freshened'; Fitzgerald, *Salt-Water Sweetned*; and below, p. 82.

[134] Below, pp. 65–6.

[135] Below, pp. 69–70.

5. Wotton's *Life*, I: its Progress and Demise

[136] Evelyn to Pepys, 7 July 1694, in Pepys, *Letters and Second Diary*, p. 241. See also Levine, *Battle of the Books*, pp. 31–3 and *passim*.

[137] Hall, 'William Wotton', p. 1049 and *passim*. See also Levine, *Battle of the Books*, ch. 1.

[138] On Burnet's patronage of Wotton see Gascoigne, *Cambridge in the Age of the Enlightenment*, pp. 148–9, and Clarke's Life of Wotton in Add. MS 4224, fol. 153. For the attribution of this text to Clarke rather than Birch (as stated by Levine, *Battle of the Books*, pp. 32n, 350n, 408n, 412–13) see Clarke to Birch, 10 September, 4 October 1739, Add. MS 4302, fols 250, 252. See also Clarke to Bowyer, 12 November 1739, ibid., fol. 254.

[139] See below, p. xxxviii.

[140] Levine, *Battle of the Books*, pp. 338f. On Evelyn's earlier literary career, see Hunter, 'John Evelyn in the 1650s'.

[141] Wotton, *Reflections*, pp. 290–307; see also Evelyn to Wotton, 28 October 1696, Evelyn, *Diary and Correspondence*, vol. iii, pp. 363–5.

[142] Pepys, *Letters and Second Diary*, pp. 241–2; Wotton to Evelyn, 11 July 1694, Evelyn MS 3.3.104; Evelyn to Wotton, 11 July 1694, Letterbook no. 709. The letters also reveal that Wotton gave Evelyn a copy of the *Reflections* and that a copy of Hooke's *Micrographia* changed hands between the two men.

[143] Bentley, *Correspondence*, vol. i, p. 117, and below, p. 84. For an overlapping account, see Evelyn to Williams, 15 June 1696, Letterbook no. 759.

[144] Bentley, *Correspondence*, vol. i, p. 118.

[145] E.g. Maddison, 'A Summary', p. 92. See especially Evelyn, *Diary and Correspondence*, vol. iii, pp. 353, 370, 388–90. Cf. Evelyn MS 3.3.111; Letterbook no. 759 (in this, Evelyn's letter to Williams of 15 June 1696, on which see further below, Evelyn went particularly far in implying that the initiative was his rather than Wotton's, but this was presumably part of his strategy to discourage Williams from his overlapping venture).

[146] Evelyn to Wotton, 26 June 1697, Letterbook no. 792.

[147] Add. MS 4276, fol. 211.

[148] See Evelyn to Tenison, 5 August 1694, Letterbook no. 711; Evelyn to Wotton, 20 September 1694, Add. MS 28104, fol. 19; Evelyn to Pepys, 2 September 1694, Pepys, *Letters and Second Diary*, pp. 249–50. See also Wotton to Evelyn, 17 September 1694, Add. MS 28104, fol. 18.

[149] Below, p. 85; Evelyn to Wotton, 12 May 1696, Letterbook no. 757.

[150] Wotton to Evelyn, 13 August 1703, *Diary and Correspondence*, vol. iii, p. 389.

[151] Below, p. 87 and 84–90, *passim*.

[152] Letterbook no. 757: *aside* replaces *downe* deleted, and *Panegyric* replaces *Oration* deleted.

[153] Evelyn to Williams, 15 June 1696, Letterbook no. 759. Cf. Evelyn, *Diary and Correspondence*, vol. iii, p. 353 (a letter from Wotton of 24 May stating that Burnet's fiat was still awaited).

[154] *Works*, vol. vi, p. 1.

[155] Williams to Evelyn, 19 June 1696, Evelyn, *Diary and Correspondence*, vol. iii, pp. 359–60. On Charlett, see Carter, *History of Oxford University Press*, pp. 150–1. Williams' original letter is now in the Pierpont Morgan Library, and is printed in Buehler,'A Projected but Unpublished Edition' (though it should be pointed out that Buehler is mistaken both in dating the letter to April rather than June, and in believing it to be previously unpublished; he is also under the misapprehension that Burnet's account of Boyle first appeared in the 1737 edition of Budgell's *Memoirs*: see p. lvi, below).

[156] Evelyn to Williams, 15 June 1696, Letterbook no. 759.

[157] Evelyn, *Diary and Correspondence*, vol. iii, pp. 359–60.

[158] Below, p. 77.

[159] Evelyn, *Diary and Correspondence*, vol. iii, p. 353.

[160] Wotton to Evelyn, 2 January 1698, Evelyn, *Diary and Correspondence*, vol. iii, p. 370.

[161] Evelyn to Wotton, 24 December 1697, Letterbook no. 802: *weare* replaces *have* deleted.

[162] Evelyn, *Diary and Correspondence*, vol. iii, p. 370.

[163] For references to this letter see Miles to Millar, 24 September 1741, 4 January, 9, 15 February 1741/2, Add. MS 4229, fols 110ᵛ, 130, 133, 134; Miles to Ward, 10 February 1741/2, Add. MS 6210, fol. 248. See also Maddison, 'A Summary', pp. 101–4.

[164] Evelyn, *Diary and Correspondence*, vol. iii, p. 370. Cf. Wotton to Evelyn, 20 January 1698, ibid., pp. 371–2.

[165] Maddison, 'A Summary', p. 93. On Wotton's work on the MSS, see Hunter, *Letters and Papers*, pp. xii-xiii.

[166] See Burtchaell and Sadleir, *Alumni Dublinenses*, p. 224; Foster, *Alumni Oxonienses*, vol. i, p. 395. See also 'A Letter from the Reverend Mr *Thomas Dent*, to Sir *Edmund King*... Concerning a sort of Worms found in the Tongue, and other Parts of the Body', *Phil.Trans.*, 18 (1694), pp. 219–21.

[167] Letterbook no. 757.

[168] *Life*, p. cxli. See also below, p. lxx.

[169] Wotton to Evelyn, 8 August 1699, 22 January 1702, Evelyn MS 3.3.112, Evelyn, *Diary and Correspondence*, vol. iii, pp. 385–6. On Boulton see further below, sect. 7.

[170] Wotton, *History of Rome*, sigs. A2–7 and *passim*; Levine, *Battle of the Books*, pp. 343f.

[171] 31 October 1701, Corr. no. 1683. This is a draft, and it is slightly garbled.

[172] Evelyn, *Diary and Correspondence*, vol. iii, p. 385.

[173] These are listed at the end of Kirkwood's letter, and most of them survive among the Boyle Letters. See below, pp. 107, 109–10. On this venture see also Johnston, 'Notices of a Collection of MSS' and Maddison, 'Boyle and the Irish Bible', pp. 97–101 (ibid., plates 4–5, reproduce the title-pages of the editions of the Bible produced in connection with it). Further information on the episode, derived from hitherto unknown MSS at New College, Edinburgh, will be provided in a forthcoming study by Janelle Evans.

[174] *Works*, vol. i, p. cci. See Bahlman, *Moral Revolution of 1688*; Curtis and Speck, 'Societies for the Reformation of Manners'. It might be added that in the high priority that he gave to education as a means of achieving this, Boyle had more in common with the Society for Promoting Christian Knowledge than with the Societies for the Reformation of Manners, though the two overlapped: Bahlman, *Moral Revolution*, esp. pp. 70f.

[175] Corr. no. 1685. See also below, pp. xlviii–ix, li. The letter opens: 'You overwhelme me Worthy Sir, with Retributions & Civilitys [followed by *and* deleted] Upon which account alone [replacing *onely* deleted] (how Conscious soever of my owne Demerits, taken in the genuine Sense) I suffer all the fine things you are pleas'd to say of me; and to passe to the Obligation, which the publique as well as my selfe, must ever acknowledge for what you have already produc't of Excellent and Instructive, [sic] and what you further promise in the Work now under your hand'. It is, of course, just conceivable that Evelyn never sent the letter.

[176] Followed by *by* deleted.

[177] Followed by an illegible deleted word. Later in the sentence, *Genius &* is written is pencil.

[178] Followed by *(that the sowreness* deleted: the text is slightly garbled at this point.

[179] A letter from Evelyn to Boyle of 9 May 1657 (which refers to Wilkins' school) is preserved in Add. MS 4229, fol. 52. Other Evelyn letters in the Birch collection, which presumably came from Wotton, comprise Add. MS 4229, fol. 54; Add. MS 4279, fols 25–6; and Add. MS 4293, fols 69–72.

[180] Evelyn, *Diary and Correspondence*, vol. iii, pp. 389–90. On the verso of the original, now Evelyn MS 3.3.114, Evelyn has made some notes relating to his reply. In addition, he noted in the margin adjacent to the passage in Wotton's postscript about Fenton: 'Felton it should be'.

[181] Below, p. 98.

[182] Document 6c. Evelyn's copy of his letter to Stanhope asking him to transcribe the monument, dated 22 August 1703, is now Corr. no. 1704.

[183] Below, pp. 94–6, 97–8. He also told the story of Cork's meeting his future wife as an infant, the plausibility of which is discussed in Birch, *Life*, p. ixn; see also Miles to Birch, 21 November 1742 and 10 August 1743, Add. MS 4314, fols 69, 88.

[184] Hearne, *Remarks and Collections*, vol. i, p. 239: no related reference appears in either Council Minutes or the Copy Journal Books of the Royal Society between April 1705 and May 1706. Add. MS 4041, fol. 317.

[185] Add. MS 4041, fol. 317. See also below, p. li. Though Wotton requests von Guericke's *Experimenta Nova Magdeburgica de Vacuo Spatio* (1672), Boyle's knowledge of von Guericke's experiments derived from their publication in Schott's *Mechanica hydraulico-pneumatica* (1657). The Galileo titles which Wotton gives do

not relate very clearly to any of Galileo's writings; he may have been referring to a quarto edition of Galileo's works such as that published at Bologna in 1655–6, each work within which had a separate title-page. His bookseller was presumably Timothy Goodwin, d. 1720: see Plomer, *Dictionary of Printers*, pp. 129–30.

[186] See Guerlac, 'Sir Isaac and the Ingenious Mr Hauksbee', pp. 228, 232, 240–2 and *passim*, and Hall, *Promoting Experimental Learning*, p. 118.

[187] Evelyn, *Diary and Correspondence*, vol. iii, pp. 388–9. But see also above, p. xliii.

[188] Levine, *Battle of the Books*, pp. 403–4. Wotton's *Letter to Eusebia* contains a reference to Boyle and the air-pump: p. 33.

[189] Hearne, *Remarks and Collections*, vol. ix, p. 286. Cf. ibid., vol. iii, p. 236, and Levine, *Battle of the Books*, p. 404 (who quotes a later account attributing the affair to Wotton's lack of 'common Decency in Respect to Wine and Women', and stating that he fled to his son-in-law's house at Buxted, Sussex, before going to Wales).

[190] Clarke to Birch, 7 September 1741, Add. MS 4302, fol. 256.

[191] Add. MS 4224, fols 165–6 (for the date, see above, n. 138); *to have* replaces *to thinke that he had*, deleted. The claim made in this text that Wotton began the task 'About the year 1701' is obviously incorrect, however. This account is slightly abbreviated in the printed version of Clarke's life in Bayle, *General Dictionary*, vol. x, pp. 203–8.

6. Wotton's *Life*, II: From Panegyric to Intellectual Biography

[192] Add. MS 4224, fol. 165. Elsewhere, Clarke stated 'I have no Papers of Dr Wotton's relating to Mr Boyle's Life that are of any Moment' (Clarke to Birch, 7 September 1741, Add. 4302, fol. 256), but this may have been aimed to get Birch and Miles off his back: see sect. 7, below. See also above, sect. 5, n. 3.

[193] Add. MS 4229, fol. 70; Miles to Birch, 21 October 1742, Add. MS 4314, fol. 69; BP 36, fol. 128.

[194] *Life*, p. liv: he there describes it as 'Dr *Wotton*'s papers'.

[195] BP 36, fol. 128. Most of the points he quotes from Wotton are purely factual – e.g. 'came home in 1645' – which Miles then embellished with queries or further information (not always accurate) of his own. Virtually all of the information that he used derived from letters which were in his possession or from the biographical sources published here; on the verso of the paper are some notes by Miles from these. In addition, Miles refers to 'Dr Beales Letter to Mr Hartlib' – evidently the document dealt with in section 3, above, in connection with the Boyle family's Herefordshire origins – which is here cited concerning Beale's acquaintance with Boyle as a schoolboy at Eton. A further instance where Miles seems to have had a source of information now missing concerns the actions of the Earl of Cork's agent, Perkins, during Boyle's travels abroad. Wotton evidently cited the Burnet Memorandum on this (see below, p. 27), and Miles added: 'he was a Taylor but reputed an Honest man, had made Clothes for the Family & under false pretence that one of the Brothers owd him money, withheld what the E[arl] put into his hands to return to his Sons abroad & never repaid it'. For a parallel account, based on the Lismore Papers, see Maddison, *Life*, pp. 45–6, 50, but Perkins' excuse as given by Miles is not there directly referred to.

[196] See above, p. xxvi. Note also Miles's reference to Wotton 'crossing' the story concerning Cork's meeting his future wife as an infant: Miles to Birch, 21 October 1742, Add. MS 4314, fol. 69.

[197] Loc. cit.

[198] Letterbook nos. 757, 759.

[199] Wotton, *History of Rome*, sig. A5. Cf. Levine, *Battle of the Books*, p. 350.

[200] Corr. no. 1685 (*bright* replaces *Shining* deleted); Wotton, *Defence of Reflections*, p. 58 and pp. 47f., *passim* (in connection with Swift's *A Tale of a Tub*). See also his *Letter to Eusebia*, p. 73 and *passim*, and note Evelyn's letter to Wotton of 27 November 1703, Corr. no. 1705, on the effects of the great storm which had recently occurred, which he speculated was not due to solely natural causes (see Evelyn, *Diary*, vol. v, p. 550, Bahlman, *Moral Revolution*, pp. 12–13).

[201] *Works*, vol. vi, p. 1. For Boyle's own sense of the need for a Reformation of Manners, see above, p. xlii. For a later echo of this didactic aim, see Miles to Birch, 10 August 1743, Add. MS 4314, fol. 89.

[202] See Korshin, 'Development of Intellectual Biography'. For an example of a life of an intellectual which makes almost no attempt to deal with his ideas, see Walton's life of Hooker in his *Lives*, pp. 153–249.

[203] Below, pp. 62, 84–5; Burnet, *Life and Death of Hale*, sig. A5v; Corr. no. 1685.

[204] Gassendi, *Life of Peiresc*, sig. a2v; Joy, *Gassendi the Atomist*, p. 53; Evelyn, *Miscellaneous Writings*, p. 552; Corr. no. 1685: this is the draft referred to on p. xlii, and the text is again slightly garbled.

[205] Replacing *Retired & Contemplative &* deleted. *Sedate* was written once and deleted but then reinstated.

[206] Followed by *might* deleted.

[207] Replacing *passages* deleted.

[208] Replacing *noble* deleted.

[209] Followed by an illegible deleted word.

[210] Followed by *Examples & Emp* deleted.

[211] Followed by *Beneficien* inserted above by line but deleted.

[212] Followed by *affusive* [?] deleted.

[213] Joy, *Gassendi the Atomist*, p. 54; Gassendi, *Life of Peiresc*, *passim*, esp. pp. 159f. (second pagination).

[214] Baillet, *Vie de Descartes*, *passim*. See below, p. 88; there is also a deleted reference to 'the Author of Des Carteses Life (which Book I remember you first shew'd me at Wotton)' in Letterbook no. 1685. See also below, p. 85.

[215] Gassendi, *Life of Peiresc*, pp. 241–96 (second pagination).

[216] On the documentary impulse, see Stauffer, *English Biography before 1700*, pp. 78–80, 258.

[217] BP 36, fols 180–9. I hope to publish a text of this, with a commentary, as part of the study referred to in n. 274, below.

[218] Wotton to Evelyn, 8 August 1699, Evelyn MS 3.3.112.

[219] See Hunter, *Letters and Papers*, pp. xii, xxv.

[220] Add. MS 4229, fol. 69 (and Miles's copy in BP 36, fol. 202). See also what are evidently Miles's notes on the MSS he received from Clarke, BP 36, fols 163 (dated 10 February 1742/3), 141–3, 164, 147–8. Add. MS 4229, fol. 70, shows that Wotton also had copies of virtually all Boyle's published scientific works.

221 Miles to Birch, 17 November 1742, Add. MS 4314, fol. 73.

222 Korshin, 'Development of Intellectual Biography', pp. 519f.

223 See e.g., the letters of John Beale to Boyle in *Works*, vol. vi, pp. 404f. The idea of systematisation also underlay the publications of Boulton and Shaw. This matter will be dealt with at greater length in the Introduction to the Pickering Masters Boyle.

224 Below, p. 86.

225 Hall, 'William Wotton', p. 1049; Wotton, *Reflections*, p. 262. Cf. ibid., pp. 194–6, 205–6, 260, 371.

226 Evelyn, *Diary and Correspondence*, vol. iii, pp. 385–6. For a parallel usage of 'matters of fact' by Birch, see Korshin, 'Development of Intellectual Biography', p. 516.

227 Corr. no. 1685: *set forth* replaces *Epitomiz'd* deleted. Illegible deletions follow *hence* and *which*. Again, the draft is garbled.

228 Add. MS 4041, fol. 317.

229 Below, p. 135. See also Miles's description of what can only be this chapter in a letter to Birch of 4 November 1742, Add. MS 4314, fol. 66, quoted below.

230 Cf. below, p. 29; *Works*, vol. i, p. 6. In addition, his summary of the air's component parts draws on Boyle's *General History of the Air*, *Works*, vol. v, pp. 612–15.

231 *Works*, vol. i, pp. 5–7. See above, pp. xliv–v.

232 Wotton, *Reflections*, pp. 194–6; below, p. 112.

233 For modern editions of the latter two, see Middleton, *The Experimenters*; Pascal, *Physical Treatises*: the original preface to the latter work (by F. Périer) includes quite a lucid ' History of the Experiments with a Vacuum' with something in common with Wotton's (pp. xv–xx).

234 Gascoigne, *Cambridge in the Age of the Enlightenment*, pp. 148–9; Bentley, *Works*, vol. iii, pp. 27f., *passim*.

235 Below, p. 135. See also p. 143. On the ethos of the *Reflections*, see Levine, *Battle of the Books*, ch. 1.

236 See Shapin and Schaffer, *Leviathan and the Air-Pump*, *passim*.

237 Wotton, *Reflections*, p. 196.

238 Below, pp. 126f. Pascal, whose work he notes but does not expound in detail in the surviving section of the text, could be seen as falling into a similar category.

239 See, e.g., Boyle's references to contemporary authors in *Works*, vol. i, pp. 12–13, 18, 22, 43–4, 50, 52, 62, and his comments in ibid., pp. 36f., 41f., 73f.

240 Below, p. 146; *Works*, vol. i, pp. 73f. See also Shapin and Schaffer, *Leviathan and the Air-Pump*, pp. 40f.

241 Below, pp. 136–7. Cf. Wotton, *Reflections*, pp. 194–5; Middleton, *History of the Barometer*, pp. 65f.

242 Below, p. 145. Cf. Shapin and Schaffer, *Leviathan and the Air-Pump*, pp. 45f., 185f. Wotton had evidently forgiven Keill for his earlier criticism of him in connection with his *Reflections*: *Defence*, p. 4.

243 See Hunter, 'Dilemma of Biography'.

244 Below, pp. 120–1.

245 Below, p. 120. It is, of course, impossible to know what was in the section which is missing at the end.

7. From Wotton to Birch

[246] Cook, 'Sir John Coltbach', pp. 498–501; *Decline of the Old Medical Regime*, pp. 242–3.

[247] Boulton, *Theological Works*, vol. i, p. xxiii.

[248] Letterbook no. 759. See also no. 757.

[249] Boulton, 'Life', p. 355.

[250] Ibid., p. 6; Wood, *Fasti Oxenienses*, vol. ii, pp. 286–7. This itself came from John Aubrey's *Brief Life* of Boyle (vol. i, p. 120), though this is otherwise quite reliable. For accounts which repeat this, see below, pp. lvi–vii.

[251] Boulton, *Life*, pp. 234–353 and *passim*.

[252] Ibid., pp. 59, 167f.

[253] Ibid., pp. 16–17, 73.

[254] Budgell, *Memoirs*, pp. 118f., Appendix (separately paginated).

[255] Budgell, *Memoirs*, pp. 135f. Cf. Boulton, *Life*, p. 167f. Budgell's evaluation of Boyle as a scientist probably owes something to Shaw's edition of Boyle's *Philosophical Works* (see below, p. lvii), but this debt is not explicitly acknowledged.

[256] Budgell, *Memoirs*, p. 145. Cf. below, p. 47. These were both matters on which Birch quoted Budgell in his *Life*: pp. cxliii, cxlv. Budgell claimed that a Mr Collier, as well as Burnet, was 'intimately acquainted' with Boyle (Budgell, *Memoirs*, p. 148). He probably refers to Jeremy Collier, whose name appears on the title-page of the 1701 English version of Moreri's *Dictionary*, in which an account of Boyle appears: however, this account had already appeared in the 1694 edition, with which Collier does not appear to have been associated. See above, p. xxxi.

[257] Budgell, *Memoirs*, pp. 132–3. Swift's *Meditation on a Broomstick* still rankled with Boyle's supporters a century later: see John Weyland's introduction to his edition of *Occasional Meditations*, pp. xviii–xxii, xxvii–viii. Birch also disapproved: *Life*, p. lxxii. Cf. Campbell in *Biographia Britannica*, vol. ii, p. 920, who claimed, however, that Swift 'borrowed the first hint of his *Gulliver's Travels*' from *Occasional Reflections*! (ibid., note).

[258] Bayle, *General Dictionary*, vol. i, sigs. a1–2 and *passim*; Osborn, 'Thomas Birch and the General Dictionary', p. 25 and *passim*.

[259] Osborn, art. cit., pp. 33, 36; Bayle, *General Dictionary*, vol. iii, pp. 541–60.

[260] See Shaw, *Philosophical Works*, vol. i, pp. i–ii and *passim*; Golinski, 'Peter Shaw', esp. p. 23.

[261] Bayle, *General Dictionary*, vol. iii, pp. 548–54n.

[262] Bayle, *General Dictionary*, vol. ii, pp. 549–74, vol. vii, pp. 776–802.

[263] Bayle, *General Dictionary*, vol. iii, pp. 555–9n.

[264] Boswell, *Life of Johnson*, vol. i, p. 159 (Johnson alludes to the torpedo fish). Cf. for instance Stauffer, *Art of Biography*, p. 254; Korshin, 'Development of Intellectual Biography', pp. 516–17.

[265] Hunter, *Letters and Papers*, pp. xiii–iv. See also Maddison, 'A Summary' and Hall, 'Henry Miles and Thomas Birch'.

[266] Maddison, 'A Summary', pp. 97–8.

[267] Maddison, 'A Summary', pp. 101–2, though he fails to note the letters from Knight to Birch of 27 July 1738 and 1 July 1742 in Add. MS 4312, fols 49, 52.

These offered help in obtaining the MSS, which he had himself tried unsuccessfully to obtain from Clarke; he does not, however, seem to have been planning a life of Boyle, as Maddison speculated.

[268] See Maddison, 'A Summary', pp. 101–5. See also above, p. l.

[269] Add. MS 4229, fol. 70. This includes items which are clearly to be identified with Documents 2, 3, 5, 6, 7 and 8 below; it does not, however, include Document 9 (or 1). In addition, it includes reference to the 'Extract' from Beale's papers on Boyle's ancestors (see above, p. xxiii), and Wotton's 'Sketch' (see above, p. xlvii), and to 'Other papers relating to Mr B[oyle]'s Life'.

[270] Viz., to Document 2 on p. xxvii; to Document 3 on pp. xxvi, lx, lxviii–ix, cxxxviii, cxlii–iii and cxliv; to Document 5 on pp. cxxxviii, cxl and cxli; to Document 6A on pp. cxxxvi–viii, 6B on pp. x and lvii, and 6c on p. x; and to Document 7 on pp. xxvi and cxli. He also cited 'Mr *Warr*'s MSS' concerning the exact time of Boyle's death (p. cxxxv) and 'Dr *Wotton*'s papers' concerning where Boyle lodged at Oxford (p. liv: see also above, p. xlvii) and concerning Barlow and casuistry (p. lvi: see also below, p. lxi). For Miles's notes, see above, n. 195.

[271] Add. MS 4229, fol. 70; Miles to Miller, 15 February 1742, Add. MS 4229, fol. 134; Maddison, 'A Summary', p. 104.

[272] Miles to Millar, 29 September 1741, Add. MS 4229, fol. 112; see above, p. xl.

[273] See Add. MS 4228 (the manuscript of Birch's *Life*), fols 16 and 14–33, *passim*. Cf. *Life*, pp. xii–xxvi. Miles's transcript is closer to the original than Birch's published version: though he partially modernised the text, while it was he who introduced the paragraphing which appears in the printed text, he retained more of the original capitalisation than does the printed version, as also the section headings (which Birch deleted), together with quite a number of Boyle's duplicated words and phrases, which Birch either joined together or ignored. For the date when Miles supplied the transcript to Birch see Maddison, 'A Summary', p. 97. In addition, the text was evidently shown to John Ward, and annotations in his hand appear on the blank versos of the manuscript; some of these are elucidations which appear in the printed text.

[274] See Hunter, *Letters and Papers*, pp. xv, xviii-xix, xxv-vi; Maddison, 'A Summary', esp. p. 99. I will be publishing an account of Miles's work on the archive, with particular reference to Boyle's correspondence, at a later date.

[275] Add. MS 4314, fol. 64; Maddison, 'A Summary', p. 99.

[276] BP 36, fol. 167; see also above, p. xlvii.

[277] Add. MS 4314, fol. 66. See also Miles's queries on Birch's draft in Add. MS 4314, fol. 88.

[278] Maddison, 'A Summary', pp. 98–9; Add. MS 4228, *passim*.

[279] *Life*, p. xlii. See also Webster, 'New Light on the Invisible College', esp. pp. 21–2.

[280] *Life*, p. iii. See also above, sect. 1.

[281] For example, the account of *Hydrostatical Paradoxes* is truncated, while those of *Forms and Qualities* and *Spring of the Air* are rewritten more briefly. The account of *Cold* is, however, rewritten in longer form. Virtually all the other descriptions are identical. *Life*, pp. lx-lxi, lxxii-iii, lxxxvii-viii and *passim*: cf. Bayle, *General Dictionary*, vol. iii, pp. 548–54n.

[282] See above, n. 270.

[283] *Life*, pp. cxlvi, cxlix-cl: as already noted, Birch's account is much indebted to Shaw at this point. Cf. Stewart, *Rise of Public Science*, pp. xix-xx. Birch did give a brief account of *Salt-Water Sweetned* in *Life*, p. cxxi.

[284] See my forthcoming study referred to in n. 274.

[285] Add. MS 4314, fol. 88ᵛ.

[286] *Life*, p. lvi. Cf. Hunter, 'Casuistry in Action', p. 86 and *passim*.

[287] Add. MS 4314, fol. 70.

[288] *Life*, p. cxlvif.

[289] See the forthcoming study referred to in n. 274.

[290] *Biographia Britannica*, vol. ii, pp. 913–34.

[291] All Campbell's references to manuscript sources come straight from Birch's *Life*, except for two early letters in the possession of the Boyle family (ibid., pp. 914–15n.), and a work entitled 'Memoirs of the Hon. Robert Boyle, by J.C. MS', which was clearly by Campbell himself. The footnotes to Campbell's account give prominence to Boyle's alchemical interests, presumably because of Campbell's known interest in related matters: see his translation of Cohausen's *Hermippus Redivivus* (1744).

[292] Even Wilson's account of Boyle in his *Religio Chemici* is dependent on Birch for biographical information, though he made unusually imaginative use of it, especially the *Account of Philaretus*: see below, n. 396. An incomplete list of, mainly nineteenth-century, biographical accounts of Boyle will be found in Fulton, *Bibliography*, section D. A number of others which had accompanying illustrations may be traced through Maddison, 'Portraiture', *passim*.

[293] See, e.g., Oster, 'Biography, Culture and Science', pp. 196–7, where Pett's notes are cited as 'anonymous materials'; see also ibid., p. 219 n. 52. Piecemeal references to the documents published here will be found in Maddison, 'A Summary', pp. 91, 93, 101; Maddison, *Life*, pp. 108, 204, 207n; Harwood, *Essays and Ethics*, pp. xliii, xlix, li, 84n, 194n and 217n. See also above, n. 70, and below, nn. 294, 298.

8. The Boyle we have lost

[294] Boas [Hall], *Boyle and Seventeenth-century Chemistry*, ch. 1 *passim*, esp. pp. 11–12.

[295] See [Boas] Hall, *Robert Boyle on Natural Philosophy*; id., 'Boyle, Robert', in DSB, vol. ii, pp. 377–82; More, *Life and Works*.

[296] See, e.g., Jacob, *Robert Boyle*, pp. 3, 113, 127.

[297] Ibid., p. 5.

[298] The only exceptions to Jacob's reliance on Birch's *Life* for this material are two citations of Add. MS 4229, fol. 39: Jacob, *Robert Boyle*, pp. 196 n. 47, 205 n. 1.

[299] North, *Lives of the Norths*, vol. iii, p. 146. For a commentary on this passage, in conjunction with North's scepticism regarding the biographical labours of Burnet and his ilk, see Hunter, 'Dilemma of Biography'.

[300] Below, pp. 28, 32.

[301] Below, Documents 4, 6A. See also Paul, *Science and Immortality*, and above, sect. 4.

[302] See Hunter, 'How Boyle Became a Scientist'.

303 Below, pp. 27, 64–6. See also above, n. 85: it is the deleted clause at the end of no. 23 which indicates Boyle's awareness that he had in some way changed between the time when he initially wrote *Seraphic Love* and *Occasional Reflections* and the 1650s.

304 Below, p. 27.

305 Below, pp. 28–9, 58.

306 Below, pp. 28, 86–7.

307 Below, pp. 28–9.

308 Below, pp. 67, 42, 48.

309 Below, p. 29. See Clericuzio, 'Carneades and the Chemists', p. 82 and *passim*.

310 See Hunter, 'Science and Heterodoxy', esp. p. 451. See also id., *Science and Society*, pp. 178–9, and Shapin and Schaffer, *Leviathan and the Air-Pump*.

311 Below, pp. 28–9. See Harwood, 'Science Writing and Writing Science', pp. 48–9.

312 'General Heads for a *Natural History of a Countrey*, Great or small', *Phil. Trans.*, 1 (1666), pp. 186–9, reprinted in *Works*, vol. v, pp. 733–43. Cf. p. 27.

313 Below, p. 29. In fact, it seems to have been *here* that the enquiries published in *Philosophical Transactions* originated: see Boyle's covering letter to Oldenburg when he sent him these queries: *Oldenburg*, vol. iii, p. 65. See also my forthcoming study referred to in n. 63, above.

314 Jacob, *Robert Boyle*, p. 146.

315 Below, pp. 59–60, 81–2.

316 Below, pp. 82–3, 105, 29.

317 Below, p. 29; Hunter, *Robert Boyle Reconsidered*, pp. 14–15; see also my forthcoming study of Boyle's medical agenda.

318 Below, p. 27; *Works*, vol. v, p. 575.

319 See further above, sect. 2.

320 Jacob, *Robert Boyle*, pp. 144f.; below, p. 33.

321 Below, pp. 33, 58. It is unclear from Pett's text whether Boyle said this while they knew each other at Oxford or later, since it seems to merge with the following section concerning milled lead. It should be noted that what he gives as Boyle's view is very similar to that expressed in Hale, *Account of New Inventions*, p. vi: but there is no reason to doubt that these were Boyle's sentiments. Compare his views in his 'Medica Praescripta Communicata R.B.', BP 17, fols 1f.

322 Jacob, *Robert Boyle*, p. 147: this opinion is there attributed to 'Pett, and Boyle through him'. Cf. ibid., pp. 34, 44, 87, 140–2, 178–9. Jacob does acknowledge that in practice the two sometimes conflicted (ibid., pp. 44, 95, 147), but the repeated thrust of his argument is encapsulated in the quotation given in the text.

323 See further Hunter, 'Conscience of Robert Boyle', pp. 152–3.

324 Jacob, *Robert Boyle*, pp. 133f.; Birch, *Life*, pp. cxli–ii; below, pp. 72–4.

325 Pett, *A Discourse*, p. 4.

326 Pett, *A Discourse, passim.* See also above, pp. xxxii–iii. See further the complementary treatise produced at Boyle's behest at this time, Barlow's 'The Case of a Toleration in Matters of Religion' in his *Cases of Conscience*, which, although rather legalistic in tone, does deal with such issues as the biblical mandate on such matters. See also below, p. lxxi.

327 Below, pp. 49, 72.

328 Below, pp. 68–9.

329 Below, p. 33.

330 See particularly Spurr, 'Latitudinarianism and the Restoration Church'.

331 Below, pp. 33–4, 71.

332 See Spurr, 'Latitudinarianism', pp. 63, 66, 73–4; Gascoigne, *Cambridge in the Age of the Enlightenment*, ch. 2.

333 Kenyon, *Revolution Principles*, p. 22; Spurr, *Restoration Church*, pp. 152f.; Beddard, 'Vincent Alsop', pp. 163–5; Marshall, 'Ecclesiology of the Latitude-men'.

334 *Life*, p. cxli. See below, pp. 105–6. From the context, it is clear that by 'this subject' Dent meant conscience in the sense of casuistry, since he then goes on to refer to Stillingfleet's help to Boyle 'upon some nice & criticall case' and to Barlow's advice to Boyle concerning 'things of this nature'.

335 See Hunter, 'Casuistry in Action', pp. 82–6, 91–3. On Barlow, see esp. Trott, 'Prelude to Restoration', *passim*, and below, p. 74.

336 See Spurr, *Restoration Church*, esp. pp. 311f. See also Tyacke, 'Religious Controversy'. Barlow may even have expressed reservations about certain implications of the new philosophy: see his *Genuine Remains*, pp. 151–9, though see also Hunter, *Science and Society*, p. 138n.

337 Barlow, 'The Case of a Toleration', in *Cases of Conscience, passim*.

338 See Hunter, 'Casuistry in Action', pp. 84–6.

339 Ibid., pp. 82–4; Hunter, 'Conscience of Robert Boyle', pp. 154–5. See also below, pp. 74–5, 79–80. On these men, see Trott, 'Prelude to Restoration', *passim*.

340 Below, pp. 78–9.

341 Below, pp. 77, 80; Wojcik, 'Theological Context of Boyle's *Things above Reason*', p. 148.

342 Ashcraft, 'Latitudinarianism and Toleration'. See also Wojcik, 'Theological Context of Boyle's *Things above Reason*', esp. p. 150.

343 Collinson, *Religion of Protestants*, ch. 6.

344 Below, pp. 68–9.

345 See Birch, *History of the Royal Society*, vol. iii, pp. 351, 363, 366–7, 371; Bodleian Library, Oxford, MS Rawlinson D 833, fol. 63ᵛ. See also Hall, *Promoting Experimental Learning*, pp. 155–6.

346 Compare Jacob, *Robert Boyle*, p. 87 and *passim*, with Hunter, 'How Boyle Became a Scientist'.

347 Below, pp. 63–4. Compare Jacob, *Robert Boyle*, pp. 128–9.

348 Below, pp. 48, 67.

349 *Spectator*, vol. iv, pp. 394–5. The source of the quotation is identified only as 'an excellent Sermon, preached at the Funeral of a Gentleman who was an Honour to his Country, and a more diligent as well as successful Enquirer into the Works of Nature, than any other our Nation has ever produced': but the reference to Boyle is unmistakeable (and is picked up in all annotated editions of *The Spectator*).

350 Below, p. 67.

351 For a full account, see Hunter, 'Conscience of Robert Boyle', pp. 153–7.

352 Printed in Hunter, 'Casuistry in Action', pp. 94–8.

353 Below, pp. 17–18.

354 Below, pp 49, 17–18.

355 Below, p. 78; Hunter, 'Casuistry in Action', pp. 83–4. Pett gives information

about the role of Sir James Shaen in connection with the grant which is not otherwise available.

[356] Below, pp. 74–5; Hunter, 'Casuistry in Action', pp. 82–3. It is perhaps worth noting that Boyle's original intention had been to provide support Sanderson for more than one year, and the episode recounted by Pett possibly accounts for its curtailment.

[357] See Canny, *Upstart Earl*, p. 13; cf. ibid., chs. 2–3 *passim*. See also below, p. 75.

[358] Below, pp. 75–7.

[359] Maddison, *Life*, pp. 257–82.

[360] Below, pp. 29–33. See Hunter, 'Alchemy, Magic and Moralism'.

[361] See Principe, 'Boyle's Alchemical Pursuits', esp. pp. 100–1, and Hunter, 'Alchemy, Magic and Moralism'.

[362] Below, p. 28.

[363] Hunter, 'Conscience of Robert Boyle', p. 158.

[364] See Hunter, *Robert Boyle Reconsidered*, pp. 10–11 and *passim*.

[365] Below, p. 89.

[366] Westfall, 'Newton and the Hermetic Tradition', p. 187.

[367] Shapin, 'Pump and Circumstance', esp. pp. 490f.

[368] See Harwood, 'Science Writing and Writing Science', pp. 46–8, and Hunter, 'Introduction' in *Robert Boyle Reconsidered*, p. 12. This matter will be considered more fully in the Introduction to the Pickering Masters Boyle.

[369] *Works*, vol. i, pp. ccxxii-iv.

[370] Canny, *Upstart Earl*. A parallel account of certain of the themes of Canny's book appears in Oster, 'Biography, Culture and Science', pp. 180f.

[371] Aubrey, *Brief Lives*, vol. i, p. 120. Cf. Canny, *Upstart Earl*, pp. 97, 109, 122, 127.

[372] Birch, *Life*, p. xlix; Canny, *Upstart Earl*, ch. 7, esp. pp. 144f.

[373] Canny, *Upstart Earl*, pp. 91, 107–8, 114; Mendelson, *Mental World of Stuart Women*, ch. 2.

[374] Below, pp. 9–10, 26, 104–5. See also Canny, *Upstart Earl*, p. 113.

[375] Canny, *Upstart Earl*, chs 4–5; below, pp. 5f.

[376] Middleton, *Lorenzo Magalotti at the Court of Charles II*, p. 135. See also below, pp. 3–4, 89.

[377] Harwood, *Essays and Ethics*, pp. xlii and, e.g., 37–8, 44; BP 3, fol. 91$^\mathrm{v}$.

[378] Oster, 'Biography, Culture and Science', pp. 198f.

[379] See the critique of claims to this effect in Oster, 'Biography, Culture and Science', in Hunter, 'How Boyle Became a Scientist'.

[380] Below, p. 20. Curiously, this episode is mentioned in Oster, 'Biography, Culture and Science' only in two digressive footnotes: pp. 219 n. 67, 222 n. 111.

[381] Below, pp. 88, 27, 32.

[382] This is set out in Maddison, *Life*, pp. 54–6, 73. It is perhaps also worth noting here that Miles's comment in his notes on Wotton's lost 'Sketch' (see above, n. 195), that 'It shoud seem by other accounts he never *Courted* any Lady See what Bishop Burnet sais', is based on this passage: there is no reason to believe that Miles had access to any information on this question beyond what is currently known. See also Oster, 'Biography, Culture and Science', pp. 203–5, though on his speculation about the 'betrothal ring' see Maddison, *Life*, p. 258n.

[383] See below, p. 27, for Miles's speculation that this referred to the offer made by Wallis in 1669 (see Maddison, *Life*, p. 56): but this seems unlikely.

[384] Below, pp. 70–1. Pett also associated Boyle's rejection of romances and plays with this change of heart. It is worth pointing out, however, that in this passage Pett may himself have been engaging in retrospective reconstruction from the text of *Seraphic Love*: cf. his citation of passages from it in his notes, pp. 63, 82, below. For Evelyn's speculation that *Seraphic Love* was linked with a love affair, see below, p. 88. See also *Works*, vol. i, p. 248, and Oster, 'Biography, Culture and Science', p. 224 n. 169.

[385] Evelyn to Boyle, 1 December 1659, Letterbook no. 158. See also Evelyn, *Diary and Correspondence*, vol. iii, pp. 121–6, and below, p. 93, which suggests that these critical remarks in Boyle's lost letter were balanced by compliments.

[386] See Hunter, 'How Boyle Became a Scientist'.

[387] See Principe, 'Early Boyle'.

[388] Hunter, How Boyle Became a Scientist'.

[389] See below, pp. 20, 89; see also Middleton, *Lorenzo Magalotti*, p. 135. On Boyle's early complexion, see esp. Maddison, 'Portraiture', p. 178. It is also worth noting that the passage concerning Boyle's complexion in Aubrey's *Brief Lives* was misleadingly transcribed by Andrew Clark. What Aubrey actually said, by way of analogy to John Dee's 'curious faire cleare, rosie complexion', was: 'so had E[arl] of Rochester exceeding. Sir John Denham was unpolished of the smallpox otherwise fine complexion. Mr R. Boyle when a Boy at Eaton: from Dr Wood who was his schoolefellow. now sickly & pale'. Bodleian Library MS Aubrey 8, fol. 6aᵛ; cf. Aubrey, *Brief Lives*, vol. i, p. 120.

[390] Boyle to [Worsley, February 1647], *Works*, vol. vi, p. 40; see also Boyle to Hartlib, 8 May 1647, ibid., vol. i, p. xli, and Birch's comment in ibid., p. xliv. On the ague, see below, pp. 8–9. In a letter to Lady Ranelagh of 13 May 1648, Boyle refers to 'the fickleness of my health' in relation to his melancholy in 1648 (ibid., vol. vi, p. 45; see also his letter to Kinalmeaky, 1 August 1642, Grosart, *Lismore Papers*, 2nd series, vol. v, p. 96): but such complaints were probably trivialised by what followed.

[391] See his letter to Lady Ranelagh, 2 August 1649, *Works*, vol. vi, p. 48. It is interesting that Boyle kept a prescription relating to this illness, by Dr Davies, dated 30 July 1649: BP 18, fols 101–2.

[392] See below, p. 27; *Works*, vol. v, pp. 315–16. See also above, n. 89.

[393] Below, pp. 51, 89.

[394] Maddison, *Life*, pp. 80–1; North, *Lives of the Norths*, vol. iii, p. 146.

[395] North, loc. cit.; Maddison, *Life*, pp. 221–2 (I am grateful to Richard Evans for his advice on this passage). For a comparable comment see Kirkwood to Boyle, 13 July 1689, *Works*, vol. i, p. cc.

[396] Cf. Hunter, 'Conscience of Robert Boyle'. As an *envoi*, it is perhaps worth pointing out that this view is not entirely new. For an interesting mid nineteenth-century reading of Boyle as a melancholy man of genius, based mainly on the *Account of Philaretus*, see Wilson, *Religio Chemici*, esp. pp. 233f. This is to be seen against the background of nineteenth-century views of scientific creativity surveyed in Yeo, 'Genius, Method, and Morality'.

PRINCIPAL EVENTS
IN THE LIFE OF ROBERT BOYLE

1627 25 January. Born at Lismore, Ireland, seventh son of Richard Boyle, 1st Earl of Cork, by his second wife, Catherine.

1635 2 October. Enters Eton College with his brother, Francis, later Viscount Shannon.

1638 23 November. Leaves Eton College.

1639 Travels to France and Switzerland with Francis, under the tutelage of Isaac Marcombes; spends several months in Geneva.

1641 Travels to Italy.

1642 Stranded at Marseilles; returns to Geneva.

1644 Returns to England.

1645 Settles at Stalbridge, Dorset, where he spends much of the next decade; possibly briefly visits France in this year.

1648 February–April. Visits Netherlands.

1649 July. Quotidian Ague.

1652 June. Travels to Ireland.

1653 June–September. Returns to England.

1654 Anasarka and ancillary ailments; leaves Ireland.

1655 *Invitation to Free Communication* published in *Addresses made to Samuel Hartlib*.

Late 1655 or early 1656 Settles at Oxford.

1659 *Seraphic Love* published.

Temporarily resident in Chelsea.

1660 *New Experiments Physico-Mechanical, Touching the Spring of the Air and its Effects* published.

28 November. Attends inaugural meeting of Royal Society.

1 December. Appointed to Council for Foreign Plantations.

1661 *Certain Physiological Essays*, *The Sceptical Chymist* and *Some Considerations touching the Style of the Holy Scriptures* published.

1662 7 February. Appointed first Governor of the Corporation for Propagation of the Gospel in New England.

Defence of the Doctrine Touching the Spring and Weight of the Air published.

1663 First volume of *The Usefulness of Experimental Natural Philosophy* published.

1664 *Experiments and Considerations touching Colours* published.

1665 *Occasional Reflections* and *New Experiments and Observations touching Cold* published.

8 September. Created Doctor of Physic at Oxford.

1666 *Hydrostatical Paradoxes* and *The Origin of Forms and Qualities* published.

1668 Settles in London, living for the rest of his life with Lady Ranelagh in Pall Mall.

1671 Second volume of *The Usefulness of Experimental Natural Philosophy*, *Cosmical Qualities of Things* and other tracts published.

1672 *An Essay about the Origin and Virtues of Gems* and other tracts published.

1673 *Essays of Effluviums* published.

1674 Tracts on *Saltness of the Sea*, etc., and *The Excellency of Theology, Compar'd with Natural Philosophy* published.

1675 *Some Considerations about the Reconcileableness of Reason and Religion* and *Experiments, Notes, &c. about the Mechanical Origin or Production of Divers Particular Qualities* published.

1678 *Of a Degradation of Gold Made by an Anti-Elixir* published.

1680 *The Aerial Noctiluca* published.

18 December. Declines presidency of the Royal Society.

1681 *A Discourse of Things above Reason* published.

1682 *The Icy Noctiluca* published.

1684 *Memoirs for the Natural History Of Humane Blood* and *Experiments and Considerations about the Porosity of Bodies* published.

1685 *Of the High Veneration Man's Intellect owes to God, An Essay Of the Great Effects of Even Languid and Unheeded Motion, Of the Reconcileableness of Specifick Medicines to the Corpuscular Philosophy* and treatise on mineral waters published.

1686 *Free Enquiry into the Vulgarly Receiv'd Notion of Nature* published.

1687 *The Martyrdom of Theodora and of Didymus* published.

1688 *A Disquisition about the Final Causes of Natural Things* published; recipe collection printed for private circulation.

1689   22 August. Resigns Governorship of Corporation for Propagation of the Gospel in New England due to illhealth.

1690   *Medicina Hydrostatica* and *The Christian Virtuoso* published.

1691   *Experimenta & Observationes Physicae* published.

       18 July Will signed and sealed.

       23 December. Lady Ranelagh dies.

       31 December. Boyle dies.

1692   7 January. Buried at St Martin's in the Fields. Burnet's funeral sermon delivered.

       First volume of *Medicinal Experiments* and *The General History of the Air* published.

1695   *Free Discourse against Customary Swearing* published.

BIOGRAPHICAL GUIDE

The following are mentioned frequently both in the Introduction and in the texts which follow; those mentioned only once or twice are identified in footnotes. Further information about many of these figures, particularly members of the Boyle family, will be found in Maddison, *Life*, and piecemeal reference to that work has not therefore been made. Entries relating to members of the Boyle family are placed alphabetically by title. 'Boyle' throughout means Robert Boyle.

Barlow, Thomas (1607–91). Bodley's Librarian, 1642–60; thereafter Lady Margaret Professor of Divinity, and, from 1675, Bishop of Lincoln. One of Boyle's closest friends among churchmen.

Broghill, Roger Boyle, Baron (1621–79). Fifth son of the Earl of Cork. Saw active service during and after the Civil War. Later a member of Cromwell's council; created Earl of Orrery in 1660. Author of the romance, *Parthenissa* (1651–69), and of various successful plays.

Burlington, Richard Boyle, 1st Earl of (1612–98). Second son of the Earl of Cork, whom he succeeded to the title in 1643. Active in the Civil War, and Lord Treasurer of Ireland from 1660 to 1695. Created Earl of Burlington in 1664. Inherits Boyle's estate at his death by virtue of the provisions of the will of the Earl of Cork.

Burnet, Gilbert (1643–1715). Churchman and politician of Scottish descent. Chaplain of the Rolls, 1675–84; travelled abroad 1685–8; Bishop of Salisbury, 1689. Close *confidant* of Boyle in his later years.

Clarendon, Edward Hyde, 1st Earl of (1609–74). Charles I's closest adviser during the Civil War, and chief minister to Charles II from 1660 to 1667.

Cork, Catherine, Countess of (d. 1630). Boyle's mother. Daughter of Sir Geoffrey Fenton, Principal Secretary of State for Ireland; became the second wife of Richard Boyle, 1st Earl of Cork, in 1603.

Cork, Richard Boyle, 1st Earl of (1566–1643). Boyle's father. The greatest adventurer in Elizabethan and Stuart Ireland, of which he became Lord High Treasurer in 1631; said to be the richest man in England. Known as the Great Earl of Cork. See the stimulating account by Nicholas Canny, *The Upstart Earl*.

Dungarvan, Richard Boyle, Viscount: see Burlington, Earl of.

Fell, John (1625–86), Dean of Christ Church and Bishop of Oxford; high churchman and scholarly publisher.

Fenton, Catherine: see Cork, Countess of.

Harrison, John (d. 1642). Head Master of Eton College, 1630–6.

Hooke, Robert (1635–1703). Assistant to Boyle at Oxford in the late 1650s; from 1662, curator of experiments to the Royal Society. Evolves from the role of Boyle's *protegé* to being one of the leading natural philosophers in Restoration England.

Hyde, Thomas (1636–1703). Orientalist; Bodley's Librarian, 1665–1701.

Jones, Lady Katherine: see Ranelagh, Lady.

Killigrew, Elizabeth (1622–81). Wife of Francis Boyle, Viscount Shannon (q.v.).

Kinalmeaky, Lewis Boyle, Viscount Boyle of (1619–42). Tenth child of the first Earl of Cork. Killed at the Battle of Liscarrol, 3 September 1642.

Marcombes, Isaac. Boyle's governor on his European tour, having previously acted in a similar capacity for his elder brothers. Marcombes came from the Auvergne; in 1637 he became nephew by marriage of the Genevan divine, Jean Diodati. Maddison, 'Grand Tour', pp. 53–5.

Orrery, Earl of: see Broghill, Baron.

Ranelagh, Katherine Jones, Lady (1615–91). Fifth daughter of the Earl of Cork, and sister to Robert. Married Arthur Jones, heir to the 1st Viscount Ranelagh, in 1630; he became 2nd Viscount in 1643. Boyle lived with her in Pall Mall from 1668 to his death. A remarkable women in her own right, the closeness of whose relationship to Boyle is indicated by Burnet's funeral sermon.

Rich, Mary: see Warwick, Countess of.

Salisbury, Bishop of: see Burnet, Gilbert.

Sarum, Bishop of: see Burnet, Gilbert.

Shannon, Francis Boyle, 1st Viscount (1623–99). Sixth son of the Earl of Cork. Attended Eton and toured Europe with Boyle.

Southwell, Sir Robert (1635–1702). Diplomat, virtuoso and Principal Secretary of State for Ireland, 1690–1702.

Tenison, Thomas (1636–1715). Rector of St Martin's in the Fields, 1680; Bishop of Lincoln, 1691; Archbishop of Canterbury, 1694.

Ussher, James (1581–1656). Scholar, divine and Archbishop of Armagh from 1625 to his death. A friend of the Earl of Cork, and an important influence on Boyle. Insightful account available in Trevor-Roper, *Catholics, Anglicans and Puritans*, pp. 120–65.

Warwick, Mary Rich, Countess of (1624–78). Seventh daughter of the Earl of Cork, and sister to Robert. Married Charles Rich, heir to the 3rd Earl of Warwick, in 1641; widowed 1673. A deeply religious figure in her mature years. See Mendelson, *The Mental World of Stuart Women*, ch. 2.

Wilkins, John (1614–72). Warden of Wadham College, Oxford, 1648–59; Dean of Ripon, 1663, and Bishop of Chester, 1668. The convenor of the scientific group in Interregnum Oxford which Boyle joined in 1655–6; after 1660, a leading light of the Royal Society and a strong protagonist of ecclesiastical comprehension.

Wotton, Sir Henry (1568–1639). Diplomat, author and Provost of Eton College from 1629 to 1639.

NOTE ON TEXTS

Manuscript texts have been transcribed literally, retaining original spelling, capitalisation and punctuation; the ampersand has also been retained. Words or phrases inserted above the line in the original have been denoted ‹thus›. Words or passages deleted in the original are recorded in endnotes. Editorial additions are indicated in square brackets: these include punctuation added (or, very occasionally, altered) to assist the reader, and words or letters obscured by damage to the manuscript. When it is unclear whether or not a word is capitalised, the reading consistent with modern usage has been adopted. Standard abbreviations have been silently expanded, with square brackets being used in doubtful cases; the thorn has throughout been expanded to 'th', and u/v and i/j have been modernised. Catchwords in the original have been ignored unless they fail to tally with the text that follows. In general, paragraphing reflects the original, but in some cases additional spacing has been added for clarity (where this occurs, this is indicated in an accompanying note). Marginal references in the original (e.g., to the Bible) have been placed in footnotes, but insertions which appear in the margin with the intention that they should be incorporated in the text have been placed there (with an accompanying endnote stating this). Original foliation has been indicated by the insertion of 'fol. 2' between soliduses at the point where each recto or verso of the manuscript text begins.

DOCUMENT 1

An Account of Philaretus during his Minority
by Robert Boyle

BP 37, fols. 170–84. The manuscript is in Boyle's own youthful hand, with original pagination. Fols. 171–2 are misbound, and should come at the end. Whereas the remainder of the text is evidently a fair copy, these represent the residue of an earlier recension, in smaller format. At the point of transition from the fair copy to the earlier draft, there is a brief repetition of the text, and the two versions are not quite identical. However, the decision to effect the transition from one version to the other in this way was apparently Boyle's own, since the original pagination of the manuscript continues to include these leaves. The text ends part of the way down a page, so it may never have been completed. All blanks in the text are Boyle's own. Throughout the text, Boyle repeatedly leaves a choice of phrasing, either by giving two versions each enclosed by soliduses (or, in some cases, one version followed by a blank), or by adding an alternative above the line without deleting the original word or phrase. The former have been reproduced in the text exactly as in the original (it should be noted that these soliduses are Boyle's, in contrast to the editorial ones used – here as throughout this volume – to indicate the foliation of the manuscript); the latter have been recorded in endnotes, using the formula 'duplicated by' to denote the word inserted above the line. In the manuscript, the first word on each page is sometimes capitalised, whereas in the catchword it is not; for this reason, these capitals have here been ignored. There is some variation in ink and handwriting, suggesting that the manuscript was completed in stages. On the background to the work, see above, pp. xv–xxi; on the history of the manuscript, see p. lix; on previous editions, see p. xii.

An
Account
of
Philaretus
During his Minority

Not needlesly to confound the Herald with the Historian, & begin a Relation by a Pedigree, I shall content my selfe to informe yow, that the immediate Parents of our Philaretus, were, of the Female Sex, the Lady [Catherine] Fenton[1], (a Woman that wanted not Buty, & was rich in Vertue;) & on the Father's side, that Richard Boyle Earle of Corke, who by God's blessing on his prosperous Industry, ‹upon›[2] very inconsiderable Beginnings built so plentifull & so eminent a Fortune, that his prosperity has found many Admirers, but fewe Parallels.

He was borne the 14 Child of his Father (of which 5 women & 4 men do yet survive[a]) in the Yeare 1626 upon St. Paul's Conversion-day, being the 25th of January; at a Cuntry-house of his Father's cal'd Lismore, then one of the noblest Seates & greatest Ornaments of the Province of Munster, in which it stood; but now so ruin'd by the sad Fate of warre, that it serves only for an Instance & a Lecture; of the Instability of that Happinesse, that is built upon the ‹incertin› Possession of such ‹fleeting Goods›[3] as it selfe was. /fol. 170v/

To be such Parent's Son, & not their Eldest,[4] was a Happinesse that our Philaretus would mention with greate expressions of Gratitude; his birth so suiting his Inclinations & his Desseins, that had he been permitted an Election,[b] his Choice would scarce have alter'd God's Assignement. For as on the one side a Lower[5] Birth wud haue too much expoz'd him to the Inconveniences of a meane Discent; which are too notorious to need specifying; soe on the other side, to a Person whose Humor indisposes him to the ‹distracting› Hurry of the World; the being borne Heire to a Greate Family, is but a Glittering kind of Slavery; whilst obliging him to a Publick & en‹tangled›[6] course of Life to support the Credit of his Family, & tying

[a] Alice, Joan (d. 1656), Katherine, Dorothy and Mary; Richard, Roger, Francis and Robert. Lewis had died in 1642.

[b] Evidently an astrological usage on Boyle's part, referring to the choice of a propritious time.

him from[7] satisfying his dearest Inclinations, it often forces him to build the Advantages of his ‹House›,[8] upon the Ruines of his owne Contentment. A man of meane Extraction, ‹(Tho never so advantag'd by Greate meritts)›[9] is seldome admitted to the Privacy & the secrets[10] of greate ones promiscuously; & scarce dares pretend to it, for feare of being censur'd sawcy, or an Intruder. And titular Greatnesse is ever an impediment to the[11] Knowledge of many retir'd /   / Truths, that cannot be attain'd without[12] Familiarity with meaner Persons, & such other Condiscentions, as fond opinion in greate men disapproves & makes Disgracefull. But now our Philaretus was borne in a Condition, that neither was high enuf to prove a Temptation to Lazinesse; nor low enuf to discourage him from aspiring.

And certinly to a Person that so much affected an Universall /   / Knowledge, & arbitrary vicissitudes of Quiet & Employments; it could not be unwelcome to be born to a Quality, that was a handsom stirrop to Preferrment without (being) an Obligation to court it: & which might at once, both protect his higher Pretensions from the Guilt /   / of Ambition, & (secure) his Retirednesse from Contempt /   /.

When once Philaretus was able without Danger to support the Incommoditys of a Remove, his Father, who had a perfect aversion for their Fondnesse[13], who use to breed their Children so nice & tenderly, that a hot ‹Sun›,[14] or a good Showre of Raine as much ‹endangers them as› if they were made of Butter [or][15] of Sugar; sends him away from home; & commits him to the Care of a Cuntry-nurse; who by ‹early› inuring him by slow degrees to a Course (but cleanly)[16] Dyet and to the Usuall Passions / / of the Aire[,] gave him so vigorous a Complexion, that both Hardships were made easy to him by Custome, & the Delights of conveniencys & ease, were endear'd to him by their Rarity./fol. 173/

Some few yeares after this, two greate Disasters befell Philaretus; the one was the Decease of his Mother; whose Death would questionless have excessively afflicted him, had ‹but› his Age[17] permitted him to know the Value of his Losse. For he wud ever reckon it amongst the Cheefe Misfortunes of his Life, that he did ne're know her that gave it him: her free & Noble Spirit[18] (which had a handsome Mansion to reside in) added to her kindnesse and sweete carriage to hir owne, making her[19] hugely regretted by her Children, & so lamented by her Husband; that not only he annually[20] dedicated the Day of hir Death to solemne Mourning for it; but burying in her Grave all thoughts of Aftermarriage, he rejected all Motions of any other Match; continuing a constant Widdower till his Death;

The second Misfortune that befell Philaretus, was his Acquaintance with some Children of his owne Age, whose stuttring Habitude he so long Counterfeited that he at last contracted it. Possibly a just Judgment upon his Derision, & turning the Effects of God's Anger into the Matter[21] of his

3

Sport. Divers Experiments, beleev'd the probablest meanes of Cure, were try'd with as much successlesnesse as Diligence; so Contagious & catching are men's Faults, & so Dangerous is ‹the› commerce[22] of those (condemnable) Customes, that by being imitated but in jeast, come to be learn'd (& acquir'd) in earnest.

But to show that these Afflictions made him not lesse the Object of Hev'n's Care; he much about this time escap't a Danger, from which he ow'd his Deliverance wholly to Providence; being so far from contributing to it himselfe, that he did his Endevor to oppose it. For waiting upon his Father[23] up to Dubline, there to expect the Returne of his Eldest Brother (then landed out of England with his new Wife the Earle of Cumberland's Heire;[a]) as they were to passe over a Brooke,[24] at that time suddenly by immoderate showers swell'd to a Torrent; he was left alone in a Coach only with a Footboy to attend him: where a Gentleman of his Father's[25] very well hors't accidentally espying him, in spite of[26] some other's & his owne Unwillingnesse & Resistance, (they not beleeving his stay Dangerous) carry'd him in his Armes over the Rapid Water; which prov'd so much beyond expectation, both swift & Deepe, that[27] horses with their Riders were ‹violently› hurry'd /fol. 173v/ downe the Streame, which easily overturn'd the ‹unloaded›[28] Coach; the horses; (after by long struggling; they had broke their harnesse) with much adoe, saving themselves by swimming.

As soone as his Age made him capable of (admitting) Instruction, his Father (by a Frenchman & by one of his Chaplaines) had him taught both to write a Faire hand, & to speake French & Latin;[b] in which (especially the first) he prov'd no ill proficient; adding to a reasonable Forwardnesse in study, a more than usuall Inclination to it.

This Studiousnesse observ'd in Philaretus, endear'd him very much unto his Father; who us'd (highly) to commend him both for it & his Veracity: of which (latter) he would often give[29] him this Testimony; that he never found him in a Lye in all his Life time. And indeed Lying was a Vice both so contrary to his nature & so inconsistent with his Principles, that as there was scarce any thing he more greedily desir'd then to know the Truth, so was there scarce any thing he more perfectly detested, then not to speake it. Which brings into my Mind a foolish Story I have heard him Jeer'd with, by (his Sister,) my Lady Ranalagh; how she having given ‹strict› order to have a Fruit-tree preserv'd for his sister in Law, the Lady Dungarvan, then big with Childe; he accidentally comming into the Garden ‹, &› ignoring the Prohibition, ‹did› eat ‹even›[30] halfe a score of them: for which being chidden by his sister Ranalagh; (for he was yet a Childe:) & being told by way of aggravation, that he had eaten halfe a dozen Plumbs; Nay truly Sister

[a] Elizabeth, Baroness Clifford (1613–91).
[b] Evidently 'Mounsieur Frances de Carey': Grosart, *Lismore Papers*, 1st series, vol. iii, p. 41.

4

(answers he simply to her) I have eaten halfe a Score. So perfect an enemy was he to a Ly, that he had rather accuse himselfe of another fault, then be suspected to be guilty of that. This triviall Passage I have mention'd now; not that I thinke ‹that in it selfe› it deserves a Relation /to be recorded/ but because as the Sun is seene best at his Rising &[31] his Setting, so men's ‹native› Dispositions are clearyest perceiv'd, whilst they are Children & when they are Dying. And certainly these little sudden Accidents are the greatest Discoverers of men's true Humors. For whilst the Inconsiderable-nesse / / of the thing affords no temptation to dissemble; & the sudden-nesse of the Time / /[32] allowes /Fol. 174/ no Leasure to put Disguizes on; men's Dispositions do appeare in their true, genuine, shape: whereas most of those[33] actions that are done before others, are so much done for others; I meane most[34] /solemne/ Actions are so personated; that we may much more probably guesse from thence, what men desire to seeme, then what they are: such Publicke, formall Acts, much[35] rather being ajusted /fitted, acc[ording]/ to men's Desseins, then flowing from their Inclinations.

Philaretus had now attain'd & /or/ somewhat past the Eighth yeere of his Age; when his Father; (who supply'd what he wanted in Schollership himselfe by being both a Passionate Affecter & eminent Patron of it)[36] ambitious to improve his early Studiousnesse; & considering[37] that greate men's Children's breeding up at home, tempts them to Nicety, to Pride, & Idlenesse; & contributes much more to give them a Good opinion of themselves, then to make them deserve it; resolves to send over Philaretus in the company of[38] (Mr F.B.) his elder brother;[a] to be bred up at Eaton Colledge (neere Windsor) whose Provost at that time was (one) Sir Henry Wotton, a Person, that was not only a fine Gentleman himselfe, but very well skill'd in th'Art of making others so: betwixt whom & the Earle of Corke an Ancient Friendship had been constantly cultivated by Recipro-call Civilitys. To him therefore the good old Earle recommends Philaretus: who having lay'd a weeke at Youghall for a wind, when he first put[39] to sea, was by a storme beaten backe againe, (not only a Tast but an Omen (& Earnest) of his future Fortune: /after Destiny./) But after eight Dayes further stay; upon the second summons of a ‹promising›[40] Gale, they went aboord once more; & (tho the Irish Coasts were then sufficiently infested with Turkish Gallyes) having by the way tuch't[41] at Ilford-combe, & Myn-head, at last they happily[42] arriv'd at Bristoll.

The Second Section /fol. 175/

Philaretus, in the Company of his Brother, after a short stay to repose & refresh themselues at Bristoll; shap't his Journey directly for Eaton-

---

[a] Francis Boyle, later Viscount Shannon.

colledge: where a Gentleman of his Father's,[a] sent to conveigh him[43] thither, departing, recommended him to the especiall Care of Sir Henry Wotton; & left with him, partly to instruct & partly to attend him one R.C. one that wanted neither vices nor cunnning to dissemble them.[b] For tho his Primitive fault was onely a dotage upon Play; yet the excessive love of thatt, goes seldome unattended with[44] a Traine of Criminall /culpable/ retainers: for fondnesse upon[45] Gaming is the seducingst Lure to Ill Company; & that, the subtlest Pander to the worst Excesses /Debauches.[/] Wherefore our Philaretus deservedly reckon'd it ‹both›[46] amongst the Greatest & the unlikelyest Deliverances[47] he ow'd Providence, that he was protected from the Contagion of such Presidents. For thogh the man wanted not a Competency of Parts; yet perverted Abilitys make men but like those wandring Fires Filosofers call Ignes fatui; whose light[48] serves not to direct but to seduce the[49] credulous ‹Traveller› & allure him[50] to ‹followe them in their›[51] Deviations. And it is very true, that during the[52] Minority of Judgment, Imitation is the Regent in the Soule; &[53] Those that are at least capable of Reason, are most sway'd by Example. A blind man will ‹suffer›[54] himselfe to be lead, tho by a Dog /or Child./

Not long our Philaretus stay'd at schoole,[55] ere its Master (Mr Harrison) taking notice of some aptnesse & much Willingnesse in him to learne; resolv'd to improve them both by all the gentlest wayes of Encouragement. For he would often dispense from[56] his attendance at schoole at the accustom'd howres; to instruct him privately & familiarly in his Chamber. He would often as it were cloy him with fruit & Sweetmeats & those little Daintys that age is greedy of; that by preventing the want[57], he may lessen both the value of them & the Desire. He would sometimes give him unask't Playdayes, & ‹oft› bestow upon him such Bals & Tops & other implements of Idlenesse, as he had taken away from others that had unduely us'd them. /fol. 175/ He would sometimes commend others before him to rowze his Emulation, & oftentimes give him Commendations before others, to engage his Endeavors to deserve them. ‹Not› to be tedious, he was carefull to instruct him in such an affable, kind, & gentle way, that he[58] easily prevail'd with him to consider /affect/ Study not so much as a Duty of Obedience to his Superiors /Parents/ but as the Way to ‹purchase for himselfe›[59] a most delightsome & invaluable Good. In Effect, he soone created in Philaretus so strong a Passion to acquire Knowledge, that what time he could spare from a Schollar's taskes (which his retentive Memory made him not find uneasy;) he would usually employ so greedily in Reading; that his Master would be sometimes necessitated to[60] force him out to Play; on which, & upon Study, he look't as if their natures were

---

[a] Thomas Badnedge: Maddison, *Life*, p. 7.

[b] Robert Carew, who carried out various services for the Earl of Cork despite the negative assessment of his character given here: Maddison, *Life*, p. 8n.

inverted. But that which he related to be the first occasion that made him
so passionate a Friend to Reading; was the accidentall Perusall of Quintus
Curtius;[a] which first made him in Love with other then Pedanticke[61]
Bookes, & conjur'd up in [him] that unsatisfy'd Curiosity[62] of Knowledge,
that is yet as greedy, as when it first was rays'd. In gratitude to this Booke
I have heard him ‹hyperbolically›[63] say, that not onely he ow'd more to
Quintus Curtius then Alexander did; but deriv'd more advantage from the /
/ Conquests themselves.

    Whilst our Youth was thus busy'd about his Studyes, there happened to
him an Accident that Silence must not cover. For being one Night gone to
bed somewhat early; whilst his Brother was conversing with some Company by the fire side;[64] without giving them the least warning ‹or summons
to take heed›, a greate part of the wall of their chamber, with the Bed,
Chaires,[65] Bookes & furniture of the next chamber over it, fell downe ‹at
unawares› upon their Heads. His brother had his band torn about his
Neck & his Coate upon his Backe, & his Chaire crush't & broken under
him; but by a lusty Youth then accidentally in the roome, was snatch't
‹from›[66] out the Ruines, by which Philaretus had (in all probability) been
remedilesly oppres't; had not his Bed been curtain'd by a watchfull Providence; which kept all heavy things from falling on it: but the Dust the
Crumbled rubbish raised, was so thicke /fol. 175v/ That he might there
have stifled, had not he remembred to wrap his head in the sheet; which
serv'd him as a strainer, through which none but the purer Aire could find
a passage. So sudden a Danger & happy[67] an Escape, Philaretus would
sometimes mention with expressions both of Gratitude & Wonder. To
which he would adde the Relation of divers other almost contemporary
Deliverances; Of these, one was, that being fallen from his horse, the beast
run over him, & trod so neare his throate, as within lesse then two inches
of it, to make a hole in his band;[68] which he long after reserv'd for a
Remembrancer. An other was, that riding thorou a Towne upon a Nagge
of his owne whose starting quality he never observ'd before; his horse
upon a sudden fright rise [sic] bolt upright upon his hinder feet, & falling
rudely backwards with all his Weight against a Wall had infallibly crush't
his Rider into Peeces, if[69] by a strange instinct, he had not cast himselfe off
at first, & ‹was quit of it›[70] for a slight Bruise. The last was ‹an Apothecary's mistake›[71]. Philaretus being newly recover'd of a Fluxe, the doctor
had prescrib'd ‹him a›[72] refreshing Drinke; the fellow that should adminis-
ter it, instead of it brings him a very strong vomit prepared & intended for
another. Philaretus was that morning visited by some of his schoole-
fellowes, who; (as we [sic] was not Ill belov'd amongst them) presented

---

[a] Quintus Curtius Rufus, the 1st-century author of *Historiae Alexandri Magni*.

him with some sweetmeats, which[73] having eaten, when afterwards he would have eaten his Breakfast, his stomacke whither out of Squeamish-nesse or Divination, forc't him to render it againe. To which[74] lucky accident, the Physitian ascrib'd his Escape from the[75] Apothecary's Error; for in the Absence of those that tended him, his[76] Fisick cast him into hideous Torments, the true Cause of which, by [them] never dream'd of, remain'd long unconjectur'd; untill at last the Effects betray'd it; for after a long struggling, at last the Drinke wrought with such violence, that they fear'd ‹that› his Life would be disgorg'd together with his Potion. This Accident made him long after apprehend more the Fisitian[77] then the Disease & was possibly the occasion that made him afterwards so inquisi-tively apply himselfe to the study of Fisicke, that he may have the lesse need of them that professe it. But Philaretus wud not ascribe[78] any of these Rescues unto Chance, but would[79] be still industrious to perceive[80] the hand of Hev'n /fol. 176/ in all these Accidents: & indeed he would professe that in the Passages of his Life he had observ'd so gratious & so peculiar a Conduct of Providence, that he should be equally blind & ungratefull, shud he not both Discerne & Acknowledge it.

Philaretus had[81] now for some 2 yeares been a constant Resident at Eaton; (if yow except[82] a few visits which during the long Vacations, he made his sister My Lady Goring, at Lewes in Sussex;[a]) when about Easter he was sent for up to London, to see his Eldest Brother the Lord Dungar-van; where being visited with a Tertian Ague, after the Queen's & other's Doctors Remedyes had been succeslesly essay['d];[b] at last he return'd againe to Eaten, to derive that Health[83] from a Good Aire & Dyet, which Fisick could not give him. Here to divert his Melancholy they made him read the stale Adventures[84] Amadis de Gaule;[c] & other Fabulous & wandring Storys; which[85] much more prejudic'd him by unsettling his Thoughts, then they could have advantag'd him; had they effected his Recovery; for meeting in him with a restlesse Fancy, then made more susceptible of any Impressions by an unemploy'd Pensivenesse; they[86] accustom'd his Thoughts to such a Habitude of Raving, that he has scarce ever been their quiet Master since, but ‹they› would take all occasions to flinch[87] away, & go a gadding to Objects then unseasonable & imperti-nent. so great an Unhappinesse it is, for Persons that[88] are borne with such Busy Thoughts, not to have congruent Objects propos'd to them at First. Tis true that long time after Philaretus did in a reasonable[89] measure fixe his Volatile Fancy & Reclaime his Ramage thoughts;[90] by the use of all those Expedients he thought likelyest to fetter (or at least, to curbe /bridle/)[91] the

---

[a] Lettice (1610–43), wife of Lord George Goring (1608–57).
[b] The Queen's doctor was Sir Theodore Turquet de Mayerne (1573–1655).
[c] The hero of a popular romance.

roving wildness of his wandring /hagard/ Thoughts. Amongst all which the most effectuall Way he found to be, the Extractions of the Square & Cubits Rootes, & specially those more laborious Operations of Algebra, which both accustome &[92] necessitate the Mind to attention, by so entirely exacting the whole Man; that the least[93] Distraction, or heedlessnesse, constraines us to renew our (Taske &) Trouble, & rebegin the Operation. Six weekes had Filaretus been troubled /fol. 176v/ with his Ague, when he was free'd from it by an accident, whic[h] is no slender Instance of the force /power/ of Imagination; for the Fisitian having sent him a Purge to give (as he say'd) the[94] fatall Blow to the Disease; our Philaretus had so perfect an aversion to all Fisicke, & had newly essay'd it so unsuccesfully, that his Complaints induc't the Maydes[95] of the House he lodg'd in, (partly out of complaysance to him, & partly out of a beleefe, that Fisick did but exasperate his Disease;)[96] unknowne to him to poure out the Potion, & fill the Viall with Syrup of stew'd Prunes; a Liquor so resembling it, that Philaretus[97] (see the force of Fancy) swallow'd it with the same Reluctancy; & found the Taste as loathsome as if't had been the Purge: but being after acquainted with the Cuzenage, whither 'twas that his sicknesse[98] (as having already reach't it's Period,) would have expir'd of it selfe, or that his Mirth dispatch't it; I pretend not to determine; but certaine 'tis, that from that Howre to this; Agues & he have still been perfect stranger. And he had much adoo to refraine from laughter, when going to thanke & reward the Doctor for his Recovery, he found it wholly ascrib'd to the Efficacy of a Potion, he had never swallow'd[99] but in Imagination. /His fancy only swallow'd./

He had now serv'd well neere halfe an Appentice-ship [sic] at Schoole, when there arriv'd Intelligence of his Father's being landed in England, & gone to Stalbridge, (a Place in Dorsetshire then newly purchas'd by him). Thither Philaretus accompanys[100] his sister the Countesse of Kildare,[a] to waite upon him. The good old Earle welcom'd him very kindly; for whether it were to the Custome of old People[101] (as Jacob doted most on Benjamin & Joseph) to give their Eldest Children the largest Proportions of their Fortunes, but the youngest the greatest shares of their Affections; or to[102] a resemblance observ'd in Philaretus both to his Fathers Outside / body/ & his mind; or to both an exact & affectionate Obedience his Duty made him pay his Commands;[103] or, as it seemes most likely, to his never having liv'd with his Father to an Age that might much tempt him to run in debt & take /fol. 177/ such other Courses to provoke his Dislike, as in his elder Children he severely disrellish't: To which of these Causes the Effect is to be ascribe[d], it[104] is not my Taske to resolve; but certine it is,

---

[a] Joan Boyle (1611–56).

9

that from Philaretus his Birth,[105] untill his Father's Death, he ever continu'd very much his Favorite. But after some weeke's Enjoyment of the
‹summer› Diversions[106] at Stalbridge, when his Father remov'd to London,
he left him by the way at Eaton-Colledge, from whence at his Returne into
the West some few months after; he tooke him absolutely away: after
Philaretus had spent[107] in that Schoole (then very much throng'd with
young Nobility) not much beneath Foure Yeares; in the last[108] of which he
forgot much of that Lattin he had gott: for he was so addicted to more
reall[109] Parts of Knowledge, that he hated the study of Bare words, naturally; as something that relish't too much of Pedantry to consort with his
Disposition & Desseins: so that by the change of his old Courteous
Schoolemaster, for a new rigid Fellow;[a] loosing those encouragements that
had formerly subdu'd his Aversion to Verball studys; he quickly quitted
his Terence & his Grammer, to read in History their ‹Gallant›[110] Acts, that
were the Glory of their owne & the Wonder of our Times. And[111] indeed,
'tis a much nobler   for Ambition, to learne to do[112] things, that may
deserve a roome in History; then onely to learne, how ‹congruously› to
write such actions, in the Gowne-men's Language.

As soone as Philaretus was arriv'd at Stalbridge, his Father assign'd the
care of teachi[n]g him to one Mr W. Douch, then Parson of that Place, ‹&
one of his Chaplaines›[b] &, (to avoid the temptations to Idlenesse, that
Home[113] might afford) made him both lodge & dyett where he was
taught; tho it were not ‹distant› from his Father's house above twice
musket-shot. This old Divine instructing our Youth both with care &
Civility, soone brought him to renew his first acquaintance with the
Roman Tongue, & to improve it so farre that in that Language he could
readily enuf expresse himselfe in Prose, & began to be no dull Proficient in
the Poeticke straine; which latter he was naturally[114] addicted to, resenting
a greate deale of Delight in the /fol. 177v/ Conversation of the Muses,
which neverthelesse he ever since that time forbore to cultivate; not out of
any Dislike or Undervaluing of Poetry; but because[115] in his Travells
having by discontinuance forgot much of the Latin Tongue, he afterwards
never ‹could find›[116] [occasion] to redeeme his Losses by a serious study of
the Ancient Poets; & then for English Verses, he say'd they[117] could not be
certaine of a lasting applause, the changes of our Language being so greate
& nimble /sudden/ that the rarest Poems within few yeares[118] will passe for
obsolete; & therefore he us'd to liken[119] Writers in English Verse to Ladys
that have their Pictures drawne[120] with the Clothes now worn; which tho
at present never so rich & never so much in Fashion, within a few yeares
hence will make them looke like Antickes. Yet did he at Idle howres write

---

[a] Probably Charles Faldoe, Usher from 1637 to 1642: Maddison, *Life*, p. 19.
[b] William Douch, Rector of Stalbridge since 1621; d. 1648.

some few verses both in French &[121] Latin: & many Copys both of amorous, Merry, & Devout ones in English; most of which, uncommunicated[,] the Day he came of Age he sacrific't to Vulcan, with a Dessein to ‹make›[122] the rest perish by the same Fate, when they came within his Power; tho amongst them were many serious Copys, & one long one amongst the rest, against Wit Profanely or Wantonly employ'd; those two Vices being ever perfectly detested by him in others, & religiously declin'd in all his Writings.[123]

About this Time also Philaretus began ‹to be taught›[124] some skill in the Musick both of the Voyce/tong/& hand; but the Discouragement of a Bad voyce quickly[125] persuaded him to Desist.

It was now the Spring of the Yeare when[126] newes was brought to Stalbridge, of the Approach of his Sister the Lady Goring, & in her Company two of his Brothers (the Lords of[127] Kinalmeakye & of Broghill:) then newly return'd from their Three Yeares Travells. In their Company arriv'd one Mr Marcombes, a French Gentleman who had been their Governor; & behav'd himselfe so handsomly in that Relation; that the Old Earle remov'd[128] Philaretus (his brother lying sick at a Doctor's house) to his owne House againe, &[129] entrusted his ‹whole› Education with this Gentleman. He was a man whose Garbe, his Mine [sic] & outside, had very much of his Nation: having been divers Yeares[130] a Traveller /fol. 178/ And a souldier; he was well fashion'd, & very well knew[131] what belong'd to a Gentleman. His Naturall were much better then his acquired Parts; tho divers of the Latter he possess't, tho not in an Eminent, yet in a very competent Degree: Schollership he wanted not[,] having in his[132] greener Yeares been a profess'd Student in Divinity; but he was much lesse Read in Bookes then men. And[133] hated Pedantry as much as any of the seaven Deadly sins. Thrifty he was extreamely, & very skilfull in the Slights of Thrift; but[134] Lesse out of Avarice then a just Ambition, & not so much out of Love to Mony, as a Desire to live Handsomely at last. His practicall Sentiments /Op[inion]s/ in Divinity, were most of them very sound; & if he were giv'n to any Vice himselfe, he was carefull by ‹sharply› condemning it, to render it uninfectious; being industrious, whatsoever he were himselfe, to make his Charges, Vertuous. Before Company he was ‹always› very[135] Civill to his Pupills; apt to Eclipse their Failings, & sett off their Good qualitys to the best advantage: but in his Private Conversation he was Stoically[136] Dispos'd; & a very nice Critick both of Words & Men; which Humor he us'd[137] so freely with Filaretus; that at last he forc'd him to a very cautious & considerate way of expressing himselfe; which after turn'd to his no small advantage. The Worst Quality he had, was his Choller, to[138] Excesses of which he was excessively[139] prone: & that being the onely Passion to which Philaretus was (much) observ'd to be inclin'd: his Desire to shunne clashing with his Governor; & his accustomednesse

11

to beare the sudden sallys of his impetuous humor, taught our Youth, so to subdue that Passion[140] in himselfe, that he ‹was soone able›[141] to governe it, habitually & with ease. The Continuance[142] of which Conquest he much acknowledg'd to that Passage of St James; For the Wrath of[143] Man worketh not the righteousnesse of God.[a] And he was ever a strict observer of that Precept of the Apostles, Let not the Sun go downe upon Your Wrath:[b] for Continu'd Anger turnes easily to Malice; which made him, upon occasion of this sentence of St. Paull; to say that[144] Anger was like the Jewish Manna, which might be wholesome for a Day or so; but if 'twere kept [at] all,[145] 'twud presently breed Wormes, & corrupt.

With this new Governor our Philaretus /fol. 178v/ spent the Greatest Part of the Summer, partly in reading[146] & interpreting the Universall history written in Latin;[c] & partly in a familiar Kind of Conversation[147] in French, which Philaretus found equally Diverting & Instructive, & which was as well consonant to the humor of his Tutor, as his owne. About this time his eldest Brother the Lord ‹of›[148] Dungarvan, having ‹at›[148] his owne Charges rais'd a Gallant Troope of Horse for the King's service in the Scotch Expedition; His father sent[149] two other of his sons, (Kinalmeaky[150] & Broghill) to accompany him in that service, & design'd Philaretus for the same Employment, but[151] the Sicknesse of his Elder Brother, (Mr F.) whom he was to go along with in that Voyage, Defeated all our Young man's greedy hopes. During his stay at Stalbridge, all that Summer, His Father, to oblige him to [be] Temperate, by freely giving him the Opportunity to be otherwise; trusted[152] him with the Key of ‹all› his[153] Garden & Orchards. And indeed Philaretus; was little given to greedinesse either in Fruite or sweetmeats; the latter he had no fancy to at all[154]; & in the former he was very Moderate: so valuing such Nicetys & Daintys, that tho he enjoy'd them with Delight; /Ple[asure]/ he could want them without the least Disquiet /Regret./ During this[155] Pleasing season, when the intermission of his Studys, allow'd Philaretus Leasure for Recreations; he would very often steale away from all Company, & spend 4 or 5 howres alone in the fields, to [walk] about,[156] & thinke at Random;[157] making his delighted Imagination[158] the ‹busy› Scene, where some Romance or other was dayly acted: which ‹tho imputed›[159] to his Melancholy, was in ‹effect›[160] but an usuall ‹Excursion› of his yet untam'd[161] Habitude of Raving; a Custome (as his[162] owne Experience often & sadly taught him) much more easily contracted, then Depos'd.

Towards the End of this Summer, ‹the Kingdome›[163] having now attain'd a seeming settlement by the King's Pacification with the Scotts,

[a] In margin: *Jam. 1. 20.*
[b] In margin: *Eph. 4. 26.*
[c] Possibly Giovanni Botero, *Relazioni universale* (1596), of which a Latin translation had been published.

there arriv'd at Stalbridge Sir Thomas Stafford, Gentleman-usher to the Queen, with his Lady[a,164] to visit their Old Friend the Earle of Corke, with whom, ere they departed, they concluded a Match betwixt his Fourth sonne, Mr F.B: & Mrs E.K.[165] Daughter to my L.S. by Sir   K: & then a Maid of Honor[166] both Young & Handsome. To make his Addresses to this Lady, Mr F. was[167] sent (& Ph. in his Company) before up to /fol. 179/ London: whither within few weekes they were follow'd by the[168] Earle & his Family; of which a greate part lived at (the Lady Staffords house),[169] the Savoy; the rest (for his Family was much encreased by the Accession of his Daughters the Countesse of Barrymore & the Lady Ranalagh, with their Lords & Children;) were lodg'd in the Adjacent[170] Houses; but ‹tooke their meales›[171] in the Savoy; where the Old Earle kept so plentifull a house that in   months his accompts for bare house-keeping exceeded   pound.[b]

Not long after his arrivall, Philaretus his Brother having been succesfull /having prosper'd/ in his Addresses to his Mistris, at last in the Presence of the King & Queen, publickely marry'd her at Court; with all that Solemnity that usually attends Matches with Maids of Honor. But to render this Joy as Short as it was Greate,[172] P. & his Brother, were within 4 dayes after commanded away for France & after having Kiss't their Majesties hands; they tooke a ‹differing›[173] farewell of all their Friends; the Bridegroome extreamely afflicted to[174] be so soone deprived of a Joy; which he had tasted but just enuf of, to encrease his Regrets by the Knowledge of what he was forc't from; but Philaretus as much satisfy'd; to ‹see›[175] himselfe in a Condition to content a Curiosity to which his Inclinations did passionately addict him. With these differing resentments of their Father's Commands; accompany'd by their Governor, two French Servants, & a Lacquay of the same Cuntry;[176] upon the End of October 1638,[c] they tooke Post for Rye in Sussex; where the next day hiring a Ship, tho the Sea were not very smooth, a Prosperous ‹Puffe of Wind›[177] did safely by the next morning, blow them into France.

The Third Section /fol. 179v/

After a short refreshment at Deepe, our Philaretus travell'd ‹through› Normandy to the cheefe Citty of it,[178] Rouën; but by the way[179] receiv'd advertisement of a Robbery freshly committed in a Wood, he must[180] traverse by Night; but judging the feare of being apprehended would

---

[a] Sir Thomas Stafford's wife was Lady Killigrew; her previous husband had been Sir Robert Killigrew (1579–1633), and Elizabeth Killigrew was the product of this marriage, as Boyle states.

[b] Throughout this paragraph, as throughout the text, the gaps are Boyle's.

[c] Sic: in fact, this must have been a slip of the pen for 1639: Maddison, *Life*, pp. 25–6.

deterre the Robbers from a suddaine return to the same Place after so recent a Crime, the Company quietly continu'd on their Journy to Roüen, & arriv'd safely at it: where amongst other singularitys, Philaretus tooke much Notice of a Greate Floating Bridge, which rising & falling as the Tyde-water[181] dos, he us'd to resemble to the vaine Amorists of[182] outward Greatnesse, whose spirits resent all the floods & Ebbes of that Fortune it is built on. From Roane they pass't to Paris; & having spent some time in visiting that vast Chaos of a Citty; they shap't their Course for Lyons; where after 9 dayes unintermitted travell they arriv'd: having by the way (besides divers considerable Places) passed by the Towne of Moulins; (here fam'd enuf for the fine Tweezes it supply's us with;) a part of the ‹French Arcadia, the› pleasant Pays de Forest; where the Marquis d'Urfé was pleas'd to lay the Scene of the Adventures & Amours of that Astrea, with whom so many Gallants are still in love so long after both his & her Decease:[a] being also by the way[183] usefully diverted by the Company of two Polonian Princes, who had[184] as well a right unto that Title, by their Vertu & their Education, as their Birth.

After some stay at Lyons, (a Towne of great Resort & no less Trading, but fitter for the Residence of Marchands then of Gentlemen,) they travers't those lofty mountaines that fortmerly [sic] belonging to the Duke of Savoy, were now by stipulation (in exchange of the Marquisat of Saluzzo) devolv'd to the French King;[b] & having by the Way, beheld that famous Place, where the Swiftest & one of the Noblest Rivers of Europe, the Rhosne, is so streightned betwixt two /fol. 180/ neighbour Rockes, that ‹'tis› no ‹such› large stride to stand on both his Bankes; the third Day after their Departure from Lyons, they arriv'd safely at Geneva; a little Common-wealth which their early embracing & constant Profession of the Reformed Religion,[185] together with that peculiar Care of Providence in so long & so unlikely a Preservation, has rendred very much the Theame ‹not only› of Discourse, but some Degree of Wonder.

Philaretus his Instructions[186] commanding him a long stay in this Place, his Governor, (having both[187] Wife & Children in the Towne) provided him Lodgings & Entertainment in his owne house; & at set howres taught him[188] both Rhethoricke & Logicke, whose Elements (not the Expositions) Philaretus wrote out with his owne hand; tho afterwards he esteem'd both those[189] Arts (as they are vulgarly ‹handled›) not only unseasonably taught, but[190] obnoxious to those (other) Inconveniences & guilty of those Defects, he dos fully particularize in his Essays. After these slighter studys he fell to learne the Mathematickes, & in a few months

---

[a] Honoré d'Urfé (1567–1625), Comte de Chateauneuf, Marquis de Valromery, author of the romance, *L'Astrée.*

[b] By the Treaty of Lyons, 1601. The *famous Place* next referred to is La Perte du Rhône.

14

grew very well acquainted with[191] the most usefull[192] Part of Arith-metick[,] Geometry (with it's subordinades [sic]) the[193] Doctrine of the Sphere, ‹that of the Globe›[194], & Fortification, in all which being in-structed by a Person that[195] had more[196] of his Scholler's Proficiency, then the Gaines he might derive from the common tedious & dilatory way of teaching, he quickly grew so enamor'd of those delightfull studys, that they were often prov'd both his Bisness & his Diversion in his Travels; & he afterwards improv'd his Opportunitys to the attainment of a more then ordinary skill in divers of them. He also frequently convert with a /fol. 180v/ Voluminous, but excellent French Booke, that call'd, Le Monde, which so judiciously informes its Readers, both of the Past & present Condition of all those States that now possesse[197] our Globe & by a Delectable & instructive Variety not only ‹satisfys men's Curiosity›,[198] but so copiously supplys them with matter, both of Curious & serious Dis-course, that he us'd to say of[199] that Booke, that it was worth it's Title (which meanes, The World).[a,200]

But to employ his Body as well as his Minde, because P. his Age was yet unripe[201] for so rude & Violent an Exercise as the Greate horse, he spent some Months in Fencing, & 10 or 12 in learning to Dance; the former of which Exercises he ever as much affected as he contemn'd the latter.

His ‹Recreations›[202] during his Stay at Geneva were sometimes Maill, Tennis (a Sport he ever passionately lov'd;) & above all the Reading of Romances; whose Perusall did not only extreamely divert him; but (assisted by a Totall Discontinuance of the English tongue[203]) in a short time taught him a Skill in French[204] (somewhat) unusuall to Strangers. In effect, before he quitted France, he attain'd a readinesse in the language of that Country, which enabled him, (when he made Concealement his Dessein) to passe for a Native of it, both amongst them that were so; & amongst Forreiners also: & in all his Writings whilst he was abroad; he ‹still› made use of the French Tongue, not out of any intention[205] to improve his Knowledge in it; but because it was that he could expresse himselfe best in.

But during Ph. his Residence at Geneva there hapned to him an Accident which he always[206] us'd to mention as the Considerablest of his whole Life. /fol. 181/ To frame a right apprehension of this, yow must understand; that tho Philaretus his Inclinations were ever Vertuus, & his Life ‹free from›[207] Scandall & Inoffensive; yet had the Piety he was master of allready, so diverted[208] him from aspiring unto more; that Christ who long had layne asleepe in his Conscience (as he once did in the Ship upon   ,[b]) must now (as then)[209] be wak't upon[210] a Storme. For

---

[a] Pierre Davity, *Le Monde ou la Description Générale de ses Quatre Parties avec tous ses Empires, Royaumes*, 7 vols. (1637).
[b] The gap – about five words in length – is Boyle's; the allusion is unclear.

on a[211] night which (being in the[212] very heart of summer) promis'd nothing lesse; about that time[213] of[214] Night that adds most terror to such accidents; P. was suddenly waked in a Fright with such loud Claps of Thunder (which are oftentimes very terrible in those hot Climes & Seasons;) that he thought the Earth would owe an ague to the Aire: & every clap was ‹both› preceded & attended with Flashes of lightning so numerous /fr[equent]/ & so dazling, that P. began to imagine them the Sallyes of that Fire that must consume the World. The long Continuance of that /a/ dismall Tempest where the Winds were so loud as almost drown'd the noise of the very thunder; & the showres so hideous, as almost quench't the Lightning ere it could reach his Eyes; confirm'd P. in his Apprehensions, of the Day of Judgment's being at hand /come:/ Whereupon the Consideration of his Unpreparednesse to welcome it; & the hideousnesse of being surprized by it in an unfitt Condition, make him[215] Resolve & Vow, that if his Feares were that night disappointed, all further additions to his life shud be more Religiously & watchfully[216] employ'd. The Morning come, & ‹a›[217] serener, cloudlesse sky return'd he[218] ratify'd his Determination so solemnly that from that Day he dated his Conversion; renewing now he was /fol. 181v/ past Danger, the vow he had made whilst he beleev'd[219] himselfe to be in it: that tho his feare was (& he blush't it was so) the Rise /occasion/ of his Resolution of Amendment; yet at least he might not owe his more deliberate consecration /ded[ication]/ of himselfe to Piety, to any lesse noble Motive then that of it's owne Excellence. Thus[220] had this happy storme an operation upon P. resembling that it had upon the Ground: for the Thunder did but terrify /fright/ & Blasted not: but with it fell such kind & geniall showres, as water'd his parch't & almost wither'd Graces /vertues/ & reviving their Greenenesse, soone render'd them both Flourishing & Fruitfull. And tho his boyling /   / Youth did[221] often very earnestly sollicit to [be] employ'd in those culpable Delights that[222] are usuall in & seeme so proper for that season, & have repentance adjourn'd till /to/ Gray haires /old Age/ yet did it's importunitys meete ever with Denyalls /repulses/ P. ever esteeming that Piety was to be embrac't not onely[223] to Gaine heav'n but[224] to serve God with. And I remember that being once in company with a Crew of mad Young Fellowes; when one of them was saying to him, What a fine thing it were if men could sinne securely all their Life-time, by being sure of Leasure to Repent upon their Death-beds; P. presently reply'd, that truly for his part he shud not like /accept/ of sinning, (tho) on those Termes; & wud not ‹all that while› deprive himselfe of the satisfaction of serving God[225], to enjoy so many Yeares Fruition of the World. In effect, 'tis strange that men shud take it[226] for an Inducement to an Action, that they are confident /certin/ that they shall /must/ repent it. But P. himselfe[227] having sufficiently discuss't[228] that Point of Early Godlinesse /Piety/ in the sixth Treatise of his Christian

16

Gentleman, I shall now[229] onely adde to the ‹Arguments /reasons/›[230] yow may find There alleadg'd;[231] that he us'd to say, that /fol. 182/ 'twas a kind of Meannesse in Devotion[232], to consider the very Joyes of t'other Life, more as a Condition then a Recompense; & that true Gallant Christians (in their Dutys to their Maker) looke upon Heav'n it selfe as Gallant Lovers do on the Portion in their services /add/ to their Mistresses; for they consider it as[233] a /property/ Consequent of their Fruition,[234] /   / not the Motive of their Love.

But (as when ‹in Sommer› we take up our[235] Grasse-horses into the Stable & give them store of oates; 'tis a signe that we meane to travell them) our P. soone[236] after he had received this new strength, found a new weight to support. For spending some of the Spring in a Visit to Chambery, (the Cheefe Towne of Savoy) Aix (fam'd for its Bathes) Grenoble the head Towne of Dauphiné & Residence of a Parliament, his Curiosity at last lead him to those Wild Mountaines where the First & Cheefest[237] of the Carthusian Abbeys ‹dos› stand seated; where the Devil taking advantage of that deepe, raving Melancholy, so[238] sad a Place; his humor, & the stange [sic] storys & Pictures he found /   / there of Bruno the Father /Patriark/ of that order;[a] suggested such strange & hideous thoughts, & such distracting Doubts of some of the Fundamentals of Christianity /Religion; that tho his lookes did little betray his Thoughts, nothing but the Forbiddenesse /   / of Selfe-dispatch, hindred his acting it. But after[239] a tedious languishment of many months in this tedious perplexity; at last it pleas'd God one Day he had receiv'd the Sacrament, to[240] restore unto him the withdrawne /   / sence of his Favor. But tho since then Philaretus ever look't upon those impious suggestions, rather as Temptations to be suppress't /rejected/ /fol. 182v/ then Doubts to be resolv'd;[241] yet never after did these fleeting Clouds, cease now & then to darken /obscure/ the clearest serenity of his quiet:[242] which ‹made› him often ‹say that›[243] Injections of this Nature, were such a Disease to his Faith as the Tooth-ach is to the Body; for tho it be not mortall, 'tis very troublesome. And,[244] (as all things worke together to them that Love God,) P. deriv'd[245] from this   Anxiety the Advantage of Groundednesse in his Religion: for the Perplexity his doubts created oblig'd him ‹(to remove them)› to be seriously inquisitive of the Truth of the very fundamentals of[266] Christianyty: & to heare[247] what both Turkes, & Jewes, &[248] the cheefe Sects of Christians cud alledge for their severall opinions: that so tho he beleev'd more then he could comprehend, he might not beleeve more then he cud prove; & not owe the stedfastnesse /   / of his Fayth to so poore a Cause as the Ignorance of what might be objected against it. He say'd (speaking of those /Persons/ that want ‹not›[249] Meanes

[a] Boyle refers to St Bruno (d. 1101), founder of the Carthusian order.

17

‹to enquire› /Conv[enience]/ ‹&›[250] Abilitys to judge) that 'twas not a greater happyness to inherit a good Religion; then 'twas a Fault /   / to have it only by inheritance; & thinke it the best because 'tis generally embrac't;[251] rather then embrace it because we know it to be the best. That tho we cannot[252] often give a Reason for /of/ What we beleeve; we shud be ever able to give a Reason Why we beleeve it. That it is the greatest of Follys to neglect any diligence that may prevent the being[253] mistaken, where it is the greatest of Miserys to be deceiv'd. That how deare soever things taken up on the score are sold; there is nothing worse taken up upon Trust then Religion; in which he deserves not to meet with the True ‹one› that cares not to examine whither or no it be soe. /fol. 183/

And now P. having spent one & 20 months in Geneva, about the Middle of September (164) departed towards Italy; & having travers't Swizzerland, & by the Way seene Lausanne (an Academy[254] seated upon the Greate Genevan lake) Zurich & Saleurre, (the heads of Cantons wearing the same Name[)]; & wander'd seven or 8 dayes amongst those hideous Mountaines wh[e]re the Rhosne takes it's source & he saw the Rhine but a Brooke; he at length arriv'd at the Valtollina; a ‹spacious› Valley wall'd round with the Steepe Alpes; but so delicious, & (especially at[255] that season) so crown'd with all that Ceres & Bacchus are able to present; that it deserv'd to be the Motive, but not the Stage; of those late Warres it has occasion'd betwixt the Rivall Crownes of France & Spaine. There Ph: had the Curiosity to visit the Place on which stood Piur, a Pleasant little Towne; once esteem'd for it's Deliciousnesse; but now much more & more meritedly famous for it's Ruine;[a] which happn'd[256] some 2 dozen of yeares since; by the sudden & unexpected fall of a neighbouring Hill upon it; which strucke the whole Towne so deepe into the Ground; that no aftersearch by digging has ever prevail'd to reach it. Having visited the singularitys of this Earthly Paradice; P. & his company began to climbe that Mountaine of the Alpes; which[257] denominated from[258] a Towne that sits upon it's foote[259], is usually call'd La Montagna di Morbegno. The Hill was 8 miles in ascent & double that Number downewards. It was then free from snow; but all the Neighbouring hils (where store of Christall's digg'd) like perpetuall Penitents, do all the Yeare weare white. Upon the top of this Hill (which is[260] entirely unhabited) P. had the Pleasure to[261] see the Clouds ‹(which they[262] pass'd through in their descent) beneath them› darkning the Middle of the Mountaine; which on the Top they had /enjoy'd/ a cleare serenity. But notwithstanding the fairenesse of the Day they spent it all in Traversing this Hill,[263] at the ‹height› of which they left the Grisons Territory,[264] & at the Bottome enterr'd a Village belonging to (that of the) Venetians; but having pass'd thorough[265] such a Purgatory as the Alpes to

---

[a] Piuro had been buried by the fall of a nearby hill in 1618.

their Italian Paradice /to that Paradice call'd Italy/, I cannot but suppose them somwhat weary; & so ‹my Pen› oblig'd to[266] let them (& it selfe) take some short rest. /fol. 183v/

The [Fourth Section]

Philaretus being thus entred into the vast & delicious Plaines of Lombardy, travers't the greatest part of that rich Province, & having stay'd the Stomacke of his Curiosity with the Observables of Bergamo, Brescia, Verona, Vincenza & Padoa (a famous University, but more peculiarly devoted to Esculapius then Minerva's Arts[267]) he gave it a full meale at Venice; where the greate Concourse of forreine Nations, numerously resorting thither ‹for Trade or Nobler Bisnesse›, presents the senses with a no lesse Pleasing then constant Variety. From Venice, returning through Padoa, & passing[268] through Bologna & Ferrara (Townes, whose Names, allowe me to spare their Characters) he at last arriv'd at Florence with a[269] Determination (having dispos'd of the Horses he rid on from Geneva thither,) to passe the Winter there. Florence is a Citty to which Nature has not grudg'd a Pleasing[270] situation, & in which Architecture has been no niggard either of Cost or[271] Skill: but has so industriously & sumptuously improov'd the Advantages liberally conferr'd by Nature, that ‹both› the[272] seate & Fabrickes /buildings/ of the Towne, (abundantly) justify the Title the Italians[273] have given it of Faire. Here P. spent much of his Time in learning of his Governor (who spake it ‹perfectly›[274]) the Italian Tong, in which he quickly attain'd a native accent, & knowledge enuf[275] to understand both Bookes & Men; but to speake & expresse himselfe readily in that Language, was a skill he ever too little aspir'd to /car'd for/ to acquire.[276] The rest of[277] his spare howres he spent in reading the Moderne history in Italian, & the New Paradoxes of the greate Star-gazer Galileo;[a] whose ‹ingenious› Opinions,[278] ‹perhaps› because they could not ‹be so› otherwise ‹were›[279] /confuted/ by a Decree from Rome; the[280] Pope it seems /fol. 184[281]/ presuming ‹(and that justly)› that the Infallibility of his Chaire extended to determine[282] points in Philosophy as in Religion; & loath to have the stability of that Earth question'd, in which he had established his Kingdome; /Greatnesse/.[283] Whilst Philaretus liv'd at Florence, this famous Galileo dy'd within a league of it; his memory being honor'd with a celebrating Epitaph & a[284] faire Tomb erected for [him] at the Publicke Charges /Cost/[b] But before his Death being long growne

---

[a] Probably his *Dialogue concerning the Two World Systems* (1632). The work on modern history could be that of Giovanni Botero.

[b] Although Boyle must have heard of plans to erect such a tomb while he was in Italy, these were not implemented until 1737: Maddison, 'Galileo and Boyle', p. 350; Fahie, *Memorials of Galileo Galilei*, pp. 142–4.

Blind; to certaine Friers (a Tribe whom for their Vices & Impostures he ‹long›[285] had hated) that reproacht him with his Blindnesse as a Just Punishment of heav'n, incens'd for being so narrowly pry'd into by him, he answer'd, that he had the satisfaction of not being blind till he had seene in heav'n what never mortal Eyes beheld before.[a] But to returne to P. the Company of certaine Jewish Rabbins lodg'd under the same roofe with him, gave him the opportunity of acquainting himselfe with divers of their Arguments & Tenents, & a Rise of further Disquisitions in that Point. When Carnevall was come (the season when Madnes is so generall in Italy that Lunacy dos for that time loose it's Name) he had the Pleasure to see the Tilts maintain'd by the greate Dukes Brothers; & to be present at the Great men's Balls. Nor did he sometimes scruple, in his Governor's Company, to visit the famousest Bordellos; whither resorting out of bare Curiosity,[286] he retain'd there an unblemish't ‹Chastity›, & ‹still› return'd thence as honest as he went thither. Professing that he never found any such sermons against them, as they were against themselves. /fol. 184v/ The[287] Impudent Nakednesse of Vice, clothing it with a Difformity, Description cannot reach, & the worst of Epithetes cannot but flatter. But tho P. were noe Fewell for forbidden Flames, he prov'd the Object of unnaturall ones. For being at that Time in the Flower of Youth,[288] & the Cares of the World having not yet stain'd[289] a Complexion naturally fresh enuf; as he was once unaccompany'd diverting himselfe abroad, he was somewhat rudely storm'd[290] by the Preposterous Courtship of 2 ‹of those› Fryers, whose Lust makes no Distinction of Sexes, but that which it's Preference of their owne creates; & not without[291] Difficulty, & Danger, forc't a scape from these gown'd Sodomites. Whose Goatish Heates, serv'd not a little to[292] arme Filaretus against such Peoples specious Hypocrisy; & heightn'd & fortify'd in him an Aversenese for Opinions, which now the Religieux discredit as[293] well as the Religion.[b]

P. having thus spent the Winter in Florence, towards the End of March began his journey to Rome, & having[294] past thorough & seen the singularitys of Siena, Montefiascone & some other[295] remarkable Places in his Passage, at the End of 5 dayes he safely arriv'd at that ‹imperious›[296] Theame of Fame, which destinated to some kind or other of universall Monarchy, is now no lesse considerable by it's present superstition then formerly by it's Victorious Armes; the ‹Moderne›[297] Popes bringing it as high a Veneration as the Antient Cesars; & /fol. 171/ The[298] Barberine Bees flying as farre as did the Roman Eagles.[c] The more conveniently to see

[a] It has not proved possible otherwise to trace the story that Boyle here retails.
[b] In margin: *Place here the story of the Duell with Coates of Maile.* It is unclear what this story was.
[c] Boyle alludes to the Barberini family, whose coat of arms included the motif of bees; Maffeo Barberini (1567–1644) had become Pope Urban VIII in 1623.

the numerous raritys of this Universall Citty, P. to decline the distracting[299] Intrusions & importunitys of English Jesuits, past for A french man, which ‹neither›[300] his Habit ‹nor› Language much contradicted. Under this Notion he˙ ‹restlesly›[301] & delightfully pay'd his Visits, to what in Rome ‹& the Adjacent Villas› most deserv'd; & amongst other Curiosityes & Antiquitys, had the Fortune[302] to see the Pope: at Chapell with the Cardinalls, who severally ‹appearing›[303] mighty Princes, in that Assembly, look't like a Company of Common Fryers. Here P. cud not chuse but smile, to see a Young Churchman after the service ended, upon his knees carefully with his feet sweepe into his handkercheeefe the Dust his holynesse's (Gowty) feet, had ‹by treading on it› consecrated, as if it had been some Miraculous Relique. Nor was P. negligent to procure the Latin & Tuscan Poëms of this Pope; whose Name Urbanus, his Actions did not belye; he having ‹more›[304] of the Gentleman in him then his Pontificall Habit, would /might/ seeme to let him weare. A Poet he was & a Mat...[a,305] & tho neither of them in perfect ... much, that he delighted & to [?] ... [P. said] /fol. 171v/ that he never[306] found the Pope lesse valu'd then in Rome, [nor] his Religion fiercelyer disputed against then in Italy; &[307] sometimes added, that he ceas'd to wonder that the Pope shud forbid the sight of Rome to[308] Protestants, since nothing could more confirme them in their Religion.

P. having in a short Time survey'd the Principall Raritys, of this proud Mistris of the World, was unwillingly driven thence by his Brothers Disability to support th'encreasing Heates; which there[309] prove often insupportable to strangers, (the neighbouring Cuntry being very scorch't[310] barren & uninhabited.) Wherefore he tooke his way backe towards Florence, by that delicious Valley that ennobles Perugia, & passing by Pistoya, came to Florence; where after a short repose, they descended the river Arno unto Pisa,[b] & from thence to Livorno, where in a Felouca[c] with a good Winde, they ventur'd, for expedition sake,[311] some 15 or 16 miles into the sea; & coasting along the country, still neere the shore for feare of sudden stormes, they each night lay in some towne, drawing their Boat ashore (which was not uneasy in regard of the Inconsiderable Tides of the Meditteranean there) & [soon]e (tho not without Danger) [reac]h't proud [?] Genoa ...[d] /fol. 172/ As if she meant to give Italy a Good Deboire; & indeed[312] the Cuntry could no way have left P. a better Relish & Idea of hir, then by[313] making the last Place he stop't at[314] in hir, one of the Pleasantest that could be seene. The next day Filaretus prosecuted his Journey, & passing by Monaco (a very strong Place, then newly betray'd

---

[a] Here and in the next line, the paper has been torn away affecting about one-third of each line in the MS.
[b] In margin: *Place here the Accident about his Teeth.* It is unclear what Boyle is referring to.
[c] A felucca, a coasting vessel equipped with oars and a lateen sail (OED).
[d] Bottom of page missing.

by the Prince of it to the French,) & by Mentone, a little Principality belonging to the same Prince, & ‹stopping a While›[315] at Nizza, a Place extreamely & meritoriously famous for that strength, which Nature & Art, have emulously given it, by night they[316] landed at Antibe, one of the Townes of France[317], that most approaches Italy.

The Fifth Section

The Morning that succeeded P. his Arrivall at Antibe, he left it to crosse the Cuntry to Marseilles. But he was welcom'd into France by an Accident which was very hazardous, & might have prov'd Tragicall.[318] During the whole time of his being a Traveller or Resident in Italy he had religiously adher'd to his ...[a] /fol. 172v/ Englishmen that ... ‹...ilist [?] before Philar.›[319] thinking as much better as safer, [to] take of their hats then to venture their heads, complimented with the Crucifixe: but P. without the least Act of superstition, tho not with ill Words, & worse Menaces, ‹ventur'd & past› boldly[320] thorough them all: as ever[321] resolving that the Soule shud not more transcend the Body in it's owne Value, then in his Esteeme. This Danger thus happily scap't, P. continues his Way to Marseilles, where the third Day he arriv'd, with intent there to expect of Exchenge promis'd to be then sent thither, to enable ‹him› to prosecute his future Travells. His Detention here was shortned by his Visits of so excellent a Harbor for Gallyes & small vessels; which as 'tis compriz'd in the Towne, is every Night, like it, ‹assur'd›[322] with Locke & Key. Here P. had the pleasure to see the French King's Fleet of Galleys put to Sea, & about 2000 ‹poore› Slaves tugge at the Oare to row them.[b]

[a] Bottom of page missing.
[b] At this point the text ends, not at the bottom of the page, so there is no reason to believe that it is incomplete. On the failure of the bills of exchange to arrive due to the outbreak of hostilities in England, see Documents 3 and 7 below, and Maddison, *Life*, pp. 45f.

DOCUMENT 2

Biographical notes dictated by Boyle
to his amanuensis, Robin Bacon

Add. MS 4229, fols. 66–9. Endorsed at the top of fol. 66: 'The Autograph of Robin Bacon, who was for many years Mr Boyle's Amanuensis. H[enry] M[iles]'. Fols. 66–7 and 68–9 are both conjugate pairs of leaves, of which fols. 67 and 69 are wholly blank apart from the following endorsement on the verso of the former in Wotton's hand: 'Papers of Mr Boyles Life delivered by him to the Bishop of Sarum'. Apart from the endorsements noted above and the words in the text which are here italicised, which are in Wotton's hand, the whole of the manuscript is in the hand of Bacon. Fol. 68 is more neatly presented than fol. 66, with no deletions, perhaps due to its having been recopied. See also above, pp. xxiii–iv.

[a] was born the 25th of January commonly called St. Pauls day in the year 1626 ([1]according to the English account) at Lismore a place of some note belonging to his Father ‹in the county of Waterford[2] Part of the Province of Munster in Ireland.›[3] His Parents were Richard   and the Lady Catherine Fenton his wife, by whom her Husband had, for he was a widower & a Father too, when he marry'd her *thirteen* children, of which our *Robert Boyle* was[4] *the youngest*[5] son[;][6] his breeding in his Fathers family til he was about nine years of age, had nothing peculiar, but about that time, he was sent together with a Brother elder than he, by above 3 years, into England, to go to school at Eaton Colledg under the inspection of the famous Sir Harry Whitton[b] then Provost of it, and a Friend of his Father, who us'd him with much civility & kindnes incouraging him to pursue his studys. In this present[7] Mansion ‹of the Muses› he continu'd for almost 4 years thô with some interpos'd Recesses to visit his Relations. During his abode in that Colledg, there happen'd to him, nothing worth taking notice of, unless one should allow some room in this Place to such petty Incidents as these.

[a] A gap is left in the MS at this point, as if a name was intended to be inserted; the same occurs at a number of subsequent points in this document which have not been separately noted. In this first instance only, *N.M.* has been faintly written in it in pencil.
[b] Sic: i.e., Sir Henry Wotton.

1. His immediate schoolmaster[a] being ‹of› a somwhat rigid humor kept him so strictly to his Book (as ‹they›[8] speak) that it made him almost loath the Books his Lessons were taken out of, on which account he could never relish, even Tullys Offices for many years after he grew a man[;] but another schoolmaster ‹Mr or Doctor Harrison› he had, who having been gentilely bred & visited forraign Countrys treated him much more winingly. ‹For›[9] observing, that as God had blest this Boy with a happy memory, he was thereby inabled to get his Lesson perfectly by heart in a short time, but seem'd not at all to relish[10] Books, because he had been /fol. 66v/ allow'd none but those read in his Form: this discreet Gentleman, I say, taking notice of this put into his hands some pleasant & yet useful Books to entertain the time with, after he had got his Lessons by heart, and finding that the Boy took this kindly, & read these diverting Books with pleasure, and almost with greedines he would now & then at hours vacant from school exercises take him into his wel furnisht study[11] & leave him there for a good while together to look on the Books & tell the Owner, which of them he took most delight in, & if they were proper he would lend him sometime one, & sometime another to peruse at home[;] that piece of these that ‹our Lad›[12] was first charm'd with was Quintus Curtius which much enamor'd him ‹on›[13] reading History, and conversing with Books, for which reason he us'd to own a peculiar obligation to that Eloquent Author, as having first ingag'd him to love Reading, & he would some times relate, that being once in a Country Inn, troubled with violent pains after his journy, & not knowing in that disfurnisht place how to relieve himself, he found among ‹som›[14] Books, which one of his servants took along with him in travelling the History of Quintus Curtius, which he had not seen of a long time, so that 'twas almost new to him, upon which account he read it with great delight & attention so long that his spirits being diverted from minding his pains, when he had don reading he found they had left him.[b]

*   *   *   *   *   *

/fol. 68/ After our   was return'd into England he posted away from Dover to London, where he found by Inquiry, for he had no full address to her, where his sister the Lady Catherine Jones who was fled thither out of Ireland, was lodg'd. And coming into the house in his french Dresse he

---

[a] This might be William Norris: Maddison, *Life*, p. 19.
[b] On Boyle's enthusiasm for Quintus Curtius, see also above, p. 7.

pass'd by several of his younger Relatives, who gaz'd at him without knowing him, as he on the other side knew them but by guess, but advancing to the innermost part of the House, and meeting no servants of whom to make Enquiry, he spoke to a Person, that was going down a pair of Stairs, your [sic] Back turn'd towards him, and Enquir'd for such a Lady. This question made her come up to see who ask'd it, and at first she look'd upon him with surprize, because she knew not at all that he was in England, whence he had been absent so many years, but afterwards having upon an attentive view, discern'd who he was, she cry'd out, oh! 'tis my Brother, when with the joy & tendernes of a most affectionate sister, she saluted & imbrac'd him[;] when many questions & answers were interchang'd between them, she took care privately to have an apartment provided for him in her House, & would by no means suffer, he should lodge any where else whilst he staid in London, which was four months & a half.

This abode with so excellent a Person in a faire large House, where there liv'd also a Sister in Law of hers, marry'd to one of the chief members of the then House of Commons, who were both of them eminently Religious.[a] Our   many times with thankfulnes to God, acknowledg'd as a seasonable /fol. 68v/ Providence to him for more than one Reason, since first in the heat of his youth, it kept him constantly in a Religious family, where he heard many pious Discourses, & saw great store pious Examples [sic], whereas if a gracious Providence had not detain'd him here, he had lost the benefit of all these, & instead of them being expos'd to the manifold & great temptations of a Court & an Army, where thô there were besides the excellent King himself, and diverse eminent divines, many worthy Persons of several ranks, yet the generality of those he would have been oblig'd to converse with were very debaucht & apt, as well as inclinable to make others so. Besides by this means our   grew acquainted with several Persons of power & interest in the Parliament and their party, which being then very great, & afterwards the prevailing one, prov'd of good use, & advantage to him, in reference to his estate and concerns both in England and Ireland.

---

[a] At this point Lady Ranelagh was evidently living with the leading Presbyterian politician Sir John Clotworthy, to whom and his wife Margaret she was sister-in-law: Maddison, *Life*, p. 53n.

DOCUMENT 3

The 'Burnet Memorandum': notes by Gilbert Burnet on his biographical interview(s) with Boyle

Add. MS 4229, fols. 60–3; fol. 62v is wholly blank, as is fol. 63 except for the following endorsement on the verso in Wotton's hand: 'Mr Boyle's papers dictated by him, copied by Dr Burnet, relating to his Life'. Endorsements by Miles on fols. 60 and 60v are recorded in footnotes; otherwise, the entire manuscript is in the hand of Burnet. The pages of the manuscript are numbered 1 to 5 in brackets at the outer top corner. The content of the first page is set out in two parallel columns, with items 1 to 20 in the first and 21 to 37 in the second. See also above, pp. xxiv–viii.

1 Age January [sic] 1627
2 Loving trueth of a child
3 beloved of his father
4 sent to Eton School for 4 years at 9 year old
5 Sir H Wotton kind but no Inclination to Latin
6 A love to read History begun with Sir W Rawleigh[a]
7 Travells at 14
8 A Gouvernour honest but Cholerick more a Gentleman then a Schollar[b]
9 at Geneva acquainted with Diodati his Gov[ernor's] Uncle by his wife
10 There he acquired the Latin
11 he had Philosophy and Mathematicks taught him
12 he fell on Senecas ‹Naturall questions›[1c] which first set him on Naturall Philosophy
13 Travells into Italy there he read the lives of the old Philosophers[d] and became[2] in love with the Stoicall Philosophy and endured a long fit of the toothach as a Stoick

[a] Sir Walter Ralegh, *The History of the World* (1614 and subsequent editions).
[b] Isaac Marcombes, who was nephew by marriage of the famous divine, Jean Diodati: Maddison, *Life*, p. 21n.
[c] Lucius Annaeus Seneca, *Naturales Quaestiones*, in which issues relating to natural phenomena and their explanation are discussed.
[d] Presumably the *Lives and Opinions of the Eminent Philosophers* of Diogenes Laertius, referred to by Boyle in *Works*, vol. i, p. 355.

26

14 On May day observed often Inauspicious hears of the Irish Rebellion[3]

15 Returns to Geneva wher by the dishonesty of a merchant he lay 2 years for want of money[;] his father dies and he in watches and rings had credit to take as much as brought him to England near the end of the warre he paied his debts and followed his studies rather reading every thing then choosing well

16 Abstained from purposes of marriage at first out of Policy afterwards more Philosophically and upon a Generall proposition with many advantages he would not know the persons name[a]

17 Upon Ar[chbishop] Ushers reproaching him that was so studious for his ignorance of the Greek he studied it and read the N[ew] Test[ament] in that Language so much that he could have quoted it as readily in Greek as in English

18 From this he went to the study of Hebrew till he both Mastered the text [bu]t could also read a Commentary[.] The Caldaick and Syriack he also acquired [;] he writ a Grammar of Hebrew for his own use

19 He set about the Arabick and had mastered it if an Infirmity in his eyes had not taken him off from reading

20 By a fall off a horse on May day and travelling too soon after he fell in an Anasarka[b] and recovering ‹of› that in the winter after into a feaver which ended in an Ague that made him keep within all winter

21 his eyes became so ill that he could not read much himselfe after that

22 he had been an Experimentall Philosopher before but now he join[ed] The Study of Phisick to it

23 He writ Seraphick Love at 22 and his Occasionall Meditations before[4]

24 His Stile of the Scriptures a part of a larger designe for raising the Scriptures in his brothers esteem[c]

25 He was of the Committee of forreigne Plantations and in that set himselfe much to know the Natural History of those Countries for which he drew a Paper of Queries

26 He pursued the same designe in the East India Company

27 His name in an Act of Parliament for the ward of a lady but his was only borrowed in trust for he made no advantage by it[d]

28 A Grant begged in which he was to be a sharer without his knowledge

---

[a] After *the persons name*, Miles has inserted in tiny script: *see Dr Wallis Letters where the name of a lady is eras'd*. Cf. Wallis to Boyle, 17 July 1669, *Works*, vol. vi, p. 459; the original of this letter is BL 5, fols. 174–5, where the name is indeed erased, though it is highly questionable whether this is the episode to which Boyle refers: see above, p. lxxvii.

[b] A dropsical complaint (OED). Cf. Boyle's preface to *Medicinal Experiments*, *Works*, vol. v, pp. 315–6 and above, p. lxxviii.

[c] Presumably Broghill.

[d] This probably refers to the wardship of Jane Itchingham, which Boyle was granted in November 1660. However, the link with an Act of Parliament is otherwise unknown. See Maddison, *Life*, p. 101.

for he often said he would ask nothing but would take what should be given by the King for the Charge he was at in his laboratory yet he never would make any advantage of his share but resolved to apply it to a good use[a]

29 He made no vowes not knowing how his Circumstances might change but usually gave the 20th

30 He generally put large Charities in other peoples hands to distribute for them

31 He had never any temptations to Popery nor was he even joined to the sectaries ‹he kept to› the Communion of the Church of England ‹yet he›[5] was not keen in any of our debates and loved vertue and goodnes in all people[;] if he sharp [sic] in any point it was against persecution which he thought Immorall

32 Ever Modest sparing to censure never speaking evill[6] of any[;] never engaged in any great mans concerns or in any faction or Intrigue but Civill to all and particular with few in busines

33 He made Conscience of great exactnes in Experiments

34 was much Visited by all forreigners Ambassadours and others

35 His ill health brought him under very strict rules of diet which he observed like a Philosopher

36 Never addicted to Vanity in his Apparell only found it necessary to shew a litle finer [sic] among his Tennants to have his land well let

37 Inclined to be be [sic] Cholerick but Governed it so that it never appeared except in his looks a litle when much provoked

/Fol. 60v/

Inclination to play but overcome becaus too earnest to win more for victories sake then for money[;] never given to Appelise nor stealing fruit or robbing Orchards when a boy[;] When he lived in Florence ther was a Jew whom the Inquisition had driven out of Spain wher he had the Government of a towne lodged in the same house wher they had many discourses about the Scriptures this led him first to enquire into them.

Deliverances when a child being left in a Coach his Father with the rest in it got on horseback to crosse a torrent somewhat Impetuous but a Gentleman took him out and set him before himselfe on his horse neck and[7] so he got over but the Coach was overturned in the water so that very probably he had drowned. another time at Eaton a horse flung him over his head and ride over him without hurting him tho he tread so near his neck that his band was torn by the nails in his shoe.[b]

His first designe in Philosophy was to recommend the Usefulnes of

---

[a] See below, p. 78, and Hunter, 'Casuistry in Action', pp. 83–4.

[b] Miles has here added in the same small hand as above: *Compare his own Narrative under the name Philaretus.* See above, pp. 4, 7.

Experimentall Philosophy to the Gentry who being disgusted with the dry Notions of that taught in schools and knowing no other were prejudiced against all Philosophy[;] he pursued it in that chiefly in relation to Phisick but designed other parts of its usefullnes in relation to trade and Mechanicall performances[;] in order to this he was gathering all the observations of the Naturall History of other Countries to try how farre it was feasible to bring the product of them to thrive in England but want of health made [sic] he could never finish this and many loose sheets he had writ of it were either losst or stolln[;] this was written before the King came in.[a] His next book was the Scepticall Chimist writ to take those ‹Artists› off ther excessive Confidence in their principles and to make them a litle more Philosophy [sic] with their Art[.] Then he fell on his Experiments about Air to examine the qualities of it and what pure Air above the Atmcsphere is which he hoped might direct him in many usefull things for the Regiment of our health and took up after many Experiments that Hypothesis of the pressure and Elasticity of the air[;] against him writ Linus a Jesuit and Professor of the Mathematicks with great Modesty and good Judgment upon the old Principles of the fuga Vacui and conamen atque motum[;][b] Hobbs as he said himselfe being resolved to quarrell with the society singled him out but answered him both Ignorantly and as he used to doe Insolently[;] in one grosse errour he fell of which Mr Boile gave him notice thinking it was the fault of the transcriber or Printer but he justified it so he resolved to expose him having otherwise a great dislike of the mans temper and Principles but having thus confuted these Answerers he resolved never to engage further into Controversy about Philosophy.[c] The Affinity of Air and Colour led him to write of that and he next writ of Cold which was a great prejudice to his health for the cold steams of ice and snow were too severe for so weak a body.

He was long very distrustfull of what is comonly related concerning the Philosophers Stone. One brought him once a grain pretended to be the powder of projection but ‹that person› putting it unskillfully into the lead the flame scorched his hand so that the shaking it made him lose it[;] somewhat of it stuck to the ‹paper›[8] but so litle that the eye without the help of a Microscope could not discern it[;] that he carefully gathered and put to lead in fusion which being so litle could not have great Operation but it stopt the fusion of the lead which made him conclude that a greater quantity might have produced some more considerable change[;] but he

---

[a] On this intended work see Glanvill, *Plus Ultra*, p. 104, and above, p. lxvii.

[b] *Avoidance of a vacuum/ exertion and thus movement*. The reference is to Francis Linus (1595–1675). See Shapin and Schaffer, *Leviathan and the Air-Pump*, pp. 155f.

[c] Thomas Hobbes (1588–1679). See Shapin and Schaffer, *Leviathan and the Air-Pump*, inc. pp. 129f., 350f. on Hobbes' professed dislike of the Royal Society (on which cf. *Works*, vol. i, pp. 188–94, 232), though Boyle's further claim is not there substantiated. See above, p. xxviii.

received afterwards a more convincing satisfaction in that matter he going to Visit a forreigner that was known to him found one with him[9] of whom he had no knowledge so the forreigner after he had whispered the other it is like to ask his leave to let him see the Experiment they were about he told him the stranger would ‹trie›[10] an Experiment of making lead ductile as wax or rather as butter but that he might not apprehend he was deceived he wisht him to send for a Crucible /fol. 61/ and some lead which being done and the lead put in fusion he put into an[a]   of lead about   of a bright powder and a litle while took the Crucible from the fire and when they Judged it was cold he expecting no such thing was not a litle surprised to find instead of lead Gold which after all the trialls that could be made was found true Gold of   finenesse.

A forreigner that was a Chimist told him that as he was going to Charenton on a hot day he being a litle faint went into a Cabaret and staying till somewhat was prepared for him[11] to eat one came in that seemed a Churchman for he had the tonsure[;] they resolved to eat together and as a French Conversation grows soon familiar the other asked him his Religion to which he only answering that he was sure he was not of his Religion the Ecclesiastick said he feared he neither beleeved God nor the Devill but he would satisfy him concerning the one and so told him he had spirits at command and could make them appear when he would and asked him if he could endure the sight of them in a terrible shape[;] the other said nothing could afright him so he spake a few words and as he thought four Wolves came into the room and run round the Table about which they were sitting for some considerable time[;] they lookt full of rage he was so terrified that he felt his hair stand on end and desired the other to deliver from that Company so upon some words spoken by the other they disappeared and then he said he would entertain him mor[e] aggreably so two ladies seemed to come beautifull and well dressed and in their behaviour had all the modes of a Courtisan[;] a banquet appeared and he thought he saw them eat and that on[e] of them reacht him a glasse of wine which as he thought he dran[k] of. they spake nothing but by many signes Invited him to come to them but he kept at a distance. he asked them questions about the Philosophers Stone and as he thought one of them writ a paper which he read and as he thought understood but within a litle they vanished so did the paper and what was writ in it went ‹so› clearly out of Memory that he could never trace it. after this he being fully convinced of the apparitions of Spirits had a great [sic] to learn Magick and what by the help of the Ecclesiastick what by a written book of Conjuring which one had losst in the high way and a poor traveller finding it and by accident

---

[a] A gap in the text has been left after *an*, as after *about* three words later. Another gap follows *Gold of*.

offered it to him for a small reward it was in French but writ in Greek Characters. he had come so farre that he could make spirits appear and offer[ed] to satisfy Mr B by shewing him in a glasse of water strange representations. but he would not accept of it. He said he had not yet the Mistery of his Art nor could he make spirits speak to him but he had a direction to use such a time of fasting to wear such clothes and such humour [?] and to use such Incantations after which he should be an Adeptus he thought it no sin for he sai[d] he made no Compact but did what he did by a rule like an Ordinar[y] Receipt and if the Evill Spirits were subject to such rules eithe[r] [by] some Compact with those who had the secret first from whom it wa[s] conveyed downe, or by the lawes of the Creator[,] he was not concerned in it[.] He was a libertine as to his pleasures and some other things yet in these things Mr Boil saw no reason to disbeleeve him especially when he offered to satisfy him in it and indeed when the thing was proposed to him he asked my[a] Opinion whether he might try whether [this][12] man spake true I said as I could not directly say it was a sin yet it was a thing justly to be doubted of and therfore to be forborn[.] He had heard another story of the like nature but from a man to whom he gave a much more undoubted credit. He gave me no other hint of him but that he was a man of quality and Estate was of the Royall society and had been upon some occasion chosen President pro tempore he was a man of a very vertuous life so that he believed him as much as he did any man.[b] This Gentleman had been much in love with a lady that did not entertain it but in his fondnes he corrupted her woman to get him /fol. 61v/ some of her hair which she gathered out of her Combs[;] this he set in a locket of Diamond and going to travell misst it at Venice at which being much concerned he went to a Priest who as he heard had a Magicall glasse which the Priest owned but said he could not look into it more then himselfe since only Virgins could see in it but the Gentleman said he should bring a Virgin with him[.] The Priest was shie for he said it was a Reserved case and he must confesse it to the Bishop who he knew would enjoin him a severe pennance but a good present removing that scruple the Gentleman brought a young girle with him[:] a glasse transparent was brought out in which the Gentleman saw nothing[;] the Priest only said this Gentleman would see who robbed him of a locket of hair so the girle that was about 9 year old looking in saw first as it were pretty boies then a boat

---

[a] I.e. Burnet's.

[b] The identity of this man is unclear. The combined reference to Italian travels and the presidency of the Royal Society might favour Boyle's friend, Sir Robert Southwell, except that his presidency occurred only in the early 1690s: see above, pp. xxv–vi. Alternatively, this could conceivably be a veiled reference to Boyle himself, who had visited Venice, Padua and Moulins, and who was elected president of the Royal Society in 1680 but turned the office down. See Hunter, 'Alchemy, Magic and Moralism', pp. 390n., 393.

such as they passe in from Padua to Venice then a chamber in the boat wher the Gentleman that brought her seemed to lie sleeping then a boy who had served him seemed to come in and take out some things out of his pocketts she described the locket and the colour of the hair and also a picture of the ladies losst at the same time with some other things one of them she could not well describe for she had never seen any of the sort only she said it was like an etisse[a] and he then an etisse [sic] of a new fashion which he had bought at Molins. by all these things the Gentleman was convinced that the Glasse told true but he had dismissed that servant a litle while before so he could not recover his locket but this did then so affect him that it cured him of his former passion[;] he came often with his Interpreter the girle who was to see for him and among things having got a processe of the Philosophers stone he brought it with him and made the Priest ask if it was a good one upon which the girle said the boies lookt very angry but she saw nothing else[;] so the question was repeated but then she said the boies lookt so on her that she was affrighted and feared they should beat her. So the Priest said he must presse it no further least the spirits should tear them to pieces.

Another Gentleman not only told him stories like this but brought him a glasse in which he did not doubt but he[13] was a Virgin and so could satisfy himselfe. He told me he lookt about it he saw it was an Ordinary double convexe glasse and had no angles in it by which tricks might have been put on those who lookt into it[;] he had the greatest Curiosity he ever felt in his life tempting him to look into it and said if a Crown had been at his feet it could not have wrought so much on him[:] but he overcame himselfe which he accounted the greatest Victory he had ever over him-selfe. The Gentleman in whom as was formerly told he had such confid-ence said to him that his over Curiosity as it cured him of his passion so it raised in him a most confounding Melancholy which those who under-stood not ‹the true cause› Imputed to his concerne in his Mistresse who was married but he said the Reflections on his curiosity and the thoughts which the satisfaction he had received raised in him drove him almost to madnes but he overcame it and it was to him the happy occasion of a great change in the whole course of his life[.]

He was also told that women with child had by vertue of the child in their belly the Priviledge of seeing into those glasses and that other Magic-ians knew how to give such sights in water of which one gave him ane account that during one of the sea fights wee had with the Dutch the roar of the Cannon being heard in London one that was with child expressed at

---

[a] This reading is slightly doubtful (partly due to an alteration during composition), and so is the meaning of this word: *estrich* was the commercial name for the fine down of an ostrich (OED). There is a lacuna in the sense at this point, presumably due to haste in composition.

randome a Curiosity of knowing how the matter went so one being expert in the Art brought a bason or pale of water in which tho the rest of the Company could see nothing yet she saw ships with English and Dutch colours engaging and beheld the issue of the battell[;] the womans name was told and he Intended to have gone and asked her about [it] but he who told him of it went suddenly out of England and Mr B. had forgot her name for depending on the person that was to conduct [him] he had not been carefull to take her name or the place where she dwelt[.] He never engaged in Correspondence by letters for he spared his eyes and so would not write himselfe so he excused himselfe to a great many who Invited him to it /fol. 62/

He was upon the Kings coming in used very kyndly by the King who said once that he deserved a statue[;] the E[arls] of Clarendon and Southampton[a] were particularly kynd to him and it was moved to him to enter into Orders and he should be made a Bishop but he said he never felt the Inward Vocation so he should be in an Imployment against the grain with him. He followed the Councill of forreigne Plantations with great application and had caried on a great undertaking with some Merchants of London for all the sagars [sic] so that it was like to be of great advantage both to the King and all concerned in it[b] but some upon private reasons or humours obstructing it he gave over further medling in that Councill but had the Kings particular thanks for what he had done. There was a Corporation with considerable gifts for propagating the Gospell in New England they bought about 700 a year in their own name ‹some of it was› of one Beningfield a Papist the Father and Son both sold it but the Corporation being dissolved by the Kings coming in Beningf[ield] the Son entered on the Estate He was desired to secure the matter with the Earl of Claren[don] who liked the thing and entred a Caveat that it might not be given away and not long after renewed the Corporation and made Mr B Governour saying that he had not a good enough name among [sic] otherwise he would have been Governour he liked the designe so well. by L[ord] Roberts[c] advice he caused see if there were previous Articles signed by those who treated with Beningf[ield] in whose name the sute was brought and the land recovered[d]

Upon a discourse the Bishop of Oxford had with him about propagating the Gospell in the East Indies in my hearing the Bishop saying £200 a year

---

[a] Thomas Wriothesley, 4th Earl of Southampton (1607–67), Lord High Treasurer 1660–7.
[b] Boyle was involved in negotiations over sugar prices: see Jacob, *Robert Boyle*, pp. 146–7. See also above, p. lxviii.
[c] Sir John Robartes, 2nd Baron Robartes, later 1st Earl of Radnor (1606–85).
[d] On the Bedingfields' attempts to recover their title to estates purchased from them by the company in 1653, see Kellaway, *New England Company*, pp. 37–8, 43–5, 53–5. For a copy of a petition concerning this case, see BP 40, fol. 9.

would serve for a nursery of 4 fellows who should be bound to goe thither he as soon as he went away told me he would lay downe £200 for setting that once on foot or if that could not be caried on to goe for translating such parts of script[u]re[14] into their langu[ages] as should be advisable[;][a] he was at the charge of putting Grotius de Veritate Religionis Christianae into Arabick in which Dr Pocock changed the 6th book in severall particulars[;] these he sent to the Indies to be given Gratis to such as would take them[;][b] He had subscribed for the Translation of the Bible into the Turkish tongue but the Company hearing that such a work was going on sent to the Undertaker that they would not have it done but upon their charge so they subscribed towards it yet he gave £70 for advancing it.[c]

[a] Boyle recounted his conversation with John Fell on this matter in a letter to Robert Thompson of 5 March 1677 printed in Birch, *Life*, pp. cviii–ix.
[b] Pococke's Arabic translation of Grotius was published in 1660; Maddison, *Life*, p. 96. On Pococke's alterations to Book 6 of Grotius' work, see Pococke to Boyle, 5 Oct. 1660, *Works*, vol. vi, pp. 323–4.
[c] William Seaman's translation of the New Testament into Turkish was published in 1666: Maddison, *Life*, p. 111. On the role of the Turkey Company, see Seaman to Boyle, 5, 19 Oct. 1664, *Works*, vol. vi, pp. 511–12.

DOCUMENT 4

Gilbert Burnet's funeral sermon, 1692

This sermon was published in quarto in 1692. Fulton, *Bibliography of Boyle*, p. 171, differentiates between two versions, distinguished by the width of the border on the title-page and the line endings on pp. 33 and 37. He designates these issues A and B, though it would be more proper to call them editions A and B, since they are from substantially different settings of type. He adds: 'It is not clear from internal evidence which appeared first'. The present text is taken from his version B, since, where the two differ, the readings in this one are almost invariably preferable. The facsimile of the title-page on p. 36, however, is taken from version A. The running title in the original is: *A Sermon at the Funeral of the Honourable Robert Boyle*. On the context of this work, see above, pp. xxix–xxxi.

[Half-title]

———————

THE

Bishop of *Sarum*'s SERMON,

At the Funeral

Of the Honourable

ROBERT BOYLE.

———————

A

SERMON

Preached at the

FUNERAL

OF THE

HONOURABLE

Robert Boyle;

AT

St. *MARTINS in the Fields*,
JANUARY 7. 169$\frac{1}{2}$.

By the Right Reverend Father in God,
GILBERT Lord Bishop of SARUM.

LONDON:
Printed for Ric. Chiswell, at the *Rose* and *Crown*,
and John Taylor, at the *Ship*, in St. *Paul's*
Church-Yard. MDCXCII.

[Title: see facsimile on facing page]

ECCLES.II.26.

For God giveth to a man that is good in his sight, wisdom, knowledge, and joy.

When the Author of this Book, the Wisest of Men, *applied his heart to know and to search, to seek out wisdom, and the reason* (or nature) *of things*;[a] and summed up the Account of all, Article by Article, *one by one*, to find out the thread of Nature, and the Plann of its great Author; *tho his Soul sought after it*, yet the Riddle was too dark, he, even he, could not discover it: *But one man among a thousand he did find*, and happy was he in that discovery, if among all the Thousands that he knew, he found One counting Figure for so many Cyphers, which tho they encreased the Number, yet did not swell up the Account, but were so many Nothings, or less and worse than Nothings, according to his estimate of Men and Things.

We have reason rather to think, that by *a Thousand* is to be meant a vast and indefinite number; otherwise it must be confessed that *Solomon's* Age was in-/6/-deed a Golden one, if it produced *one Man, to a Thousand* that carry only the name and figure, but that do not answer the end and excellency of their being. The different Degrees and Ranks of Men, with relation to their inward powers and excellencies, is a suprizing but melancholy Observation: Many seem to have only a Mechanical Life, as if there were a moving and speaking Spring within them, equally void both of Reason and Goodness. The whole race of men is for so many years of Life, little better than encreasing Puppits; many are Children to their Lives end: The Soul does for a large portion of Life, sink wholly into the Body in that shadow of death, Sleep, that consumes so much of our time; the several disorders of the Body, the Blood and the Spirits, do so far subdue and master the Mind, as to make it think, act, and speak according to the different ferments that are in the humours of the Body; and when these cease to play, the Soul is able to hold its tenure no longer: all these are strange and amazing speculations! and force one to cry out, Why did such a perfect Being make such feeble and imperfect Creatures? *Wherefore hast*

---

[a] In margin: *Chap. 7. 25, 27.* Subsequent quotations are from verses 27–8. The reference to *less and worse than Nothings* could allude to Isaiah, 40, 17.

thou made all men in vain?[a] The Secret is yet more astonishing, when the frowardness, the pride and ill-nature, the ignorance, folly and fury that hang upon this poor flattered Creature, are likewise brought into the Account. He that by all his observation, and *encrease of knowledge*, only *encreaseth sorrow, while he sees that what is wanting cannot be numbred, and that which is crooked cannot be made straight*, is tempted to go about, and with *Solomon*, to *make his heart to despair of all the labour wherein he has travelled.* /7/

But as there is a dark side of Humane Nature, so there is likewise a bright one, The flights and compass of awakened Souls is no less amazing. The vast croud of Figures that lie in a very narrow corner of the Brain, which a good memory, and a lively imagination, can fetch out in great order, and with much beauty: The strange reaches of the Mind in abstracted Speculations, and the amazing progress that is made from some simple Truths into Theories, that are the admiration as well as the entertainment of the thinking part of mankind; The sagacity of apprehending and judging, even at the greatest distance; The elevation that is given to Sense, and the Sensible powers, by the invention of Instruments; and which is above all, the strength that a few thoughts do spread into the mind, by which it is made capable of doing or suffering the hardest things; the Life which they give, and the Calm which they bring, are all so unaccountable, that take all together, a Man is a strange huddle, of Light and Darkness, of Good and Evil, and of Wisdom and Folly. The same Man, not to mention the difference that the several Ages of Life make upon him, feels himself in some minutes so different from what he is in the other parts of his Life, that as the one fly away with him into the transports of joy; so the other do no less sink him into the depressions of sorrow: He scarce knows himself in the one, by what he was in the other: Upon all which, when one considers a Man both within and without, he concludes that he is both *wonderfully*, and also *fearfully made*:[b] That in one side of him he is but a little lower than Angels; and in another, a little, a very little higher than Beasts. /8/

But how astonishing soever this Speculation of the medly and contrariety in our composition may be, it contributes to raise our esteem the higher, of such persons as seem to have arisen above, (if not all, yet) all the eminent frailties of humane nature; that have used their Bodies only as Engines and Instruments to their Minds, without any other care about them, but to keep them in good case, fit for the uses they put them to; that have brought their souls to a purity which can scarce appear credible to those who do not imagine that to be possible to another, which is so far

[a] Psalms, 89, 47. Subsequent quotations are from Ecclesiastes, 1, 15, 18; 2, 20, 22.
[b] Psalms, 139, 14.

out of their own reach; and whose Lives have shined in a course of many years, with no more allay nor mixture, than what just served to shew that they were of the same humane nature with others; who have lived in a constant contempt of Wealth, Pleasure, or the Greatness of this World; whose minds have been in as constant a pursuit of Knowledge, in all the several ways in which they could trace it; who have added new Regions of their own discoveries, and that in a vast variety to all that they had found made before them; who have directed all their enquiries into Nature to the Honour of its great Maker; And have joyned two things, that how much soever they may seem related, yet have been found so seldom together, that the World has been tempted to think them inconsistent; a constant looking into Nature, and a yet more constant study of Religion, and a Directing and improving of the one by the other; and who to a depth of Knowledg which often makes men morose, and to a heighth of /9/ Piety, which too often makes them severe, have added all the softness of Humanity, and all the tenderness of Charity, an obliging Civility, as well as a melting kindness: when all these do meet in the same person, and that in eminent degrees, we may justly pretend that we have also made *Solomon*'s observation of *one man*; but alas! the Age is not so fruitful of such, that we can add *one among a thousand*.

To such a man the Characters given in the words of my Text, do truly agree, That *God giveth to him that is good in his sight, Wisdom, Knowledg, and Joy:*[a] The *Text* that is here before us, does so agree to *this* that I have read, that the Application will be so easie, that it will be almost needless, after I have a little opened it.

A man that is good in the sight of God, is a Character of great extent: Goodness is the probity and purity of the mind, shewing it self in a course of sedate Tranquility, of a contented state of Life, and of Vertuous and Generous Actions. A good man is one that considers what are the best Principles of his Nature, and the highest Powers of his Soul; and what are the greatest and the best things that they are capable of; and that likewise observes what are the disorders and depressions, the inward diseases and miseries, which tend really to lessen and to corrupt him; and that therefore intends to be the purest, the wisest, and the noblest Creature that his nature can carry him to be, that renders himself as clean and innocent, as free from designs and passions, as much above appetite and pleasure, and all that sinks the Soul deeper into the body; that is as tender and compassionate, as gentle and good natured as he can possibly make himself to /10/ be. This is the *good man* in my Text; that rises as much as he can above his body, and above this world, above his senses, and the impressions that

---

[a] Burnet now quotes further from Ecclesiastes, 2, 17–26; subsequent quotations come from chs. 2 and 7.

sensible objects make upon him; that thinks the greatest and best thing he can do, is to awaken and improve the seeds and capacities to Vertue and Knowledg, that are in his nature; to raise those to the Noblest objects, to put them in the rightest method, and to keep them ever in tune and temper: and that with relation to the rest of Mankind, considers himself as a Citizen of the whole world, and as a piece of Humane Nature; that enters into the concerns of as many persons as come within his Sphere, without the narrowness or partiality of meaner regards; that thinks he ought to extend his care and kindness as far as his capacity can go; that stretches the Instances of this, to the utmost corners of the earth, if occasion is given for it; and that intends to make mankind the better, the wiser, and the happier for him in the succeeding as well as in the present Generation.

This is the truly *Good man* in *God's sight*, who does not act a part, or put on a Mask; who is not for some time in a constraint, till the design is compast, for which he put himself under that force; but is truly and uniformly good, and is really a better man in secret, than even he appears to be; since all his designs and projects are worthy and great; And *Nature, Accidents and Surprizes* may be sometimes too quick and too hard for him; yet these cannot reach his heart, nor change the setled measures of his life; which are all pure and noble. And tho the errors of this *good man*'s conduct may in some /11/ things give advantages to bad men, who are always severe censurers; yet his unspeakable comfort is, That he can make his secret Appeals to God, who knows the whole of his heart as well as the whole of his life; and tho here and there, things may be found that look not quite so well, and that do indeed appear worst of all to himself, who reflects the oftenest, and thinks the most heinously of them; yet by measuring Infinite Goodness with his own proportion of it, and by finding that he can very gently pass over many and great defects in one whose principles and designs seem to be all pure and good, he from that concludes, That those allowances must be yet infinitely greater, where the Goodness is infinite; so being assured within himself, that his vitals, his inward principles, and the scheme and course of his life are good, he from thence raises an humble confidence in himself, which tho it does not, as indeed it ought not, free him from having still low thoughts of himself, yet it delivers him from all dispiriting fear and sorrow, and gives him a firm confidence in the love and goodness of God, out of which he will often feel an incredible source of satisfaction and joy, springing up in his mind. A man who is thus *good in the sight of God*, has, as one may truly think, happiness enough within himself. But this is not all his reward, nor is it all turned over into a Reversion. We have here a fair particular given us, by one that dealt as much both in Wisdom and Folly, as ever man did; who run the whole compass of pleasure, business, and learning, with the freest range, and in the greatest variety, and who by many repeated Experiments knew the strong and the weak sides of /12/ things: He then who had found the *vanity*, the *labour*, the *sore travel*, and the *vexation of spirit*, that was in all other things; the many disappointments that were given by them, and the painful reflections that did arise out of them so sensibly, *that they made him hate life for the sake* of all the *labour* that belonged to it, and *even to make his heart despair of all the travel he had undergone*, gives us in these words another view of the effect of true Goodness, and of the happy consequences that follow it.

The first of these is *Wisdom*, not the art of craft and dissimulation; the cunning of deceiving or undermining others: not only the views that some men may have of the springs of humane nature, and the art of turning these; which is indeed a Nobler Scene of *Wisdom*, by which Societies are conducted and maintained. But the chief acts and instances of true Wisdom, are once to form right Judgments of all things; of their value, and of their solidity; to form great and noble thoughts of God, and just and proper ones of our selves; to know what we are capable of and fit for; to know what is the true good and happiness of Mankind, which makes Societies safe, and Nations flourish. This is solid Wisdom, that is not mis-led by false appearances, nor imposed on by vulgar opinions. This was the Wisdom that first brought men together, that tamed and corrected their natures; and established all the art and good Government that was once in the world; but which has been almost totally defaced by the arts of *Robbery* and *Murder*, the true names for *Conquest*; a specious colour for the two worst things that humane nature is capable of, *Injustice and Cruelty.* /13/

Wisdom in gross, is the forming true Principles, the laying good Schemes, the imploying proper Instruments, and the chusing fit seasons for doing the best and noblest things that can arise out of humane nature. This is the *defence* as well as the *glory* of Mankind: *Wisdom gives life to him that hath it, it is better than strength, and better than weapons of war*; it is, in one word, The Image of God, and the Excellency of Man. It is here called *the Gift of God*; the seed of it is laid in our Nature, but there must be a proper disposition of body, a right figure of brain, and a due temper of blood to give it scope and materials. These must also be cultivated by an exact education; so that when all these things are laid together, it is plain in how many respects *Wisdom* comes from God. There are also particular happy flights, and bright minutes, which open to men great Landskips, and give them a fuller prospect of things, which do often arise out of no previous Meditations, or chain of thought, and these are flashes of light from its Eternal sourse, which do often break in upon pure minds. They are not Enthusiasms, nor extravagant pretentions, but true views of things which appear so plain and simple, that when they come to be examined, it may be justly thought that any one could have fallen upon them, and the simplest are always the likest to be the truest. In short, a pure mind is both better prepared for an enlightning from above, and more capable of receiving it; the natural strength of mind is awakened as well as recollected; false Biasses are removed; and let prophane minds laugh at it as much as they please, there is a secret commerce between God and the Souls of good /14/ men: They feel the influences of Heaven, and become both the wiser and the better for them: Their thoughts become nobler as well as freer; and no man is of so low a composition, but that with a great deal of goodness, and a due measure of application, he may become more capable of these, than any other that is on the same level with him, as to his natural powers, could ever grow to be, if corrupted with Vice and Defilement.

Knowledge comes next: This is that which opens the mind, and fills it with great Notions; the viewing the Works of God even in a general survey, gives insensibly a greatness to the Soul. But the more extended and exact, the more minute and severe, the Enquiry be, the Soul grows to be thereby the more inlarged by the variety of Observation that is made, either on the great Orbs and Wheels that have their first motion, as well as their Law of moving, from the Author of all; or on the composition of Bodies, on the Regularities, as well as the Irregularities of Nature; and that Mimickry of its heat and motion that Artificial Fires so produce and shew. This Knowledge goes into the History of Past Times, and Remote Climates; and with those livelier Observations on Art and Nature, which give a pleasant entertainment and amusement to the mind, there are joined in some, the severer studies, the more laborious as well as the less-pleasant study of Languages, on design to understand the sense, as well as the discoveries of former Ages: and more particularly to find out the true sense of the sacred Writings. These are all the several varieties of the most useful parts of Knowledge; and these do spread over /15/ all the powers of the Soul of him that is capable of them, a sort of nobleness; that makes him become thereby another kind of Creature than otherwise he ever could have been: He has a larger size of Soul, and vaster thoughts, that can measure the Spheres, and enter into the Theories of the Heavenly Bodies; that can observe the proportion of Lines, and Numbers, the composition and mixtures of the several sorts of Beings. This World, this Life, and the mad Scene we are in, grow to be but little and inconsiderable things, to one of great views and noble Theories: and he who is upon the true scent of real and useful Knowledg, has always some great thing or other in prospect; new Scenes do open to him, and these draw after them Discoveries, which are often made before, even those who made them were either aware, or in expectation of them: These by an endless Chain are still pointing at, or leading into further Discoveries. In all those, a man feels as sensibly, and distinguishes as plainly an improvement of the strength and compass of his powers, from the feebleness which ignorance and sloth bring upon them, as a man in health of Body can distinguish between the life and strength which accompany it, and the flatness and languidness that Diseases bring with them. This enlarges a Man's Empire over the Creation, and makes it more intirely subject to him by the Engines it invents to subdue and manage it, by the dissections in which it is more opened to his view, and by the observation of what is profitable or hurtful in every part of it: from which he is led to correct the one; and exalt the other. This leads him into knowledge /16/ of the hidden Vertues that are in Plants and Minerals; this teaches him to purify these, from the Allays that are wrapped about them, and to improve them by other mixtures. In a word, this lets a man into the Mysteries of Nature: It gives him both the Keys that open it, and a Thread that will lead him further than he durst promise himself at first. We can easily apprehend the suprising joy of one born blind, that after many years of darkness, should be blest with sight, and the leaps and life of thought, that such a one should feel upon so ravishing a change; so the new Regions into which a true Son of Knowledge enters, the new Objects and the various shapes of them that do daily present themselves to him, give his mind a sight, a raisedness, and a refined joy, that is of another nature than all the soft and bewitching Pleasures of sense. And tho the highest reaches of knowledge do more clearly discover the weakness of our short-sighted powers, and shew us difficulties that gave us no pain before, because we did not apprehend them; so that in this respect, he that *increases knowledge, increases sorrow*: Yet it is a real pleasure to a Searcher after Truth, to be undeceived, to see how far he can go, and where he must make his stops: It is true, he finds he cannot compass all that he hath proposed to himself, yet he is both in view of it, and in the way to it, where he finds so many noble Entertainments, that though he cannot find out the whole work of God, which the *Preacher* tells us, that though *a wise man thinks he may know it, yet even he shall not be able to find it out*;[a] yet he has this real satisfaction in /17/ himself, that he has greater Notions, nobler Views, and finer Apprehensions then he could have ever fallen upon in any other method of life.

This *knowledge*, though it may seem to be meerly the effect of thought, of labour, and industry, yet it is really *the gift of God*. The capacity of our Powers, and the disposition of our Minds are in a great measure born with us: The circumstances and accidents of our lives depends so immediately upon Providence, that in all these respects, Knowledge comes, at least in the preparations to it, from God: There are also many happy openings of thought, which arise within the minds of Searchers after it, to which they did not lead themselves by any previous inferences, or by the comparing of things together. That which the Language of the World calls *chance*,

[a] In margin: *Eccles. 8. 17.*

43

happy accidents, or *good stars*, but is according to a more sanctified dialect *Providence*, has brought many wonderful secrets by unlookt for hits, to the knowledge of men. The use of the Loadstone, and the extent of sight by Telescopes, besides a vast variety of other things that might be named, were indeed the immediate gifts of God to those who first fell upon them. And the profoundest Inquirers into the greatest mysteries of Nature, have and still do own this, in so particular a manner, that they affirm, that things that in some hands, and at some times are successful almost to a Prodigy, when managed by others with all possible exactness do fail in the effects of them so totally, that the difference can be resolved into nothing, but a secret direction and blessing of Providence. /18/

The third gift that God bestows on the Good man, is *joy*, and how can it be otherwise, but that a good, a wise and knowing Man, should rejoyce both in God and in himself; in observing the works and ways of God, and in feeling the testimony of a good Conscience with himself. He is happy in the situation of his own mind, which he possesses in a calm contented evenues [sic] of Spirit. He has not the agitations of Passions, the ferment of Designs and Interests, nor the disorders of Appetite which darken the Mind, and create to it many imaginary Troubles, as well as it encreases the Sense of the real Ones which may lye upon ones Person or Affairs. He rejoyces in God when he sees so many of the hidden beauties of his Works, the wonderful fitness and contrivance, the curious disposition, and the vast usefulness of them, to the general good of the whole. These things afford him so great a variety of Thought, that he can dwell long on that noble exercise without flatness or weariness. He rejoyces in all that he does, his imployments are much diversified, for the newness of his discoveries which returns often, gives him as often a newness of joy. His views are great, and his designs are noble; even to know the works of God the better, and to render them the more useful to Mankind. He can discover in the most despised Plant, and the most contemptible Mineral that which may allay the miseries of humane Life, and render multitudes of men easie and happy. Now to one that loves Mankind, and that adores the Author of our Nature, every thing that may tend to Celebrate his praises, and to sweeten the lives of Mortals, affords a joy that is of an exalted and generous kind. If this at any time goes so far as to make him a little too well pleased with the /19/ discoveries he has made, and perhaps too nicely jealous of the honour of having done those Services to the World, even this which is the chief and the most observed defect, that is much magnified by the ill-natured censures of great Men, who must fix on it because they can find nothing else, yet I say even this shews the fullness of *joy* which *wisdom* and *knowledge* bring to good minds, they give them so sensible a pleasure, that it cannot be at all times governed: and if it break out in any time in less decent instances, yet certainly those who have deserved so highly of the

Age in which they have lived, and who have been the Instruments of so much good to the World, receive a very unworthy return, if the great services they have done Mankind do not cover any little imperfection, especially when that is all the Allay that can be found in them, and the only instance of humane Frailty that has appeared in them. But if the joy that wisdom and knowledge give, is of so pure and so sublime a Nature; there is yet another occasion for *joy*, that far exceeds this: it arises from their integrity and goodness which receives a vast accession from this, that it is *in the sight of God*, seen and observed by him, who accepts of it now, and will in due time reward it. The terror of mind, and the confusion of face that follows bad actions, and the calm of thought and chearfulness of look that follows good ones, are such infallible indications of the suitableness or unsuitableness that is in these things to our natures, that all the contempt with which Libertines may treat the Argument will never be able to overcome and alter the plain and simple sense that Mankind agrees in upon this head. A good Man finds that he is acting according to his nature, and to the /20/ best principles in it, that he is living to some good end, that he is an useful piece of the World and is a mean of making both himself and others wiser and happier, greater and better. These things give him a solid and lasting joy, and when he dares appeal to that God to whom he desires chiefly to approve himself, who knows his Integrity and sees how thoroughly *good* he is, even in his secretest Thoughts and Intentions, he does upon that feel a *joy* with in himself, that carries him through all the difficulties of Life; and makes most accidents that happen to him pleasant, and all the rest supportable. He believes he is in the favour of God, he hopes he has some Title to it, from the Promises of God to him, and his grace in him. He can see Clouds gather about him and threaten a Storm, and though he may be in circumstances, that render him very unfit to suffer much hardship; yet he can endure and bear all things, because he believes all God's Promises. He may sometimes from the severe Sense that he has of his duty, be too hard, and even injust to himself, and the seriousness of his Temper may give some harsher thoughts too great occasion, to raise disquiet within him; but when he takes a full view of the infinite goodness of God, of the extent of his Mercy, and of the riches of his Grace; he is forced to throw out any of those Impressions, which Melancholy may be able to make upon him: and even those when reflected on in a truer light, though they might have a little interrupted his *joy*, yet tend to encrease it, when by them he perceives, that true strictness of principles that governs him, which makes him tender of every thing that might seem to make the least breach upon his purity and holiness, even in the smallest Matters. /21/

I will go no further upon my Text, nor will I enter upon the Reverse of it, that is in the following words, *but to the Sinner he giveth travel, to*

gather and to heap up, that he may give to him that is good before God.
These I leave to your Observation: they are too foreign to my subject to be
spoke to, upon this occasion, that leads me now to the melancholy part of
this said Solemnity.

I confess I enter upon it, with the just Apprehensions that it ought to
raise in me: I know I ought here to raise my stile a little, and to triumph
upon the Honour that belongs to Religion and Virtue, and that appeared
so eminently in a life, which may be considered as a Pattern of living: and
a Pattern so perfect, that it will perhaps seem a little too far out of sight,
too much above the hopes, and by consequence above the Endeavours of
any that might pretend to draw after such an Original: which must ever be
reckoned amongst the Master pieces even of that *Great Hand* that made it.
I might here challenge the whole Tribe of *Libertines* to come and view the
usefulness, as well as the Excellence of the Christian Religion, in a Life that
was entirely Dedicated to it: and see what they can object. I ought to call
on all that were so happy, as to know him well, to observe his temper and
course of life, and charge them to sum up and lay together the many great
and good things that they saw in him, and from thence to remember
always to how vast a Sublimity the Christian Religion can raise a mind,
that does both throughly believe it, and is entirely governed by it. I might
here also call up the Multitudes, the vast Multitudes of those who have
been made both the wiser and the easier, the better and the happier by his
means; but that I might /22/ do all this with the more advantage, I ought
to bring all at once into my memory, the many happy hours that in a
course of nine and twenty years conversation have fallen to my own share,
which were very frequent and free for above half that time: that have so
often both humbled and raised me, by seeing how Exalted he was, and in
that feeling more sensibly my own Nothing and Depression, and which
have always edified, and never once, nor in any one thing been uneasie to
me. When I remember how much I saw in him, and learned, or at least
might have learned from him; When I reflect on the gravity of his very
Appearance, the elevation of his Thoughts and Discourses, the modesty of
his Temper, and the humility of his whole Deportment, which might have
served to have forced the best thoughts even upon the worst minds, when,
I say, I bring all this together into my mind; as I form upon it too bright an
Idea to be easily received by such as did not know him; so I am very
sensible that I cannot raise it, equal to the thoughts of such as did. I know,
the limits that custom gives to Discourses of this kind, and the hard
Censures which commonly follow them: These will not suffer me to say all
I think; as I perceive I cannot bring out into distinct thoughts all that of
which I have the imperfect hints and ruder draughts in my mind, which
cannot think Equal to a Subject so far above my own level. I shall now
therefore shew him only in Perspective, and give a General, a very general

View of him, reserving to more leisure and better opportunities, a farther and fuller account of him. I will be content at present to say but a Little of him; but that Little will be so very much, that I must expect that those who do never intend to imitate /23/ any part of it, will be displeased with it all. I am resolved to use great Reserves; and to manage a tenderness, which how much soever it may melt me, shall not carry me beyond the strictest measures, and I will study to keep as much within *bounds*, as he lived beyond *them*.

I will say nothing of the *Stem* from which he sprang: that watered Garden, watered with the blessings and dew of Heaven, as well as fed with the best Portions of this life, that has produced so many noble Plants, and has stocked the most Families in these Kingdoms of any in our Age. Which has so signally felt the effects of their humble and Christian Motto, *God's Providence is my Inheritance*. He was the only Brother of five,[a] that had none of these Titles that sound high in the World; but he procured one to himself, which without derogating from the dignity of Kings must be acknowledged to be beyond their Perogative. He had a great and noble Fortune; but it was chiefly so to him, because he had a great and noble Mind to imploy it to the best Uses. He began early to shew both a Probity and a Capacity; that promised great things: and he passed through the Youthful parts of life, with so little of the *Youth* in him, that in his travels while he was very young and wholly the Master of himself he seemed to be out of reach of the disorders of that Age, and those Countries through which he passed. He had a modesty and a purity laid so deep in his Nature, that those who knew him the earliest have often told me, that even then *Nature* seemed entirely sanctified in him. His piety received a vast encrease as he often owned to me from his Acquaintance with the great Primate of *Ireland*, the never enough admired *Usher*, who as he was /24/ very particularly the Friend of the whole Family, so seeing such Seed and beginnings in him, studied to cultivate them with due care. He set him chiefly to the Study of the Scriptures in their *Original Languages*, which he followed in a course of many Years, with so great exactness he could have quoted all remarkable Passages very readily in *Hebrew*: and he read the *New Testament* so diligently in the *Greek*, that there never occured to me an occasion to mention any one passage of it, that he did not readily repeat in that language. The use of this he continued to the last, for he could read it with other mens Eyes; but the weakness of his sight forced him to disuse the other, since he had none about him that could read it to him. He had studied the Scriptures to so good purpose, and with so critical a strictness, that few men whose Profession oblige them chiefly to that sort of learning

---

[a] This number includes Lewis Boyle (d. 1642), though not Roger Boyle, buried at Deptford (see below, pp. 101–2), or Geoffrey (d. 1617).

47

have gone beyond him in it: and he had so great a regard to that *Sacred Book*, that if any one in Discourse had dropped any thing that gave him a clearer view of any passage in it, he received it with great pleasure, he examined it accurately, and if it was not uneasie to him that offered it, he desired to have it in writing. He had the profoundest Veneration for the great *God of Heaven and Earth*, that I have ever observed in any Person. The very Name of God was never mentioned by him without a Pause and a visible stop in his Discourse, in which one that knew him most particularly above twenty Years, has told me that he was so exact, that he does not remember to have observed him once to fail in it. /25/

He was most constant and serious in his secret Addresses to God; and indeed it appeared to those, who conversed most with him in his Enquiries into Nature, that his main design in that, on which as he had his own Eye most constantly, so he took care to put others often in mind of it, was to raise in himself and others vaster Thoughts of the Greatness and Glory, and of the Wisdom and Goodness of God. This was so deep in his Thoughts, that he concludes the Article of his *Will*, which relates to that *Illustrious* Body, the Royal Society, in these Words, *Wishing them also a happy success in their laudable Attempts, to discover the true Nature of the Works of God; and praying that they and all other Searchers into Physical Truths, may Cordially refer their Attainments to the Glory of the Great Author of Nature, and to the Comfort of Mankind.*[a] As he was a very Devout Worshipper of God, so he was a no less Devout Christian. He had possessed himself with such an amiable view of that Holy Religion, separated from either superstitious Practices or the sourness of Parties, that as he was fully perswaded of the Truth of it, and indeed wholly possessed with it, so he rejoyced in every discovery that Nature furnisht him with, to Illustrate it, or to take off the Objections against any part of it. He always considered it as a System of Truths, which ought to purifie the Hearts, and govern the Lives of those who profess it; he loved no Practice that seemed to lessen that, nor any Nicety that occasioned Divisions amongst Christians. He thought pure and disinteressed Christianity was so Bright and so Glorious a thing, that he was much troubled /26/ at the Disputes and Divisions which had arisen about some lesser Matters, while the Great and the most Important, as well as the most universally acknowledged Truths were by all sides almost as generally neglected as they were confessed. He had therefore designed, tho' some Accidents did, upon great Considerations, divert him from settling it during his Life, but not from ordering it by his Will, that a liberal Provision should be made for one, who should in a very few well digested Sermons, every year, set forth the Truth of the Christian Religion in general, without descending to

[a] Boyle's surviving will has *Benefit* where Burnet has *Comfort*: Maddison, *Life*, p. 261.

the Subdivisions among Christians, and who should be changed every Third year, that so this Noble Study and Imployment might pass through many Hands, by which means many might become Masters of the Argument. He was at the Charge of the Translation and Impression of the New Testament into the *Malyan* [sic] Language, which he sent over all the *East-Indies*.[a] He gave a Noble Reward to him that Translated *Grotius* his incomparable Book of the *Truth of the Christian Religion* into *Arabick*,[b] and was at the Charge of a whole Impression, which he took care to order to be scattered in all the Countries where that Language is understood. He was resolved to have carried on the Impression of the New Testament in the *Turkish* Language, but the Company thought it became them to be the doers of it, and so suffered him only to give a large share towards it. He was at 700 l. Charge in the Edition of the *Irish* Bible, which he ordered to be distributed in *Ireland*; and he contributed liberally both to the Impressions of the *Welsh* Bible, and of the *Irish* Bible for *Scotland*. He /27/ gave during his Life 300 l. to advance the design of propagating the Christian Religion in *America*, and as soon as he heard that the *East-India* Company were entertaining Propositions for the like design in the *East*, he presently sent a 100 l. for a Beginning and an Example, but intended to carry it much further, when it should be set on foot to purpose. Thus was his Zeal lively and effectual in the greatest and truest concerns of Religion; but he avoided to enter far into the unhappy Breaches that have so long weakened, as well as distracted Christianity, any otherwise than to have a great aversion to all those Opinions and Practices, that seemed to him to destroy Morality and Charity. He had a most particular zeal against all Severities and Persecutions upon the account of Religion. I have seldom observ'd him to speak with more Heat and Indignation, than when that came in his way. He did throughly agree with the Doctrines of our Church, and conform to our Worship; and he approved of the main of our Constitution, but he much lamented some abuses that he thought remained still among us. He gave Eminent Instances of his value for the Clergy. Two of these I shall only mention. When he understood what a share he had in Impropriations, he ordered very large Gifts to be made to the Incumbents in those Parishes, and to the Widows of such as had died before he had resolved on this Charity.[c] The Sums that, as I have been Informed, by one that was concerned in two Distributions that were made, amounted upon those two occasions, to near 600 l. and another very liberal one is also

---

[a] This translation, executed by Thomas Hyde, was published in 1677: Birch, *Life*, pp. cix–x; Fulton, *Bibliography of Boyle*, pp. 164–5.

[b] I.e. Edward Pococke (1604–91). On the translating projects referred to here, see above, p. 34. See also Maddison, 'Boyle and the Irish Bible'.

[c] I.e. in Ireland. Burnet's informant was probably Sir Robert Southwell: see Hunter, 'Casuistry in Action', p. 84.

ordered by his Will, but in an indefinite Sum, I suppose, by reason of the /
28/ present condition of Estates in *Ireland*: So plentifully did he supply
those who served at the Altar, out of that which was once devoted to it,
though it be now converted to a Temporal Estate. Another Instance of his
sence of the Sacred Functions went much deeper. Soon after the Restoration in the Year Sixty the great Minister of that time, pressed him both by
himself and by another, who was then likewise in a high Post, to enter into
Orders.[a] He did it not meerly out of a respect to him and his Family, but
chiefly out of his regard to the Church, that he thought would receive a
great strengthening, as well as a powerful Example from one, who, if he
once entered into Holy Orders, would be quickly at the Top. This he told
me made some Impressions on him. His mind was, even then at Three and
thirty, so intirely disingaged from all the Projects and Concerns of this
World, that as the prospect of Dignity in the Church, could not move him
much, so the Probabilities of his doing good in it, was much the stronger
Motive. Two things determined him against it; one was, That his having
no other Interests, with relation to Religion, besides those of saving his
own Soul, gave him, as he thought, a more unsuspected Authority, in
writing or acting on that side: He knew the prophane Crew fortified
themselves against all that was said by Men of our Profession, with this,
That it was their Trade, and that they were paid for it: He hoped therefore,
that he might have the more Influence, the less he shared in the Patrimony
of the Church. But his main Reason was, That he had so high a sense of the
Obligations of the Pastoral care; and of such as watch over those Souls /29/
which *Christ purchased with his own blood*, and for which they must give
an Account, at the last and great day, that he durst not undertake it,
especially not having felt within himself an *Inward motion to it by the
Holy Ghost*; and the first Question that is put to those who come to be
Initiated into the Service of the Church, relating to that *Motion*, he who
had not felt it, thought he durst not make the step; least otherwise he
should have lyed to the Holy Ghost: So solemnly and seriously did he
judge of Sacred Matters. He was constant to the Church; and went to no
separated Assemblies, how charitably soever he might think of their Persons, and how plentifully soever he might have relived their Necessities.
He loved no narrow Thoughts, nor low or superstitious Opinions in
Religion, and therefore as he did not shut himself up within a Party, so
neither did he shut any Party out from him. He had brought his Mind to
such a freedom, that he was not apt to be imposed on; and his Modesty
was such, that he did not dictate to others; but proposed his own Sense,
with a due and decent distrust; and was ever very ready to hearken to what
was suggested to him by others. When he differed from any, he expressed

[a] I.e. Clarendon and the Earl of Southampton: see above, p. 33.

himself in so humble and so obliging a way, that he never treated Things or Persons with neglect, and I never heard that he offended any one Person in his whole Life by any part of his Deportment: For if at any time he saw cause to speak roundly to any, it was never in Passion,or with any reproachful or indecent Expressions. And as he was careful to give those who conversed with him, no Cause or Colour for displeasure, so he was yet more careful of those who were /30/ absent, never to speak ill of any; in which he was the exactest Man I ever knew. If the Discourse turn'd to be hard on any, he was presently silent; and if the Subject was too long dwelt on, he would at last interpose, and between Reproof and Rallery, divert it.

He was exactly civil, rather to Ceremony; and though he felt his easiness of access, and the desires of many, all Strangers in particular, to be much with him, made great wasts on his Time; yet as he was severe in that, not to be denied when he was at home, so he said he knew the Heart of a Stranger, and how much eased his own had been, while travelling, if admitted to the Conversation of those he desired to see; therefore he thought his Obligation to Strangers, was more than bare Civility, it was a piece of Religious Charity in him.

He had for almost Forty years, laboured under such a feebleness of Body, and such lowness of Strength and Spirits, that it will appear a surprizing thing to imagine, how it was possible for him to Read, to Meditate, to try Experiments, and to write as he did. He bore all his Infirmities, and some sharp Pains, with the decency and submission that became a Christian and a Philosopher. He had about him all that unaffected neglect of Pomp in Cloaths, Lodging, Furniture and Equipage, which agreed with his grave and serious course of Life. He was advised to a very ungrateful simplicity of Diet; which by all appearance was that which preserved him so long beyond all Mens expectation; this he observed so strictly, that in a course of above Thirty years, he neither eat nor drank to gratify /31/ the Varieties of Appetite, but meerly to support Nature; and was so regular in it, that he never once transgressed the Rule, Measure, and Kind, that was prescribed him. He had a feebleness in his Sight; his Eyes were so well used by him, that it will be easily imagined he was very tender of them, and very apprehensive of such Distempers as might affect them. He did also imagine, that if sickness obliged him to lie long a Bed, it might raise the Pains of the Stone in him to a degree that was above his weak Strength to bear; so that he feared that his last Minutes might be too hard for him; and this was the Root of all the caution and apprehension that he was observed to live in. But as to Life it self, he had the just indifference to it, and the weariness of it, that became so true a Christian. I mention these the rather, that I may have occasion to shew the Goodness of God to him, in the two things he feared; for his sight began not to grow dimm above four Hours before he died; and when death came

51

upon him, he had not been above Three hours a Bed, before it made an end of him, with so little uneasiness, that it was plain the Light went out, meerly for want of Oil to maintain the Flame.

But I have looked so early to this Conclusion of his Life, yet before I can come at it, I find there is still much in my way. His Charity to those that were in Want, and his Bounty to all Learned Men, that were put to wrastle with Difficulties, were so very extraordinary, and so many did partake of them, that I may spend little time on this Article. Great Summs went easily from him, without the Partialities of Sect, Country, or Relations; for he / 32/considered himself as part of the Humane Nature, and as a Debtor to the whole Race of Men. He took care to do this so secretly, that even those who knew all his other Concerns, could never find out what he did that way; and indeed he was so strict to our Saviour's Precept, that except the Persons themselves, or some one whom he trusted to convey it to them, no body ever knew how that great share of his Estate, which went away invisibly, was distributed; even he himself kept no Account of it, for that he thought might fall into other hands. I speak upon full knowledge on this Article, because I had the honour to be often made use of by him in it. If those that have fled hither from the Persecutions of *France*, or from the Calamities of *Ireland*, feel a sensible sinking of their secret Supplies, with which they were often furnished, without knowing from whence they came, they will conclude, that they have lost not only a Purse, but an Estate that went so very liberally among them, that I have reason to say, that for some years his Charity went beyond a Thousand Pound a year.

Here I thought to have gone to another Head, but the Relation he had, both in Nature and Grace, in living and dying, in friendship, and a likeness of Soul to another Person,[a] forces me for a little while to change my Subject. I have been restrain'd from it by some of her Relations; but since I was not so by her self, I must give a little vent to Nature and to Friendship; to a long Acquaintance and a vast Esteem. His Sister and he were pleasant in their Lives, and in their Death they were not divided; for as he lived with her above Fourty years, so he did not outlive her above a Week. Both died from /33/ the same Cause, Nature being quite spent in both. She lived the longest on the publickest Scene, she made the greatest Figure in all the Revolutions of these Kingdoms for above fifty Years, of any Woman of our Age. She imployed it all for doing good to others, in which she laid out her Time, her Interest, and her Estate, with the greatest Zeal and the most Success that I have ever known. She was indefatigable as well as dextrous in it: and as her great Understanding, and the vast Esteem she was in, made all Persons in their several turns of Greatness, desire and value her Friendship; so she gave her self a clear Title to imploy

[a] Lady Ranelagh.

52

her Interest with them for the Service of others, by this that she never made any use of it to any End or Design of her own. She was contented with what she had; and though she was twice stript of it, she never moved on her own account, but was the general Intercessor for all Persons of Merit, or in want:[a] This had in her the better Grace, and was both more Christian and more effectual, because it was not limited within any narrow Compass of Parties or Relations. When any Party was down, she had Credit and Zeal enough to serve them, and she imployed that so effectually, that in the next Turn she had a new stock of Credit, which she laid out wholly in the Labour of Love, in which she spent her Life: and though some particular Opinions might shut her up in a divided Communion, yet her Soul was never of a Party: She divided her Charities and Friendships both, her Esteem as well as her Bounty, with the truest Regard to Merit, and her own Obligations, without any Difference, made upon the Account of Opinion.

She had with a vast Reach both of Knowledg and Apprehensions, an universal Affability and Easiness /34/ of Access, a Humility that descended to the meanest Persons and Concerns, an obliging Kindness and Readiness to advise those who had no occasion for any further Assistance from her; and with all these and many more excellent Qualities, she had the deepest Sense of Religon, and the most constant turning of her Thoughts and Discourses that way, that has been perhaps in our Age. Such a *Sister* became such a *Brother*; and it was but suitable to both their Characters, that they should have improved the Relation under which they were born, to the more exalted and endearing one of *Friend*. At any time a Nation may very ill spare one such; but for both to go at once, and at such a time, is too melancholly a Thought; and notwithstanding the Decline of their Age, and the Waste of their Strength, yet it has too much of Cloud in it, to bear the being long dwelt on.

You have thus far seen, in a very few hints, the several Sorts and Instances of Goodness that appeared in this Life, which has now its Period; that which gives value and lustre to them all, was, that whatever he might be in the sight of Men, how pure and spotless soever, those who knew him the best, have reason to conclude, that he was much more so in the sight of God; for they had often Occasions to discover new Instances of Goodness in him; and no secret ill Inclinations did at any time shew themselves. He affected nothing that was solemn or supercilious. He used no Methods to make Multitudes run after him, or depend upon him. It never appeared that there was any thing hid under all this appearence of Goodness, that was not truly so. He hid both his Piety and Charity all he could. He lived

[a] Lady Ranelagh's circumstances were straitened in the late 1640s: see Maddison, *Life*, p. 67n. On her husband, Arthur, Viscount Ranelagh's, activities during the Interregnum, see GEC.

in the due Methods of Civility, and would never assume the Authority which all the World was /35/ ready to pay to him. He spoke of the Government even in Times which he disliked, and upon occasions which he spared not to condemn, with an exactness of respect. He allowed himself a great deal of decent chearfulness, so that he had nothing of the moroseness, to which Philosophers think they have some right; nor of the Affectations which Men of an extraordinary pitch of Devotion go into, sometimes, without being well aware of them. He was, in a word, plainly and sincerely in the sight of God, as well as in the view of Men, *a good Man, even One of a Thousand.*

That which comes next to be considered, is the share that this *good Man* had in those Gifts of God, *Wisdom, Knowledg, and Joy.* If I should speak of these, with the copiousness which the Subject affords, I should go too far even for your Patience, tho I have reason to believe it would hold out very long on this Occasion. I will only name things which may be enlarged on more fully in another way. He had too unblemish'd a candor to be capable of those Arts and Practices that a false and decitful World may call *Wisdom.* He could neither lie nor equivocate; but could well be silent, and by practising that much, he cover'd himself upon many uneasy Occasions. He made true Judgments of Men and Things. His Advices and Opinions were solid and sound; and if Caution and Modesty gave too strong a Biass, his Invention was fruitful to suggest good Expedients. He had great Notions of what Humane Nature might be brought to; but since he saw Mankind was not capable of them, he withdrew himself early from Affairs and Courts, notwithstanding the Distinction with which he was always treated by our late Princes. But he had the Principles of an English-man, as well as of a Protestant, too deep in /36/ him to be corrupted or cheated out of them; and in these he studied to fortify all that conversed much with him. He had a very particular Sagacity in observing what Men were fit for; and had so vast a Scheme of different Performances, that he could soon furnish every Man with Work that had leasure and capacity for it; and as soon as he saw him engaged in it, then a handsom Present was made to enable him to go on with it.

His Knowledg was of so vast an Extent, that if it were not for the variety of Vouchers in their several sorts, I should he afraid to say all I know. He carried the study of the Hebrew very far into the Rabbinical Writings, and the other Oriental Languages. He had read so much of the Fathers, that he had formed out of it a clear Judgment of all the eminent Ones. He had read a vast deal on the Scriptures, and had gone very nicely through the whole Controversies of Religion; and was a true Master in the whole Body of Divinity. He run the whole Compass of the Mathematical Sciences; and though he did not set himself to spring new Game, yet he knew even the abstrusest Parts of Geometry. Geography in the several parts of it, that related to Navigation or Travelling, History and Books of Travels were his Diversions. He went very nicely through all the parts of Physick, only the tenderness of his Nature made his less able to endure the exactness of Anatomical Dissections, especially of living Animals, though he knew these to be the most instructing: But for the History of Nature, Ancient and Modern, of the Productions of all Countries, of the Virtues and Improvements of Plants, of Oars and Minerals, and all the Varieties that are in them in different Climates; He was by /37/ much, by very much, the readiest and the perfectest I ever knew, in the greatest Compass, and with the truest Exactness. This put him in the way of making all that vast variety of Experiments, beyond any Man, as far as we know, that ever lived. And in these, as he made a great progress in new Discoveries, so he used so nice a strictness, and delivered them with so scrupulous a Truth, that all who have examined them, have found how safely the World may depend upon them. But his peculiar and favourite Study, was Chymistry; in which he engaged with none of those ravenous and am[b]itious Designs, that draw many into them. His Design was only to find out Nature, to see into what Principles things might be resolved, and of what they were compounded, and to prepare good Medicaments for the Bodies of Men. He spent neither his Time nor Fortune upon the vain pursuits of high Promises and Pretensions. He always kept himself within the Compass that his Estate might well bear: And as he made Chymistry much the better for his dealing in it, so he never made himself either the worse or the poorer for it. It was a Charity to others, as well as an Entertainment to himself, for the Produce of it was distributed by his Sister, and others, into whose hands he put it. I will not here amuse you with a List of his astonishing Knowledg, or of his great Performances this way. They are highly valued all the World over, and his Name is every where mentioned with most particular Characters of Respect. I will conclude this Article with this, in which I appeal to all competent Judges, that few Men (if any) have been known to have made so great a Compass, and to have been so exact in all the Parts of it as he was. /38/

As for Joy, he had indeed nothing of Frolick and Levity in him, he had no Relish for the idle and extravagant Madness of the Men of Pleasure; he did not waste his Time, nor dissipate his Spirits into foolish Mirth, but he possessed his own Soul in Patience, full of that solid Joy which his Goodness as well as his Knowledg afforded him: He who had neither Designs nor Passions, was capable of little Trouble from any Concerns of his own: He had about him all the Tenderness of good Nature, as well as the Softness of Friendship, these gave him a large share of other Mens Concerns; for he had a quick sense of the Miseries of Mankind. He had also a feeble Body, which needed to be look'd to the more, because his Mind went faster than that his Body could keep pace with it; yet his great

Thoughts of God, and his Contemplation of his Works, were to him Sources of Joy, which could never be exhausted. The Sense of his own Integrity, and of the Good he found it did, afforded him the truest of all Pleasures, since they gave him the certain Prospect of that *Fulness of Joy*, in the Sight of which he lived so long, and in the Possession of which he now lives, and shall live for ever; and this spent and exhausted Body shall then put on a new Form, and be made a fit Dwelling for that pure and exalted Mind in the final Restitution. I pass over his Death, I looked at it some time ago, but I cannot bring down my Mind from the elevating Thoughts that do now arise into that depressing one of his Death; I must look beyond it into the Regions of Light and Glory, where he now dwells.

The only thought that is now before me, is to triumph on the Behalf of Religion, to make out due Boast of it, and to be lifted up (I had al-/39/-most said proud) upon this occasion: how divine and how pure a thing must that Religion be in it self, which produced so long a Series of great Effects, thorow the whole Course of this shining Life? What a thing would Mankind become if we had many such? And how little need would there be of many Books writ for the Truth and Excellency of our Religion, if we had more such Arguments as this one Life has produced? Such single Instances have great Force in them; but when they are so very Single, they lose much of their Strength by this, that they are ascribed to Singularity and something particular in a Man's Humour and Inclinations, that makes him rise above common Measures. It were a Monopoly for any Family or Sort of Men to ingross to themselves the Honour which arises from the Memory of so great a Man. It is a Common not to be inclosed. It is large enough to make a whole Nation, as well as the Age he lived in, look big and be happy: But above all it gives a new Strength, as well as it sets a new Pattern to all that are sincerely zealous for their Religion. It shews them in the simplest and most convincing of all Arguments, what the Humane Nature is capable of, and what the Christian Religion can add to it, how far it can both exalt and reward it. I do not say that every one is capable of all he grew to; I am very sensible that few are; nor is every one under equal Obligations: for the Service of the Universe, there must be a vast Diversity in Mens Tempers, there being so great a Variety of Necessities to be answered by them: but every Man in every Imployment, and of every size of Soul, is capable of being in some Degrees *good in the sight of God*; and all such shall receive proportioned Degrees of *Wisdom, Knowledg and Joy*; even though /40/ neither their Goodness nor these Accessions to it, rise up to the Measure of him who was a while among us, indeed *one of a thousand*, and is now but one of those *ten thousand times ten thousand* that are about the Throne, where he is singing that Song which was his great Entertainment here, as it is his now endless Joy there; *Great and marvellous are thy Works, O Lord God Almighty; and just and true are thy Ways, O King of Saints.*[a] To follow him in the like Exercises here, is the sure Way to be admitted to join with him in those above; to which God of his infinite Mercy bring us all in due time, through Jesus Christ our Lord. *Amen, Amen.*

FINIS

[a] Revelation, 15, 3.

DOCUMENT 5

Sir Peter Pett's notes on Boyle

Add. MS 4229, fols 33–49. Written in sections, with fols 33–5 and 36–40 in a scribal hand, the remainder in Pett's holograph. The scribal text on fol. 40 breaks off in mid-sentence and is continued (after an intermediate leaf, fol. 41, which is blank except for Miles's endorsement: '*Sir Peter Pett*') on fol. 42 in Pett's hand, presumably because the process of retranscribing the notes was never completed. Pett also added further remarks in his own hand at the end of the scribal text on fol. 35, while many of the insertions in the scribal sections are in his hand. The MS has the following original pagination: pp. 1–5 (fols 33–5ᵛ), pp. 1–9 (fols 36–40) and pp. 13–21 (fols 42–6); the other leaves are not paginated. There is a blank leaf (unfoliated) between fols 46 and 47. Fol. 49 is a part sheet which is blank except for Miles's note on the verso: 'Sir P. Petts Papers[.] bundled separately to prevent Confusion being writt at different times'; in addition, fol. 35 is endorsed by William Clarke: 'Sir Peter Pets Papers of Mr Boyle', while fol. 48ᵛ is endorsed, 'Relating to Mr Boyles Life', in the top half of the lefthand margin by Wotton, and '*Sir Peter Pett*' by Miles. Certain of the blank pages (fols 41ᵛ, 46ᵛ [bis], 49) have impressions of the text of the facing pages on them (in the case of fol. 49ᵛ, that of the first page of Doc. 7, which follows it in the MS). See further above, pp. xxxii–v.

During the time that I had the honour and happynesse of frequently conversing with Mr Boyle in Oxford, he would sometimes discourse about the Providence of God causing men to be borne in such particular Ages or seasons of time, as seemed most agreeable to their Genius or Temper, & he said that his brother Brohil (afterwards Earle of Orrery) has sometimes wished, that it might have been Gods pleasure, to have sent him into the world to have run his race of time, either an Age sooner, or an Age later then he did, but said Mr Boyle, I have often thanked God, for being pleased to send me into the World in this particular Age when real knowledge is in so triumphant a state and Experimental Philosophy crowned with so much succes, and he esteemed it his great pleasure, and his glory, that by his various labours he had been herein usefull to Mankind, ‹&› he tooke notice to me how every great New Invention, necessarily crossing the private Interest of many particular persons, was thereby hindred in its birth and growth, by such interested persons, and that such

58

Births being of difficult parturition, severall discouraged Inventors had applied to him in the way of fer opem Lucina,[a] and that God had blessed the Experimental knowledge he had attained, by giving some of those Inventions a succesfull midwivery into the World, and that particularly in the Case of that most famous and useful Invention of the Mill'd-Lead, which was so beneficial for Covering of Houses, and many other purposes as well as for Navigation, that it merited the highest applause and encouragement from the Age; That ingenious person Mr Hale ‹the inventor› will attend the Bishop with a Book he has published of this Invention, and the transactions about it, & whereby his Lordship will find that he has demonstrated it, to be better and more durable for Covering of Churches, Houses &c then Cast Sheet-Load can be, and above 20 per Cent cheaper, As also for Sheathing of Shipps to preserve their Planke from the Worme, against which it must be a certain Security as being Metall, and that lying so thin & smoothly on, ‹it is› no hindrance to Sayling, as the Wood-sheathing is, and this also above Cent per Cent cheaper ‹then that›.[b] But as the old Corporation of Plumbers have by their numbers endeavoured to decry (tho in vaine) the usefulnes of this Invention as to Houses, their Interest tempting them so to do; So the Corporation of Shipwrights from the same Principle endeavoured to suppresse the usefulnes of this Invention as to the Sheathing of Shipps; for it is notorious that the Worme which this Invention provides against, is (as King Charles 2 said upon this occasion) one / fol. 33v/ of the best friends the Shipwrights have, bringing them gaine by their frequent Sheathings with Wood, and when all their clamours and Objections would not do (being answered by above 5 yeares experience to the satisfaction of the Navy Board who then Contracted for it) they raised new ones, as senselesse and ridiculous as any thing imaginable, namely, that this Lead-sheathing was the cause of a very extraordinary corrosion and decay of the Ruddar Irons, not in the least taken notice of in the 5 yeares before, and by that clamour gave new trouble to the Navy Board, that nothing but demonstrating that the Thing was impossible and the Judgment of so great and Oraculor a Judge of Nature, as Mr Boyle given in the case, could have preserved this ‹excellent›[1] Invention from ruin.[c] Thus usefull to the Age was the weight of this great Mans Reason. Mr Boyles name is used in pag. 21 of the Booke about the Milld Lead Invention, for the shewing that nothing in the nature of this Sheathing singly, or from

[a] *Bring help, Lucina*: the cry of mothers and midwives during labour, invoking the Roman goddess of childbirth.
[b] See Thomas Hale, *Account of Several New Inventions and Improvements*, pp. xixf., 1f. A company set up in 1670 was ordered to sheath a number of ships for trial, but the subsequent discovery of rusted ironwork in certain of the ships led to prolonged controversy, in which vested interests were involved. Hale and his partners took over the project in 1687. See MacLeod, *Inventing the Industrial Revolution*, pp. 35, 113, 231 n. 103.
[c] See Hale, *Account*, pp. 20–1. For Charles II's comment, see ibid., p. xxi.

any thing thatt can be begotten by its meeting with Salt water, can contribute ought to the decay of Iron by corrosion, Lead being a Metall so void of any disposition that way, as to subdue that very quality of corroding in other bodies the most acid & sowre; And the aforesaid Mr Hale doth averre that he having often consulted Mr Boyle about the premisses Mr Boyle told him at first that it was vaine to imagine any such corroding quality in this Sheathing, saying Lead is a principall ingredient in a certaine Oyle made in Italy which they use upon the steele beames of their finest & nicest Scales to preserve them from rust, but if he would send in a Rundlet of Sea-water, and some of the Lead and Nailes, he would provide Iron himselfe, and discover what he could in this matter, which Mr Hale having sent him in accordingly, some time after when Mr Boyle had tryed what Experiments he thought fitt, ‹Mr Boyle›[2] told him that he could find no such corroding quality in the Lead or Nayles, but that the Salt-water affected the Iron, as much singly, as in conjunction with them, and said that he had been discoursing with the King about it and ‹had› said the same to his Majesty also, so that upon his own particular Experiment he pronounced the complaint of this Corroding quality in the Lead-sheathing impossible to be true. And in fine, now in the Reigne of his present Majestie King William this Invention is practiced so much to the satisfaction of severall Merchants & Owners of Shipps in their Sheathing that the aforesaid Clamour appears to all, as vaine and ridiculous: And seeing that the course of publick Affaires is likely to require much of the Navigation of our Capital Shipps in the Streights, where they are to provide against the Worme by the best meanes possible; And whereas it is notiorious [sic] that thô sheathing of /fol. 34/ Shipps with Wood against the Worme is a great impediment ‹to the swiftnes› of such Shipps sayling, and that the Lead-Sheathing is no impediment thereto, it may be expected that the Shipps imployed in the Streights hereafter may be ordred to be Sheathed with Lead, and the Nation have further cause to celebrate the memory of Mr Boyle, for having as aforesaid patronized one of the most Usefull Inventions of this later Age:

But here occasion must be taken to observe the singular and transcendent modesty of Mr Boyle, and his being perfectly free from any thing of a magisterial or positive humour, in regard that tho the Milld Leads corroding the Ruddar Irons appeared to his Reason as a first-rate impossibility; yet he did not expect that his Ipse dixit[a] should silence Clamour in the Case, but the sheathing of the Royal Navy & of Merchants Shipps being concerned in the Affaire, and he expecting that King Charles 2 as criticall a Judge about all mechanicall matters relating to shipping & navigation as any person whatsoever would call for his opinion about the thing, as he

---

[a] *Say so*: a tag denoting an unsupported assertion.

did, he therefore made the Experiment himselfe as hath been related, and whereby he was enabled sufficiently to enlighten the World without imposing on it. And when some of his inquisitive friends on his having acquainted them with some Experiments had told him that they themselves would make their Experiments too about the things, he was so far from being displease[d] with such a temporary suspension of their beliefe in the case, as their intending to do soe implyed, that he reply'd to them that[3] it was the very highest complement they could put on him. And this brings to my mind a passage long agoe, how an intimate friend of his, presenting him with a Pindaric Ode, wherein that which Mr Boyle called the highest Complement was addrest to him, the Author had more thanks from him for it, then for all the rest of the Poem, and I then taking a Copy of[4] the second Stanza which ends with that Complement as Mr Boyle accounted it, and which some other person might have thought Satyr, shall here set it downe as followeth.[a]

> Few till old Age make Truth's discovery:
> They of the Tree of Knowledge taste, and dye.
> They find out Nature, as them Nature leaves,
> Whome like poore Old Projectors none believes;
> And (when none cares for knowing them) they know
> Something, as they themselves to nothing goe.
> The World and they e'en to each other lost,
>   Of a poore εὕρικα they boast;
> Thunder-struck with that whispering voyce of Fate

/fol.34v/
>   That then still tells them, 'tis too late
> And like a Death-watch still their Eares doth grate.
> Even by these hissings of Times Serpent stung,
>   They sigh, & wish that they were young.
>   But O you can in blooming youth
> Resolve this old great Question, What is truth?
> And like trees that in Eastern Countries grow,
> You have at once both fruit and blossomes too.
> Credulous Youth its beliefe Knowledge calls,
> And when all's lost, finds that the dice were false:
> But in Experiment where its Knowledge ends,
> Yours beginnes, and a Crowne its birth attends.
> Since neare Land the seas danger most prevailes,
> And Truth in sight of Error alwaies sailes,
> We must go sounding all the way: We must
> Not take even your Experiments on Trust.

[a] The author of the poem has not been identified.

But more of this Ode shall not be inserted, Poetry & History not looking
well together ⟨and⟩ especially in so great a Life where the Historicall relation
of so many matters of fact greater then all praise might be censured ⟨as⟩[5]
Poetry or fiction, did not the knowledge of so many Eye-witnesses yet living
enforce the beliefe of them. Yet here by the way I shall take notice that I am
informed, that shortly after the publication of Mr Boyles Funeral Sermon,
there were published in Monsieur de Croses or Mr Duntons or Rhodes his
Monthly Prints severall latine Monumentall Inscriptions for the Memory
of Mr Boyle, & writ in the way of what we call Lapidary Verse.[a] If the
Booksellers bring them me to peruse I shall give my judgement whether
any of them are fitt for the Bishop's consideration. If none of them are fitt
for it, perhapps (if it be desired) I can engage Dr Bathurst of Oxon to do
somewhat of that kind, and as he has admirably done before a Booke of
Mr Hobbs I have by me.[b] I here offer it as my humble opinion to my Lord
Bishop's consideration, that the Booksellers ought to be at the charge of
having Mr Boyles Effigies engraved[6] by White before the Book, and some
such Hexameter & Pentameter Verses to be under it, as are under the
Effigies of Grotius, Selden, & Bishop Barlow's, and of other famous men.[c]
And forasmuch as it will tend to the honour of Mr Boyle and to the
instruction of the Age and to the perfection of the Bishop's Book of his Life
to have the particular Experiments or Inventions, of which he was the
Author /fol. 35/ summarily therein mentioned, if the Booksellers who
expect proffit by the Book will attend me, I will give them directions how
with very small charge and paines, they may repaire to the Records of the
Royal Society of which I was once a Member,[d] and thence collect materialls for his Lordship's consideration. By that meanes will this designed
Booke last to all ages, and be translated into many Languages as Gassendus his Life of Peiresk hath been, and in which are contained some curious
remarks of his Experimental Philosophy. I have heard that in the Records

[a] For these elogies see above, pp. xxix–xxx, and Fulton, *Bibliography of Boyle*, pp. 172–4.
Jean Cornand de la Crose was editor of *The Works of the Learned*, in which one of these
appeared; John Dunton was publisher of *The Athenian Mercury*, where another appeared;
Henry Rhodes or Rodes was another bookseller. Plomer, *Dictionary of Printers*, p. 252.
[b] Ralph Bathurst (1620–1704), Dean of Wells, had written dedicatory verses for Hobbes'
Humane Nature (1650). Pett also referred to these in his edition of Anglesey's *Memoirs*, sig.
a4.
[c] Pett evidently refers to: (1) the engraved portrait prefacing the 'Vita Hugonis Grotii' in
the 1679 Amsterdam edition of Grotius' *Opera Omnia Theologica*; (2) the portrait frontispiece to the 1682 edition of Selden's *Janus Anglorum*, engraved by White; and (3) the portrait
frontispiece to Barlow's *Cases of Conscience*, also engraved by White. The latter has Latin
verses under it; the others do not, though Latin verses are to be found at the end of the 'Vita
Hugonis Grotii', and in the prefatory matter to *Janus Anglorum*. An engraved portrait of
Boyle by Robert White (1645–1703), based on the Kerseboom portrait, appeared as the
frontispiece to the posthumous publication of Boyle's *Customary Swearing* (1695). See
Maddison, 'Portraiture', p. 162.
[d] Pett had been expelled in 1675: Hunter, *Royal Society*, p. 143.

of the Royal Society there is to be found one Experiment of Mr Boyles, deposited there in his life time, and which he order'd not to be opened by the Society, till after his death,[a] I am likewise ready to give directions to the Booksellers how they may effect, that the most acute men in the Royal Society shal give Characters in writing, of all Mr Boyles printed Philosophical Works, and of their respective excellencies, for the Lord Bishop to consider, and make use of as he shall thinke fitt, and that without any charge to the Booksellers; Yet the smal charge of paying the Register of the Societys servant for transcribing the matter of Mr Boyles Experiments out of the Records there, they must expect to be at, But as for Mr Boyles Moral or Theologicall workes, none [is] abler to give the just characters of them then the Bishop himself:[7]

But for the saving of the Bishop's time in reading them all over, I shall observe that in Mr Boyles seraphic love, 5th edition, p.42 & 43 he shews himselfe to have then been a Reader of Romances. p.104. He shews himselfe to have then studyd the Arminian Controversy, & there laudably refuseth to give his sense therin

p.105 writes to the same purpose as in the foregoing page.

p.108.109 shews himselfe to have then studyd the Socinian controversy.

p.110. He shews that a Jewish professor of the Hebrew tongue instructed him in that language.

p.146 He saith he was not 22 yeares old, when he writt that book.

p.158 He dates the writing of that book on the 6th of August 1648.[8]

* * * * * *

Fol. 36/ 1. Tho it be most true what the Lord Bishop of Sarum observes in p. 29 of his admirable Sermon at the funeral of Mr Boyle, that he was constant to the Church and went to no separate Assemblys, yet according to what we say of the Exception's confirming the Rule, he told me that he once had the curiosity to goe to Sir Henry Vane's house & there heare him preach,[b] where in a large thronged Roome Sir Henry preached a long sermon on the text of Daniel 12.2 And many of them that sleepe in the dust of death shall awake, some to Everlasting Life, and some to ‹shame &› Everlasting[9] Contempt, And that the whole scope of Sir Henrys sermon

---

[a] Boyle deposited various sealed papers with the Royal Society, which were opened at a Council meeting on 9 February 1692: Maddison, *Life*, p. 202. Pett must have been ignorant of this event, since it seems unlikely that his notes predate it.

[b] Sir Henry Vane the younger (1613–62), one of the leading radical politicians of the Civil War and Interregnum and well-known for his heterodox religious opinions.

was to shew that many doctrines of Religion, that had long been dead and buried in the world should before the end of it, be awakend into Life, and that many false doctrines being then likwise revived, should by the power of Truth be then doomed to Shame and everlasting contempt. Mr Boyle told me that at the end of Sir Henrys Sermon he spake to this effect to Sir Henry before the people viz. That being informed that at such private Meetings it was not uncustomary for any one of the Hearers who was unsatisfied about any matters there uttered, to give in his Objections against them and to prevent any mistakes in the Speaker or Hearers, I think my selfe for the honor of Gods Truth here obliged to say that that place in Daniel being[10] the clearest one in all the Old Testament for the proof of the Resurrection, we ought not to suffer the meaning of it to evaporate into Allegory, and the rather for that reference is made by our Saviour in the New Testament by way of asserting the Resurrection, to that place of Daniel in the Old.[a] And if it be denied that the plaine & genuine meaning of those words in Daniel is to assert the Resurrection of dead bodies, I am now ready to prove it so to be, both out of the words of the Text and Context in the Originall Language, and from the best Expositors both Christian & Jewish; But if this be not denied but granted me, and your discourse of the Resurrection of Doctrines true & false be declared by you, as designed only in the way of your occasionall meditations from those words of Daniel and not to enervate the Literal Sense as the genuine one, then I have nothing further to say.

Mr Boyle then sitting downe Sir Henry rose up and sayd that his discourse on those words of Daniel was only in the way of such occasional meditations, and which he thought edifying to the /fol. 36v/ people, and he declared that he agreed that the literal sense of the words was the Resurrection of dead bodies. And so that Meeting broke up.

Mr Boyle afterwards speaking of this to me, said that Sir Henry Vane at that time being one of the first-Rate Grandees in the State and his Auditors at that Meeting consisting chiefly of dependents on him, and Expectants from him, the feare of loosing his favour would probably have kept them from contradicting any of his Interpretations of Scripture how ridiculous soever; but I (said Mr Boyle) having no little awes of that kind upon me thought my selfe bound to enter the Lists with him as I did, that the sense of the Scripture might not be depraved.

And having thus faithfully related the courage and skill of Mr Boyle ‹&› his behaviour as a Confessor in this memoire, I shall here obiter[b] take notice how Sir Henry Vane who was so childish an Allegoriser of Scripture was afterwards given up to such an injudicious minde as to deprave the

---

[a] See Matthew, 25:46, John, 5:28–9.
[b] *In passing.*

Sense of the Scotch Covenant, in the making of which he was pars magna:[a]
for in his Speech printed by his Friends in 1662, as what he intended to
speake on the Scaffold, he calls the Commons Acting without a King or
House of Lords the refined Sense of the Solemne League & Covenant.[b]

2. The Bishop in p. 36 of his Sermon does right to Mr Boyle's Talents in
the knowledge of the Orientall Languages. And for this I can referre to a
letter I lately had from Dr Hyde the Keeper of the Bodleian Library, and
one whome fame speakes as more knowing in those Languages then any
Christian in the world, in which letter the Doctor, having mentioned his
having been acquainted with Mr Boyle for about 30 yeares, thus goes on
viz Mr Boyle besides his skill in the moderne Languages, was excellently
versed in the ready and true speaking of the Latin tongue which I have
heard him do to my admiration. He was also well skilld in Greek &
Hebrew, and he hath told me that when the Chapters were reading in the
Church, he alwaies had in his hand the original, wondring to heare our
English translation so different from it.

And Mr Boyles curiosity in studying the Hebrew text was observed
likewise by myselfe on the following occasions. I once discoursing with
him about the words in Exodus, Let every man borrow of his Neighbour
Jewells of Silver & Gold &c,[c] and how I was troubled to find out the
meaning of them, for that borrowing doth vi termini[d] imply a promise of
restitution; and that I did not more firmly ‹believe› the being of a God then
that dantur rationes boni & mali æternae & indispensabiles,[e] ‹and therefore that› it seemed to me unworthy of God to bid any borrow, with an
intent of never restoring ‹and that assertion of Dantur rationes boni &
mali &c was the Thesis of Dr Cudworth on which he answered the
Doctors when he took his Degree of Doctor of Divinity at Cambridg. and
in his noble folio he tells us that Gods will is ruled by his justice and not
his justice by his Will, and therefore he cannot command what is in its own
nature unjust.[f] But that›[11] As to the words and they spoiled the Egyptians,
I told Mr Boyle I had no scruple at all, for that God as the universall Lord
of all things might alter the property of /fol. 37/ them as he pleased, In fine
Mr Boyle told me, that the same place of Scripture troubled him more then

---

[a] *A great part.*
[b] See Vane, *The Substance*, p. 3. In 1643, Vane had altered the draft covenant submitted by
the Scots by inserting the words *according to the word of God*, hence giving greater flexibility.
[c] Exodus, 11:2.
[d] *By the force of the end, inevitably.*
[e] *Reckonings concerning good and evil are laid down as things both eternal and indispensable.*
[f] Ralph Cudworth (1617–88), Cambridge Platonist. In fact the Cambridge thesis to which
Pett refers was for his B.D: see Cudworth, *True Intellectual System*, vol. i, p. viii. See also ibid.,
esp. vol. i, pp. 310f.: this work, published in 1678, is the 'noble folio' to which Pett refers.

any thing in all the Old Testament, till that his weighing the words in the original Hebrew gave him reliefe, and that thereupon he wished that ‹instead of› the word borrow, the words let them ask or demand had been used in our Translation. And Thereupon lately consulting Dr Hyde, he wrote to me that he concurred in opinion with that of Mr Boyle; that the word borrowing being in our English translation, instead of aske, or desire, or demand, an occasion of scruple was left by our Translators.

Another occasion I had to find the curiosity of Mr Boyle's observations about the Hebrew tongue was this. I discoursing with him about the Jews applying to Cromwel for their being tolerated here, I found he wished well to the same, provided that the Government would take care to give good Salarys to 2 or 3 men for their studying the Orientall Languages, and the text of the Old Testament most accurately, that so our Ministers might be the better enabled to confute their Rabbis, which few of them were then able to do. For Mr Boyle observed that the Jewish Rabbis on all occasions of discourse did continually fly to false acceptions of the Hebrew words, and particularly where it is said a Virgin shall conceive, they replyd that by a Virgin was meant a young woman. He said that the Rabbis lay much stresse on the perpetuall continuance and lasting of Moses his Law, whereas the words for ever (said Mr Boyl) do signify only a very long time. And he thereupon directed me to some Authors who writ ex professo of the various acceptions of the Hebrew words. He further observed to me the condiscention of the divine benignity to the Jewes in suffering them to have Sacrifices, since they were so much addicted to them, on their finding all the Nations of the World to use them, and as indeed (said he) they naturally do, which appeares in their useing them in the West Indies. He then further remarked that the Heathens in their Sacrifices used hony & no Salt, and that God required Salt to be used and no hony.

3. If any in company with Mr Boyle related for entertainment any comicall Jests, or blunderings of any men in mispronouncing, misapplying or misrepresenting words of scripture, with an intent to promote mirth, he would with a gentle severity reprove the doing so, and he inculcated the sinfulnes of mens diminishing thereby the constant awe that the Scriptures should have on their thoughts: and minded the company of the Words of Isaiah to him will I look, who is of a contrite heart, & trembles at my word.[a] He further urged that our memories were not such table bookes as wherein /fol. 37v/ with a spunge we could blot out what we would, and that perhapps as long as we lived, whenever we read or heard of that [b]place of Scripture againe afterwards, the former ridiculeing of it would be apt to recurre upon our thoughts and enervate its majesty.

---

[a] Isaiah, 66:2.
[b] A gap has been left in the text between *that* and *place*.

4. I here am minded according to the common connexion of thoughts of having recourse to p. 24 of the Bishop's sermon, where his Lordship saith Mr Boyle had the profoundest Veneration for the great God of Heaven & Earth, that he ever observed in any person, and that the name of God was never mentioned [by] him without a pause, and a visible stop in his discourse, in which one who knew him most particularly above 20 yeares said he was so exact that he remem[bered] not to have observed him once to faile in it. And I can averre the same thing upon neare 40 yeares acquaintance with Mr Boyle, and I once discoursing with him of that his practice, he said, that not to have an awe upon us when the name of God is spoken of in Company, is a sign of want of Grace. And Bishop Barlow who did frequently converse with Mr Boyle tooke occasion once to tell me that he was an improver in this great point of Morality by meanes of Mr Boyle. The occasion was this. The Bishop gave me a letter of his to send to Ireland thus superscribed by him, To the Right Reverend Father Henry Lord Bishop of Meath:[a] and I asking him why he did not give him the usuall Stile Right Reverend Father in God, he told me that the name of God was so sacred that we ought never to use it without a particular awe upon our Spirits, and only when it was necessary to use it, and that he therefore omitted the use of it in his Superscriptions of letters to Bishops and then he told me of the practice of Mr Boyle as exemplary herein. I here as pertinent to this head, call to mind how Mr Boyle told me that the Lord Secretary Falkland[b] in giving a visite to a great Lady of the Court, was by another person of quality who came to see her observed to sit and talk to her with his hat on, and that th[is] omission of the ceremony usually paid by men at Court to the faire Sex did afterwards occasion that Courtiers inquiry why his Lordship sat with his hat on befo[re] her, and his Lordship for his satisfaction told him, that according to the mode of the Court, he paid her the usuall civil respect of sitting bare-headed [by] her a good while, till shee at last in her discourse Swearing by God, he immediatly put on his hat, designing to shew her that shee was not worthy of civil respect from him, after shee had appeared so irreverent towards God.

Under this head I might observe how Mr Boyle would frequently complaine of the Books of Metaphysics and Schoole Divinity read in the Universities where the name of God is so much trifled with, that it looseth its awe in mens minds for ever after. Mr Boyle would not endure to heare some of their Questions named, when I went about to repeat them /fol. 38/ as they are satyrized in Erasmus his Moria encomium[c] But according to

---

[a] Henry Jones (1605–82), Bishop of Meath from 1661 till his death.
[b] Lucius Cary, 2nd Viscount Falkland (1610?-43), convenor of the 'Tew Circle' and Secretary of State, 1642. For Pett's intended life of him see above, p. xxxiv.
[c] Desiderius Erasmus (1467–1536) wrote his *Moriae encomium* (1511) during his second visit to England, 1509–14.

the saying, that he who is afraid of Leaves must not go into a Wood, so he who hath an awfull concerne against Gods name being trifled with, must have a care how he reades those Books.

5. Mr Boyle sometimes visiting me at Oxford when he resided there[12] some years before the Restoration of King Charles 2, and once calling on me on a Sunday to go to Church with him at St Marys, I told him that the Minister who was that day to preach was an extraordinary dull man and I therefore desired to be excused from accompanying him to Church, but he soone convinced me of my duty in the case, by saying to me, that whoever preacheth, and let his parts be never so dull, we ought on the Lords days to be present at the Publick Congregation, because God is there honoured by the joint and sociall Prayers of his people, and for that our being ‹then› present in the Publick Assemblies is a necessary Confessing of Christ before men, and said that if the prayer of one righteous man prevailes much, what may we not expect from the joint prayers of many of the righteous? and moreover that there is not in Scripture an instance to the contrary of the Church of God joining in any common petition & not speeding.[a] He at that time told me how he had formerly used the like freedome with a sister of his (whose name I have forgot) as with me, and whome when he would have waited on her to heare a certaine Minister, shee said he did not preach rationally enuf to encourage his hearers to keepe awake, and that his voice was too loud to permit them to sleepe.

I shall under this head observe that while he lived in Oxford, and where the University Church was his parish Church, he constantly went to that and no other, and that in the afternoone aswell as forenoone, as supposeing, that if he had not frequented the Church in the afternoone notice would have been taken of it: But when he removed from Oxford about a yeare before the Restauration of Charles the 2nd and lived in London I observed that then he went not to Church on the Lords day in the afternoone tho constantly in the forenoone; and he satisfied me with the reasonablenes of that his practice and of his being then to be found at home & at leisure to receive the visites of such of his friends who would discourse with ‹him› of religious and moral matters, and among which Visitants I had the honour then to be constantly one, till nere a yeare after the Restoration. Whether any thing of this his practice is proper to be published, I submitt it entirely to the Bishop's judgment. But considering that as Auricular Confession is the great part of /fol. 38v/ the Religion of Papists, so Auricular profession or hearing of Sermons is the total summe of that of the generality of Protestants (and many of whose greedines of Sermons making them think the Repetition of them necessary to Salvation),

---

[a] Succeeding (OED).

and considering that while many Christians among us think themselves bound in conscience to spend the whole Sabbath with a kind of Jewish strictnes, they do in a manner damne the forraine Protestant Churches, who are farr from being such Sabbatarians, I submit it humbly to his Lordship's thoughts, whether it may not be conducive to true and solid piety to have the example of Mr Boyle as to this practice set[13] out in its true light and the rather for that I had it from the relation of one who lived with him in London the last 17 yeares of his life, that ‹all that time› he went not to Church on Sundaies in the afternoone.[a]

An example of the absurdity of the rigid Sabbatarian-humour among some Protestants, here occurrs to my remembrance. Mr Abraham Cowly told me that he being with King Charles 2 in Scotland at the time of his Majestie's Coronation at Schone, his Majesty on a Sunday having been at Church in the forenoone, Commanded Mr Cowly to go with him into the Garden after dinner, and as he was there walking behind the King & discoursing with him, a scotch Minister came up to the King and told him, that he ought not to take his pleasure by walking in the Garden o[n] the Lord's day. The King replied to him that he intended when it was Sermo[n] time to go to Church, but the Minister persisting in his reproofe, and telling the King he offended the people of the Lord, the King thereupon presently went up to his chamber.[b]

6. The having mentioned the name of Mr Cowly occasionally puts me in mind of Mr Boyles being a man of admirable wit, loving ingenious men eo nomine,[c] and on that account the aforesaid Lord Falkland and Mr Waller were very deare to him;[d] And Mr Cowly and Sir William Davenant having desired me to introduce them into Mr Boyles conversation, I had his leave to do it in some afternoone where to encourage their longer stay he (tho he never drank between meales) had wyne and tabacco on the table for their refreshment, and I remember when I went away with them from him, they gave their Judgment of Mr Boyles being equall in the talents of Wit to any of the first-rate men of the Age. Such witty mens soules have the happynes that Æneas Sylvius[e] tells of the Rhodian[s] to be blest with a continual

[a] This is probably Thomas Smith (d. 1742), but it could be John Warr (d. 1715): see Hunter, *Letters and Papers*, p. xxxii. See also below, pp. 78, 82.

[b] Charles' coronation at Schone took place on 1 January 1651; Abraham Cowley (1618–67), the poet, who was at this time cipher secretary to Queen Henrietta Maria, visited the King in Scotland in February 1651 (Nethercot, *Abraham Cowley*, p. 128). This story does not appear to be otherwise recorded.

[c] Literally *by that name*: i.e., *for that reason.*

[d] On Cowley and Falkland, see above, n. b and p. 67 n. b. Edmund Waller (1606–87) and Sir William Davenant (1606–68) were also leading *litterateurs* of the day.

[e] Enea Silvio Piccolomini (1405–64), Italian humanist; later Pope Pius II. Pett evidently refers to his account of Rhodes in ch. 88 of his *Cosmographia*, where gold (a metaphor for sunshine?) is said to have rained down on the island.

Document 5

Sunshine; Judicious men in general have the Light of Reason, but the Ingenious are favoured with the Splendor of it, and as one saith well, There are Interni colores,[a] which are divine lights severally distributed to Soul[es] where some have a faire glosse set on them, a twinkling & glittering Soule, all bespangled with light, others have more sad & dark-coloured Soules. And when I have often heard Mr Boyle discourse of the magnalia[b] of the God of Nature both with depth of Reason and height of fancy, and with the fortunate resultances of thought and such as are remote from vulgar minds, I thought the glory of his Intellectuall Endowments shined out of his mouth, as Porphyry said Plotinus his soule did when he spake.[c] This excellent sprightfull vivacity of his Thoughts appeares sufficiently /fol. 39/ in his grave Writings on Moral and Theologicall Subjects, and as much as it could do without affectation, and where a nimiety[d] of wit would have been no ornament: And tho the expression of grave Authors frequent among Writers is scarce thought to signify more then dull; when I referre to him or any other as a grave Author I take gravity according to the naturall construction of a Weighty Author, and such was he, where the solid matter resembled both the weight and compactnes of gold, and the wit and politenes of his style, the same golds brightnesse or lustre. The brisknes of his fancy when he was about 16 yeares old did shew it selfe in some short poeticall Sallies, and when the beauty of the faire Sex might seeme to have made some impressions on him; & there was about that time a Lady whose name was Killegrew,[e] who had been the most celebrated beauty of the Court, and when some other Ladys who had therefore envyd her did (afterwards upon the judgment given by ‹the› Critics in faces, that hers was decaying) together with Time triumph over her, then did Mr Boyle with this ingenious Stanza think fit to give her the Poets eternity. viz.

> Tho we should grant what th' Envious say,
>   That Killegrew's past her noone day;
> Yet the Sun when it declines,
>   Is still the brightest thing that shines.

But his fancy then made a short ‹yett› everlasting turne from all impressions by Ladies Eyes, and the which appeared ‹afterward› by his book of Seraphic Love, writ by him when he was ‹not 22›[14] yeares old. By that Love was the lesse noble one in him, naturally impaired, as the Culinary

[a] *Inward colours.*
[b] Wonders (OED).
[c] See the *Life* that the ancient Greek philosopher, Porphyry, prefixed to his edition of his master, Plotinus', *Enneads*, chs 10, 13.
[d] Excess (OED).
[e] Probably Anne Killigrew: see Butler, *Theatre and Crisis*, pp. 116–17.

Fire is by the beames of the Sun, or as I may in a way more worthy of his sublime thoughts expresse myselfe, by that Love causing in him a noble heate without the trouble of desire, was the fire of his Soul raised up to its owne high Element, and where it was not to burne but purify.

And that he might ever afterwards devote his thoughts with the greater Purity to the Divine Life & Love, and the glories of the Divine holynesse, he totally forbore reading bookes of poetry & never saw any Play, nor read either Play or Romance; no not so much as those writ by his ingenious brother the Earle of Orrery, as I have reason to believe.

7. It is most true what his Lordship observes in p. 35 of his Sermon, that he spoke of the Goverment even in times which he disliked and upon occasions which he spared not to condemne, with an exactnes of respect. Therein his Lordship to my knowledge doth him right; For my acquaintance[15] beginning with him during his residence at Oxford in the time of Cromwells Usurpation, I observed, that he sufficiently avoyded all guilt /fol. 39v/ of Scandall by shewing the least approbation of it: Nor did he there give any visite to Dr Goodwin or Dr Owen who were Oliver Cromwells great Clerica[l] Supporters,[a] and placed in the highest Posts in the University. And there was at the same time there a private Meeting in an Upper roome, where when the afternoone sermon at St Marys was done severall schollars and others went to heare the Common prayer read by Dr Fell who was afterwards Bishop of Oxon, and where the Sacrament was given by him in the way of the Church of England. But Mr Boyle thought not fitt to go to that Meeting, nor by so doing to appeare as a professed Cavalier. But of his free censuring in his private discourse the irrational politics of Oliver Cromwell I remember divers instances, as for example, his assisting the King of Sweden[b] in his war against Denmark, and whereby he might have been the Sole Master or monopolizer of the Sound, and likewise his fantastic war with Spaine, and his breaking the ballance of Christendome by siding with France: and upon which Mr Boyle further observed to me that Cromwells war with Spaine, obstructing the returnes of the Spanish Galeons with Silver from the West Indies, and thereby necessarily causing a dearth of Silver both in Spaine & France & else where in Christendome, was the true and real cause of the making the Peace between Fran[ce] and Spaine, and the which had a naturall tendency to promote the Kings Restoration; and to the which on the account of the justice of the thing I alwaies judged Mr Boyle to be a hearty welwisher.

8. I being once with Mr Boyle when some in the company were ridiculing some passages in the Singing Psalmes, hee in his gentle way reproved their

---

[a] Thomas Goodwin the elder (1600–80) and John Owen (1616–83).
[b] Charles X. On Cromwell's foreign policy, see Coward, *Oliver Cromwell*, pp. 168f.

so doing, and told them there was enuff in that work of Hopkins and Sternhold[a] to serve the uses of devotion in the church ‹very› well,[16] and that there were many things there so well said, that none could wish for better and he instanced in one staffe or stanza there for that purpose and which he said his Brother Broghil an allowed Judge of Poetry did often applaud and declare, that it much elevated his thoughts in devotion, and so I declare that it hath mine said Mr Boyle, and he thereupon repeated it viz.

> The Lord is kind and mercifull
> When Sinners do him grieve:
> The Slowest to conceive a wrath,
> The readiest to forgive.

And I remember that I have since in Ireland heard the Earle of Orrery often repeate those words in a high commendatory way, and as helpful to him in his devotion.

9. The Bishop in p. 27 of his Sermon observes of Mr Boyle what is very mome[nt]ous, and where his Lordship saies he had a most particular zeale against all severities /fol. 40/ and persecutions upon the account of Religion. I have seldome heard him speak with more heat and indignation[, than] when that came in his way. And pursuant to what is here said by the Bishop I shall relate it that shortly after the King's Restoration Mr Boyle and I largely discoursing of the extravagant severitys practiced by some Bishops towards Dissenters in the Reigne of King Charles the 1st as likewise of the horrid persecutions that the Church of England Divines suffered from the following Usurpations, and we fearing that our restored Clergy might be thereby tempted to such a Vindictive retaliation as would be contrary to the true measures of Christianity & Politics, we came in fine to an agreement that it would tend to the public good to have something writ and published in print assertive of Liberty of Conscience. He undertooke to engage Dr Barlow whose Judgment therein he very well knew, to write of the Theologicall part of the Question, and was pleased then to desire me to write of the political part thereof, and which I told him I would do as well as I could, provided he would please to let me read my Manuscript to him before it went to the Presse, and give me his opinion ‹about› what was fitt to be blotted out, or altred therein: which he frankly promised me. And Mr Boyle himselfe undertooke to write of the fact of the allowance of Liberty of Conscience in forrain parts.[17] /fol. 42/ [but] that Mr John Dury a Divine famous for his being a great traveller ‹& in the

---

[a] Thomas Sternhold and John Hopkins, *Certain Psalms chosen out of the Psalter of David* (1549, and much reprinted).

Northern Countrys where Mr Boyle had never been›[18] & ‹for being an› Essayer to reconcile the Lutherans & Calvinists, would be able to write of the aforesaid fact more extensively then Mr Boyle could, he engaged Mr Dury to do it & rewarded him for it:[a] & deliverd Mr Durys treatise of the subject to me, & which I publishd in print at the end of my owne in the yeare 1660. For in that yeare it was printed, however the booksellers more suo[b] antedated in the title 1661. And in the title page both of my owne work & that of Mr Durys I thought fitt to amuse the Reader with the concluding letters of both oure names in stead of inserting the initial ones, after the example of Dr Wilkins & Dr Ward who in a[19] book then lately writt by them in concert had used the same way, for the title presenting them as incognito.[c] But however I have herewith sent that volume of[20] Mr Durys & my writing for the Lord Bishop to reade if he pleaseth, & request the returne therof, I having not another of them for the reprinting it by, ‹and for which I suppose there will shortly be an occasion, there being an honest designe among some to have[21] many of the treatises about liberty of Conscience to be printed together in folio, & Mr Durys & mine among the rest.›[22d]

I am next to speake of Bishop Barlow's having sent his[23] learned book of toleration[24] wholy writt with his owne hand to Mr Boyle: & which Mr Boyle having read & much approved he gave it me to print when I thought fitt: &[25] which he the rather desired ‹to have›[26] printed because he justly observd it was more excellent then Dr ‹Jeremy› Taylors liberty of Prophecying,[e] & did upon an accurate & scholastic state of the question prove the unlawfulnes of persecuting for Religion, in a way beyond the performance of Dr ‹Jeremy› Taylor, tho yett that book of Dr Taylors hath a lively genius of witt & christianity in it which will make it live in the World for ever, & partake naturally of the fate of bookes mentiond in that knowne verse,

Victurus genium debet habere liber.[f]

But the reason why I did not print that book of Dr Barlow's then & wherewith I afterward satisfyd /fol. 42v/ Mr Boyle for my not having

---

[a] John Dury (1596–1680), writer, ecclesiastical conciliator and close colleague of Samuel Hartlib.

[b] *In their usual manner.* Pett refers to his *Discourse concerning Liberty of Conscience*. For a modern account of this episode see Jacob, *Robert Boyle*, pp. 133–4, but see also above, p. lxix.

[c] *Vindiciae Academiarum* (1654), by John Wilkins and Seth Ward (1617–89).

[d] Nothing is otherwise known of this planned anthology of texts on liberty of conscience.

[e] *A Discourse of the Liberty of Prophesying* (1647) by the Anglican divine, Jeremy Taylor (1613–67).

[f] *The book that is going to succeed needs to have something special about it*, quoting Martial 6.61.10.

done it, ‹was because› the Restorati[on] of the Church together with the King, was attend[ed] with the restoring of the ‹old Churchmens›[27] old Mumpsimus-doctrine[a] & practice of Persecuting ‹the› Nonconformists as such,[28] & Dr Barlows proving persecution unlawful would not have had a[ny] effect in that Conjuncture but the raising a persecution against himselfe, & with which he was soone after the Restoration threatned by the clerical grandees, because he being against the necessity of baptising Infants (& which he only held lawful because the Church Commanded it) had not long before the Restoration out of his zeale for that his opinion unwarily writt a letter to Mr Tombes the famous Anabaptist,[b] & wherein he told him that he was a friend to his person, & likewise to his opinion in the maine, & which letter Mr Tombes soone after printing, Dr Barlows enemys among the high Churchmen made him not long after the Restoration[29] so uneasy that he was importuned by many of his friends to[30] secure himselfe in his station in the University by recanting that opinion: But he could not be brought to do it either by friends or foes. I have by me that letter in print, & it is[31] a very learned & argumentative one.[c]

But had his Discourse of the unlawfulness of denying liberty of Conscience been printed[32] tis probable that his /fol. 43/ enemys in that Conjuncture would have made it more fatally prejudicial to him, then they could the[33] forementiond letter. For the Kings primier ministre[d] then was observd on all occasions[34] to patronize the Great clergy men in theire opposing liberty of conscience, & I heard it from one present at his table that upon the discourse there of that subject, his Lordship having with a great copia verborum[e] commended the extraordinary witt in Dr Jeremy Taylors book of liberty of Prophecying he concluded his censure of the book with this sarcasme, that there was in it too much of the Taylor, & too litle of the Jeremy.[f]

I should have mentiond it before & respecting the series temporum inserted it after the[g]   paragraph or memoire, as being a transaction of moment that preceded the Restoration of King Charles the 2d about a yeare or two, that Mr Boyle having a very great & just value for Dr Sandersons talents in Casuistical Learning, engaged Dr Barlow to write to

---

[a] *Mumpsimus* is a term of abuse for religious conservatives dating back to the Reformation (OED).

[b] The Baptist divine, John Tombes (1603?-76). Pett evidently refers to a letter quoted anonymously by Tombes in his *Ante-Paedobaptism*, part 3, sigs. b2ᵛ-3, which expressed trenchant opposition to arguments for the necessity of child baptism.

[c] Barlow's treatise was published in his *Cases of Conscience* (1692).

[d] Clarendon.

[e] *Abundance of words.*

[f] I.e., the Jeremiah.

[g] There is a gap between *the* and *paragraph.*

him & acquainte him that he would setle on him £50 a yeare during his life, if he would[35] yearly publish a volume of Case divinity: & that if in the yeare then Current he would prepare to publish such a volume, Mr Boyle would immediatly send him the first fifty pound.[a] Dr Sanderson accepted of the proposition, & shortly after received the £50, & in the yeare 1659 printed his 10 famous Divinity lectures of the Obligation of Conscience. For tho according to the custome of booksellers ‹to antedate›, the title mentions it printed in the yeare 1660, yet it came out of the presse in 1659, & his dedication therof to Mr Boyle beares the date of the yeare 1659. /fol. 43v/ But in the Doctors epistle dedicatory to Mr Boyle (wherin he celebrates Mr Boyles piety, & learning & mentions his[36] munificence) there was one unlucky expression & capable of a good construction [sic],[37] from which the generality of malevolent Critics took occasion to renew the discourse of the towne that Mr Prynnes ‹much-read› large folio calld a compleat history of the trial & condemnation of Archbishop Laud, had formerly causd.[b] That expression of the Doctors was that wherin he desires Mr Boyle that as he had done, he would still literatos fovere, pietatem colere, fidos sibi ex infidi Mammonæ impendio amicos comparare &c.[c] Mr Prynne p.[38] 81 there setts downe a letter of Dr Bramhal to Archbishop Laud wherin tis sayd the Earl[e] of Corke holds the whole Bishopric of Lismore at the rate of 40 s[hillings] or 5 markes by the yeare, meaning his ho[l]ding it so from the Crowne, & in p. 86 there is Archbishop Lauds letter to the archbishop of Dublin mentioning that the Earle of Cork had gott into his hands no sma[ll] proportion of the churches meanes with severe reflections on the Earle relating to the bishopr[ic] of Lismore & the Colledge of Youghhal. ‹I can lend my Lord Bishop the book of Mr Prynne.›[39]

But ‹as› I am confident that Dr Sanderson was a better courtier than to reproche his benefactor ‹before the World› & more wise then by such ingratitude to[40] forfeit the donative for the future, so I believe Mr Boyle was more a Christian then to withhold from the Church any thing belonging to it if any such thing had any way come to his possession. For no

---

[a] On Boyle's patronage of Sanderson in 1659, see Hunter, 'Casuistry in Action', pp. 82–3, and above, p. lxxiii.

[b] Prynne, *Canterburies Doome*, esp. pp. 82–6 on the Earl of Cork, including Laud's correspondence with John Bramhall (1594–1663), then Bishop of Derry (later Archbishop of Armagh), and Lancelot Bulkeley (?1568–1650), Archbishop of Dublin.

[c] See Sanderson, *De Obligatione Conscientiae Praelectiones Decem*, sig. A5, where the passage is as given by Pett except that, in quoting it, he has altered *tibi* (*yourself*) to *sibi* (*himself*). In the English translation, *Ten Lectures*, sig. A2ᵛ, the following translation is given: *to favour learned men, to advance Piety, to procure you faithful friends at the charges of unfaithful Mammon.* The latter phrase is a little awkward, possibly because of the embarrassment to which Pett refers: clearly Sanderson intended to imply that that untrustworthy thing, money, could gain trustworthy friends, whereas the hostile critics exploited his ambiguity over the role of treacherous Mammon.

doubt but his conscience was more tender then that of the canonists, & yett they all agree in telling us that possessor malæ fidei nunquam præscribit.[a]

[41]For most certainly he who in his will very piously restored to the Church ‹double›[42] all the monys he receivd from forfeited impropriations ‹that came from the crowne in Harry the 8ths time, & to which as forfeited by the Irish Rebellion of 41›[43] he was entitled by[44] a claus in an ‹Irish› Act of parliament in King /fol. 44/ Charles the 2ds time, would not detaine from the protestant church of Ireland any ‹of its› lands ‹tho› left him by will.[b] I therfore believe that if the fact were true which the Archbishop of Canterbury & Dr Bramhal afterward Archbishop of Armagh affirme of the lands of the Bishopric of Lismore that Mr Boyles father possesd, yet that none of those lands were devised to Mr Boyle by his fathers will. However I submitt this memoire en secret to the Lord Bishop[45] of Sarums consideration, & if his Lordship for his private information & curiosity desires to be particularly informd in this matter further, I can serve him therin by having Mr Boyles will consulted in the Prerogative Court's registry here, & his fathers will in that Court's registry in Ireland, & which may be done without the Earle of Burlingtons knowing any thing thereof, & for the giving of no offence to whose Lordship, I doubt not but the Bishop & my selfe have the like tender regard.[46] But to his Lordships secrecy I shall further committ the relating how that I ‹as Causarum patronus›[c] proving the old Earle of Corkes will specifice & per testes[d] in the prerogative Court ‹in Ireland› I found therin mentiond that that Earle being prosecuted in the Starre-chamber commonly calld the Castle chamber because held in the Castle of Dublin, by the order of the Earle of Strafford then Lord Lieutenant or Lord Deputy of Ireland, that Earle of Corke did at the great importunity of the present Earle of Burlington & Corke after a Submission to the Court pay £8 or 10000 for a fine (& which I suppose might have been doubled, trebled or quadrupled but for that prostration of the old Earle) tho I remember not the cause of the prosecution mentiond in the will, yet I fancy to my selfe that it could be nothing else but his possessing himselfe of the lands of the Bishopric of Lismore.[47,e] And yett I thinke if that Summe of mony was not a full satisfaction[48] in value for any lands of the Bishopric supposed to be possessd by the old Earle, that quietus obtaind from the starre-chamber could be no warranty to Mr /fol. 44v/

---

[a] *He who possesses bad faith never makes a point of declaring it publicly.*

[b] For Boyle's will, see Maddison, *Life*, pp. 257–82, esp. pp. 270f. On the impropriations see Hunter 'Casuistry in Action', esp. pp. 83–4.

[c] *Advocate.*

[d] *Individually and by the testimony of witnesses.*

[e] On Cork's struggle with the Earl of Strafford (1593–1641), see Canny, *Upstart Earl*, esp. ch. 2. For his will see Townsend, *Great Earl of Cork*, Appendix III; however, Pett's memory seems to have been mistaken as to its content, since it does not refer to this fine.

Boyle's conscience before God for not restoring to the church any of its lands he was possessd of, or the remainder of the value. I presume not in my private thoughts to give judgment ‹about›[49] any actions of his father, & the rather for that I heard Mr Boyle once say that he judging his fathers life exemplary for probity, intended to write it. Yett after all this by me sayd in this paragraph, I shall further en secret acquaint the Lord Bishop that I once telling Mr Warre Mr Boyles executor & who hath the custody of all Mr Boyles papers, that his Lordship could not possibly write Mr Boyle's life to the utmost advantage unlesse he had the perusal of those papers, & many of which were (as Mr Warre told me) Bishop Barlows & Mr Baxters[a] resolutions of cases of conscience put to them by Mr Boyle, & many of them letters about experimental philosophy in the way of re-scripts from learned men, Mr Warre sayd that the papers were very many & would aske some moneths time to peruse, & that he had a caution given him by the Earle of Burlington to shew them to no body, & that he thought the will of Mr Boyle ‹likewise› Cautiond theire being kept secret. I told him it was fitt the will it selfe should be consulted as to that particular: & so it may easily be at Doctors Commons[b] (if the Bishop pleaseth.) ‹And my present belief is that there is no cautioning clause of that Nature in the will; & if upon the inspection of the will, it be found as I believe, Mr Warre (I suppose) will consent to the Bishops perusing the papers, & so will Sir Henry Ashurst another of the executors:[c] & in case both they & the Earle of Burlington the other executor should all three deny it upon a caprice, the Archbishop of Canterbury[d] hath power to monish them judi-cially to consent it to it [sic] sub pœna juris & contemptas.›[50,e] Having here mentiond Mr Warre, I shall say of him that I believe him to be a discreet & pious man & such an one as the French call tout a fait honet homme:[f] but I believe he will not shew any one Mr Boyle's will, tho yett I must acknowledge he hath been so courteous to me as to give me a copy of a great part of it that was necessary for my direction in the disposal of the mony arising from the forfeited impropriations, & for which My Boyle made my selfe together with his Brother & the Lord Massarene[51] & his

---

[a] Richard Baxter (1615–91), Presbyterian divine. No other evidence survives of his resolu-tion of cases of conscience on Boyle's behalf.

[b] The building occupied by the Association or College of Doctors of Civil Law in London, where Prerogative Court wills were registered.

[c] Sir Henry Ashurst (1645–1711), London alderman and colonialist. See also below, pp. 78–9.

[d] Thomas Tenison.

[e] *Under the penalty of law, and [on pain of] contempt [of court].*

[f] In trying to sum up Boyle's trusted amanuensis and executor, John Warr (d. 1715), Pett uses a concept made famous by Molière which it is hard to convey in as few words in England: it implies a sensible, unpretentious, generous-spirited person, something which the extra phrase *tout a fait – altogether –* underlines.

sonne Mr Clotworthy Skeffington trustees,[a] & as to the /fol. 45/ discharge of which trust, Mr Warre can tell the Lord Bishop that I am able to wash my hands in innocence, tho none of that mony so long since receivd is yett distributed.

But I shall here relate it for the entertainment of his Lordships curiosity that[52] my curiosity making me enquire how Mr Boyle came at first to obtaine that clause in the ‹Irish› Act[53] of Parliament for giving him those forfeited impropriations, I found it out from the mouth of Sir James Shane that Mr Boyle desiring him to discover somewhat to him of the forfeited lands that he might aske of King Charles the 2d & to the end that Mr Boyle might ‹for pious uses› lay out the[54] moneys that should thence accrue, Sir James therupon projected that grant for him.[b]

And if ever the Lord Bishop should obtaine the leave for the perusal of Mr Boyles papers, I thinke that Dr Hook & Dr Slare[c] two worthy members of the Royal Society who lived with Mr Boyle as his salariated dependants, & Mr Smith[d] an expert chymist who lived with him the last 17 yeares of his life, might be very assistful to his Lordship in the perusal, & furnishing him with remarkes about which of the philosophical papers of Mr Boyle were proper to be mentiond or referrd to in the history of Mr Boyle's life. And I believe they would be ready to be ministerial herein gratis, unlesse his Lordship should require the booksellers to give them some moderate requital for theire paines.

Upon my revolving in my thoughts the trusts reposed by Mr Boyle in his will, I finde that in ‹his›[55] Designation of the persons he shewd a latitude in his principles of Christian charity. For his executors were his brother Burlington a professd member of the Church of England, ‹in the way of Archbishop Tillotson,›[56,e] Sir Henry Ashurst, & Mr Warre two pious dissenters: & his trustees for laying out the mony accruing from /fol. 45v/ forfeited impropriations were the Earle of Burli[ng]ton & my selfe an unworthy member of the Communion of the Church of England ‹in the way of Archbishop Sancroft,›[57,f] & the lord Viscount Massarene & his sonne both of the presbyterian persuasion: & his trustees for the naming his lecturers against Atheisme were Mr Evelyn of the church of England, &

---

[a] John Skeffington, 2nd Viscount Massereene, was succeeded as 3rd Viscount by his son, Clotworthy Skeffington (1660?–1714) on his death in 1695. The brother to whom Pett refers is Lord Burlington. For Boyle's will see Maddison, *Life*, pp. 257–82, esp. pp. 271–2.

[b] On these impropriations, which the Irish office-holder, Sir James Shaen (d. 1695), was instrumental in helping to obtain, see Hunter 'Casuistry in Action', pp. 83–4.

[c] Frederic Slare (1648–1727), natural philosopher and London doctor.

[d] Thomas Smith (d. 1742), servant of Boyle and later an apothecary.

[e] By referring to John Tillotson (1630–94), Pett alludes to the broad church policy over which he presided as Archbishop of Canterbury after the Glorious Revolution.

[f] By referring to William Sancroft (1617–93), the Archbishop of Canterbury who was suspended in 1689 for his refusal to take the oath of allegiance to William and Mary, Pett aligns himself with the high church party.

Sir Henry Ashurst, & Sir John Rotheram a learned dissenter[a] & who was my[58] contemporary in Oxford, & was one of the Judges in Westminster hall appointed by King Jam[es] & a concurrer with the others in asserting his Majestys dispensative power, & who in the giving the charge at the Assizes in the Country pursued the Instruction given by the Lord Chancellor to him & the rest of the Judges to magnify the Kings goodness & his justice in his declaration for liberty of Conscience,[59] took occasion (as I am informd) to inveigh against the Church of England, & say[60] that the Church of England is a bloudy church. And another thing comes into my mind upon the thinking of Mr Boyle's will, namely that because his fathers will was ‹(as I thinke)› that if he dy'd without heires of his body the estate left him should descend to the Earle of Burlington & his brother Shannon, Mr Boyle left a great estate to those his brothers who needed it not, & forbore to dispose of it to public uses or to give it away to kindred who more wanted it, because he thought it a matter of conscience to obey his father's will in the case.

And from what I have before mentiond of Bishop Sandersons dedicatory epistle to Mr Boyle I may take occasion to submitt it to his Lordships Consideration to have the many dedicatory epistles to Mr Boyle before the moral & philosophical books of learned men consulted, & some philosophical Bookes likewise wherin learned men[61] do right to some of Mr Boyles philosophical notions & experiments by celebrating them & extolling theire /fol. 46/ usefulness. General praises are in a manner as insignificant as the ‹epithets that the Poets call otiosa epitheta.›[62,b] And if the ‹writings of› any pretended critics have detracted from the usefulnes of any of Mr Boyles experiments, such too I thinke may be properly consulted & animadverted upon. I have heard that a book that went under my Lord Chief Justice Hales his name calld Difficiles Nugae reflects on some of them. But I suppose that learned man was more prudent then to do so.[c] For I know of no experiment that ever Mr Boyle made but what was useful in solving the phænomena. And the noble[63] discovery of the circulation of the bloud, hath hardly had any other use but that, the materia medica & the methodus Medendi being still applyd in the same way as they were before that Discovery made by Dr Harvy, & the incomparable defence of it in latine by Sir George Ent & wherby he deservd almost as well[64] from that Grande Inventum as Dr Harvy.[d]

---

[a] Sir John Rotherham (1630–1696?), judge with low church sympathies, whose connivance with James II's religious policies Pett aptly describes.

[b] *Superfluous epithets.* The exact source that Pett has in mind is uncertain; the Roman rhetorician, Quintillian, discusses otiose expressions.

[c] Contrary to Pett's view, Sir Matthew Hale (1609–76) did indeed take issue with Boyle in his *Difficiles Nugae* (1674) and other tracts.

[d] Sir George Ent (1604–89) vindicated Harvey's discovery in his *Apologia pro curcuitione sanguinis* (1641).

As[65] for dedications of bookes of Matters Religionary to Mr Boyle, I have by me one of Bishop Gaudens, & wherin the dedication is very amply laudatory.[a] ‹I have heard of some[66] such bookes dedicated to him by Dr Hyde & Mr How.›[67,b] Whether Bishop Sprats history of the Royal Society mentions any particular things of Mr Boyle, I have forgott.[c] But if any of the booksellers attends me to know the names of ‹learned› authors both of our owne Country & forrainers who have celebrated Mr Boyles experiments particularly, I can direct him whether to go to finde out theire names & titles of theire bookes.

But having here mentiond the name of Dr Hyde & having so much discoursd of Mr Boyle's will, I shall further[68] observe how the Dr in the forementiond letter to me relates that Mr Boyle propounded to him the[d]

*   *   *   *   *   *

Fol. 47/ In the reighne of King James the 2d a french protestant refugee who had[69] devoted his life to the study of Navigation, was growne so great a Master of the English tongue, as to publish in it a very useful treatise of Navigation, & which he dedicated to the Commissioners of the Navy, & of which my brother Sir Phinehas Pett was then one, & who approved much of the book, & moved the other commissioners ‹(tho in vaine)› to recommend the Author to his Majesty for some beneficial employment. A book was by the author presented to each of them. But I finding by my brother that the rest of the Navy board had not a just value of the excellence of the book, I gave the book to Mr Boyle, & told him ‹what›[70] the fate of the book & its author were, & he having read it told me he was so well satisfyd with the Authors performance therin that he would recommend the care of the promotion of the Author most earnestly to the Kings ministers.[e]

[a] John Gauden (1605–62) dedicated his *Discourse Concerning Publick Oaths* (1662) to Boyle.

[b] Thomas Hyde dedicated his translation of the Gospels into Malayan to Boyle in 1677 (see Birch, *Life*, pp. cix–x; Fulton, *Bibliography of Boyle*, pp. 164–5); John Howe (1630–1705) wrote his *Reconcileableness of God's Prescience of the Sins of Men, with the Wisdom and Sincerity of.... Whatsoever Means He uses to prevent them* (1677) in the form of a letter to Boyle.

[c] Thomas Sprat's *History of the Royal Society* (1667) does not make specific reference to Boyle; on Glanvill's *Plus Ultra*, however, which does, see above, p. xxii.

[d] The text ends in mid-sentence at the bottom of fol. 46: *to him the* is the catchword leading on to the subsequent missing section. Fol. 46ᵛ is blank, as is an unfoliated conjugate leaf. The incomplete passage perhaps dealt with some intended missionary work which Boyle intended to fund.

[e] The book in question, and its Huguenot author, have not been identified.

Mr Hale's book of New inventions publishd neare a yeare before[71] Mr Boyle's death was (as I found by a message he sent me) read over by him with great delight.[a] But the Gentleman who was a common acquaintance to us both & brought me the message told me that Mr Boyle made very melancholy reflections on one passage in the book namely that in p. 12. ‹that›[72] the Invention of oure English frigats of which the Constant Warwic was the first & built in the yeare 1646, was likely to come shortly among Pancirolls Res deperditæ,[b] for that King Charles the 2d observd that the fabrics of oure english shipps did for several yeares degenerate more & more from the Frigat way in which the Constant Warwic was /fol. 47v/ built, to ‹that of› oure[73] sluggish old built shipps & not at all adapted for swiftnes of sailing:[c] & he told me that Mr Boyle desired me as I was a lover of my Country to take care of the preservation of the draught of that ship, & the which he knew it was in my power to do. And that being the last Command I had the honoure to receive from him I have effectually obeyd it, & ‹I believe that›[74] but for that desire of Mr Boyle's to me the draught of that ship & the invention had been for ever lost. And since *Primum in unoquoque genere est mensura reliquorum,*[d] ‹with›[75] the helpe of that draught any ordinary artist may build another ‹such› shipp by it, & with the care of the public the invention may be transmitted safe to posterity. And it is a remarkable thing that that shipp the Constant Warwic ‹being›[76] taken by the french in this warre, it hath fallne out well for us that that ship having been rebuilt at Portsmouth by a Master builder who not having the original draught of its first building before him, it was ‹by him› spoild in its rebuilding: & otherwise the french might have wrought off the curious lines of its fabric by the body of the ship taken. Mr Hale's book p. 13 mentions that shipp's being spoyld in its rebuilding.[e]

My brother[77,f] coming hither out of the country & causing several of his draughts of shipps to be brought with him in order to his building new shipps for king William, was by my cautioning him kept from bringing among them, the original draught of the Constant Warwic: & all those draughts being now not to be found where before his death he left them,

[a] For Hale's *Account of New Inventions,* see above.

[b] Guido Pancirolli's *Rerum memorabilium jam olim deperditae libri duo* first appeared in 1599.

[c] Cf. Hale, *Account of New Inventions,* pp. ix–x, xii–iii, which Pett quotes almost exactly concerning the *Constant Warwick* and Charles II's solicitousness for her. The mutual friend referred to might conceivably have been John Daniel, referred to in Pett's letter to Pepys of 3 May 1696: see Pepys, *Private Correspondence,* vol. i, p. 115. On the *Constant Warwick* see Pepys, *Naval Minutes,* pp. 15–16, 18; Evelyn, *Diary,* vol. v, p. 10 and n.

[d] *The first example, in any kind of field, is also the measure for the rest.* The source of Pett's quotation has not been identified.

[e] Hale does not, however, mention the ship's capture by the French, on which see *CSP Dom* 1691–2, p. 106.

[f] Sir Phineas Pett, master shipwright at Chatham, knighted 1680.

the draught of the Constant Warwic had been lost too, but for Mr Boyle's cautioning me about it, as aforesaid. And I am ready to shew it as likewise many /fol. 48/ other more curious draughts of capital shipps then can elsewhere be found in Christendome to any who desires to see them. But Mr Boyle being dead, whether any surviving virtuoso thinkes it tanti[a] to come to me & looke on them, is more then I know.

Another invention highly useful to all oure shipping both men of warre & merchant-men here occurres to my remembrance, that had never been perfected & preservd but by the skill & paines of Mr Boyle, & that is the dulcoration of salt water, so as to be as grateful to the palate as any fresh water what so ever. Colonel Robert Fitsgerald a kinsman of Mr Boyle's designing to passe a patent for the benefit of that invention, applyd to Mr Boyle about the perfecting of it, & a pamphlet by[78] the Collonel publishd in English & French doth particularly sett forth Mr Boyles experiments about the same.[b]

I knowing that there was a great friendship & much conversation between Mr Boyle, & Mr Pepys of the Royal Society &[79] formerly Secretary of the Admiralty, I wrote to him lately to informe me whether by the recollection of his memory he might know of any experiments that Mr Boyle had acquainted him with relating to[80] Navigation, or the better sailing of the shipps of the navy royall, or the improvement of theire powder & gunnes. And Mr Pepys in his answer to my letter promised me to do so, & gave me hopes of his being able to[81] recollect some things of that nature.[c]

‹His very many ‹excellent› similitudes taken from the loadstone in his book of Seraphic Love, & those too not vulgar ones, but such as penetrate into the nature of it, do shew him a critic in some points about Navigation.›[82]

The Bishop in p. 37 of his sermon ‹shews›[83] how Mr Boyles charity was helpful to others in the distribution of his chymical preparations, & if[84] his Lordship thought it tanti[d] to have particular instances of the cures by Mr boyle wrought on several diseasd persons, I might perhaps /fol. 48v/ be able to furnish him with them from the mouth of Mr Smith an ingenious chymist & who servd Mr Boyle in officiating about his elaboratory during the last 17 yeares of his life.[85,e] But Mr Boyle having made the many moral

---

[a] *Worth it.*

[b] Fitzgerald's project for making saltwater drinkable is described in his *Salt-Water Sweetn'd*; see esp. pp. 1–3 and 13f. on Boyle's active role in it. See also Maddison, 'Salt water Freshened'. On the 'numerous' French and Latin translations of the treatise, see Fulton, *Bibliography of Boyle*, p. 142.

[c] On Pett's approach to Pepys, see above, p. xxxiii. The intimacy between Boyle and Pepys to which Pett here refers is otherwise undocumented.

[d] *Worth it.*

[e] On Smith, see above, p. 78.

vertues that were in him, the more conspicuous & useful to the World by his skill in physic, & in his free communication of his excellent & successful ‹medicaments› mindes me of some things sayd by Signor Ciampoli a late celebrated witt of Italy,[a] in his book calld Prose di Monsignor Giovanni Ciampoli, & printed at Rome & in which book he tells his Readers that he was Secretary to three popes successively, & where Della Potenza, Discorso nono, capo settimo[b] his contents of that chapter are, Christo, per fare gli Apostoli Principi volontariamente obbediti, die[de] loro potenza di risanare gl'infermi,[c] & there saith how in effect he made them Doctors of physic, & in the end of the chapter[86] referring to the Acts of the Apostles, he tells us that the light of the Sun was not to be compared[87] to the shaddow of St Peter, & he hath there many chapters tending to the same scope. In his 6th the contents were, that No power is greater then that of Medicine, because no humane good is greater then health; & afterwards those of the 8th are, A comparison between the Caesars triumphing, & the Apostles healing, & those of the 13th are, Vespasian establishd himselfe in the Roman empire more by the reputation of his being a[88] Physician, then by the glory of his being a Conqueror.[d]

In fine, it was Mr Boyle's constant practice to endeare his Conversation to such of his visitan[t] friends as labourd under chronical diseases,[89] by communicating to them such medicinal directions as were highly useful to them, &[90] what by that meanes & the charmes of his many vertues, & of his rational discourses, it may be affirmd that the age hath produced no one[e]

---

[a] Giovanni Ciampoli (c.1590–1643), Italian cleric and intellectual, who played a crucial role in the Galileo affair. See Redondi, *Galileo Heretic*, pp. 97–9, 265f.

[b] *On Power, Discourse 9, ch. 7*; Ciampoli, *Prose*, pp. 207f.

[c] *Christ, in order to make the apostles rulers who were voluntarily obeyed, gave them power to heal the sick.* Cf. Ciampoli, *Prose*, pp. 214–15.

[d] Cf. Ciampoli, *Prose*, pp. 213–16, 222–4. Ciampoli cites as his source for this Suetonius' *Life* of Vespasian, ch. 7, in which Suetonius reported cures that Vespasian effected as emperor. He also refers to Tacitus' lengthy account of Vespasian in the *Histories*.

[e] The text breaks off incomplete at the bottom of fol. 48[v]; *one* is the catchword leading on to the subsequent missing section.

DOCUMENT 6A

John Evelyn's letter to William Wotton,
29 March 1696

Add. MS 4229, fols 58–9. Holograph. Endorsed by Wotton on fol. 59ᵛ: 'Mr Evelyns Letter with an Account of Mr Boyles Life', to which is added in a different ink: '& the Order of his Treatises, with the Marks of lost No.' [?]. Published in Evelyn, *Memoirs* (1818), vol. ii, pp. 302–8 (and from thence, in varying degrees of modernisation, in many subsequent editions of Evelyn's diary and correspondence). The source of the printed text appears to have been the copy of the letter in Evelyn's Letterbook, now Christ Church, Oxford, Evelyn MS 39, no. 754 [in fact, '854', due to an error of numeration at this point in the manuscript], though see below, n. 34. Minor differences of phrasing, etc., between the two versions have here been ignored, but more significant variants are recorded in endnotes. In addition, whereas the copy sent to Wotton has a single marginal note, the Letterbook version has three more, which appear to have been added by Evelyn at a later date, perhaps on a rereading of the text: these have here been included in footnotes. See also above, pp. xxxvii–iii.

Wotton 29o March – 96[1]

Worthy Sir,

I most heartily beg your pardon for detaining your Books so unreasonably long, after I had read them; which I did, with greate satisfaction; especially the Life of Descartes:[a] The truth is, I had some hopes I might have Enjoy'd you longer againe here: for methought (or at least I flatter'd my-selfe with it) you said[2] you would do us that favour, before my journey to London: whither I am (God willing) going tomorrow or next day for some stay; not without Regret, unlesse I receive your Commands, if I may be any way serviceable to you, in order to that noble, & universaly-obliging Undertaking you lately mentiond to me; I meane, your generous offer, & Inclination to Write the Life of our Late Illustrious Philosopher Mr. Boyle; and to honor the Memorie of a Gentleman of that singular Worth & Virtue. I am sure if you persist in that Designe; England shall never envy France, or

---

[a] By Adrian Baillet: see above, p. xxxvii.

neede a Gassendus, or a Baillet to perpetuate, & transmit the Memory of One, not onely equaling; but in many things Transcending either of those excellent, & indeede extraordinary Persons, whom their penns have render'd Immortal.[a] I wish myselfe was furnish'd to give you any Considerable supplys,[3] after my so Long Acquaintance with Mr. *Boyle*, who had honor'd me with his particular Esteeme, now very neere fourty-yeares; & which I might have don, by more duely cultivating the frequent Opportunities he was pleas'd to allow me: But so it is, That his Life & Virtues have ben so Conspicuous, as you will neede no other Light to direct you, or Subject Matter to worke on, than what is universaly known, and by what he has don & publish'd in his Books: —

You may neede perhaps, some particulars as to his Birth, Family, Education & other lesse necessary Circumstances for Introduction; and such other passages of his Life, as are not so distinctly known but by[4] his neere Relations: If in this I may serve you, I shall do it with greate readynesse, & I hope successe; having some pretence by my Wife, in whose Grand-fathers house (which is now mine at Deptford) the Father of this Gentleman was so Conversant; that (contracting an Affinitie there) he left his ⟨then⟩ Eldest son with him (whilst himselfe went into Ireland) who in his absence died, and now lies buried in our Parish-Church with a remarkeable Monument.[b]

I mention this; because my Wifes Relation to that Family, giving me Accesse to divers of his neerest Kindred; the Countesse Dowager of Clancarty (living now in an House of my sons in Dover-streete) & the Countess of Thannet; both his Niepces, will, I question not, be able to Informe me[5] what they cannot but know of those, & other Circumstances of their Unkle, which may not be unworthy your taking notice of; Especialy my Lady Thannet, who is a greate Virtuosa, & uses to speake much of her Unkle.[c] You know she lives in one of my Lord of Notinghams Houses at St. James's; and therefore will neede no Introduction there: I will waite upon ⟨my⟩ Lord[6] Burlington likewise if there be occasion; if in the meane time (⟨&⟩ after all this Officiousnesse of mine) it be not the Proffer of a very Use-lesse Service; since my Lord Bishop of Salisbery (who made us expect, what he is now devolving on you) cannot but be fully Instructed[7] in all particulars.

It is now (as I said) almost fourty-yeares, since first I had the honor of

---

[a] Peiresc and Descartes. See above, pp. xxxiv, xxxvii, xlviii–ix.

[b] The Letterbook version has a marginal note by Evelyn at this point: *a Tent & [repeated] Map of Ireland in Relievo.* On the tomb of Roger Boyle (1606–15) at Deptford, see below, pp. 100–2. Evelyn's maternal grandfather was Sir Richard Browne (c. 1539–1604).

[c] Elizabeth Maccarty, Countess of Clancarty; Elizabeth Tufton (née Boyle), Countess of Thanet (?1638–1725): on her learned pretensions see Evelyn, *Diary*, vol. iv, p. 505; she was related to Wotton's patron, the Earl of Nottingham, through his second wife, Anne, daughter of Viscount Hatton by his wife Cicely, 4th daughter of the 2nd Earl of Thanet (GEC s.v. Winchilsea).

being acquainted with Mr. Boyle; both of us newly Return'd from Abroad; tho', I know not how, never meeting there: Whether he Travell'd more than France & Italy, I cannot say: but he had so universal an Esteeme in Foraine Parts; that not any Stranger of note or quality; Learn'd, or Curious coming into England, but us'd to Visite him, with the greatest respect & satisfaction.[8]

Now as he had an early Inclination to Learning (so especialy to that part of Philosophy he so happily succeeded in) he often honord Oxford and those Gentlemen there with his Company, who more particularly apply'd themselves to the Examination of the long domineering Methods & Jargon of the Scholes: You have the Names of this Learned Junto (most of them since deserv'dly dignify'd) in that elegant History of the *Royal Society*, which must ever owne its Rise from that Assembly, as dos the preservation of that famous University, from the phanatic Rage, & Avarice of those Melancholy times.[a] These, with some others (whereof Mr. Boyle, the Lord Viscount Brounchard [sic], Sir Robert Murrey &c. ‹were most Active›) spirited with the same zeale, and under a more propitious Influence, were the Persons to whom the world is Oblig'd, for ‹the› promoting of that generous & Reale Knowledge, which gave the ferment that has ever since Obtain'd, and surmounted all those many Discouragements, which ‹it› at first Incountered.[9] But by no man more has [sic] the Territories of the most /fol. 58v/ Usefull Philosophy ben inlarg'd, than by our Hero, to whom there are many Trophes due; And accordingly his Fame was quickly spread, not onely among us here in England; but thrô all the Learned World besides: It must be confess'd that he had a marvelous sagacity in finding out many Usefull, & noble Experiments; Never did stubborn Matter come under his Inquiry; but he extorted a Confession of all that was in her most intimate recesses; & what he discover'd, he as faithfully Registerd & freely communicated; in this exceeding my Lord Verulam;[b] who (tho never to be mention'd without honor & admiration) was us'd to take all that came to hand, without much Examination: His was probability, Mr. Boyls suspicion of successe. Sir, You will heare have ample Field, and infinitely Gratifie the Curious, with a[10] fresh Survey of the progresse he has made in these Discoveries, freed from those incumbrances, which render the Way somewhat tedious, tho' abundantly recompencing the pursuit; especialy, those noble Achivments ‹of his;› The Spring & Weight of[11] the two so necessary Elements of Life, Aer & Water, & their Effects: The Origin of Formes, qualities, & Principles of Matter:

---

[a] Sprat, *History of the Royal Society*, esp. pp. 53f. Evelyn echoes Sprat's stress on the 'Oxford' origins of the Society, whereas modern scholarship has illuminated its more varied roots. On this issue and on the founding Fellows mentioned in the next sentence see Webster, *Great Instauration*, pp. 92f. By *Brounchard* Evelyn means *Brouncker*.

[b] Francis Bacon.

Histories of Cold, Light, Colours, Gemms, Effluvias, and other his Works so well establish'd on Experiments; Polycrests[a] & of universal Use to Real Philosophy; besides other beneficial Inventions peculiarly his; such as the Dulcifying Sea-Water with that ease & plenty, together with many Medicinal Remedys, Cautions, directions; Curiosities & Arcana which owe their Birth or Illustration to his indefatigable Recherches.[b] He brought the Phosphorus[12] to the cleerest light that ever any did, after innumerable attempts; It were needlesse to reckon[13] particulars to one who knows ‹them› better than, myselfe: Nor will you omitt those many other Treatises relating to Religion, which indeede runs thro' all his Writings upon Occasion, and shew ‹how› unjustly that Aspersion has ben cast on Philosophie, that it disposes Men to Atheisme: Neither did his severer studys yet soure his conversation in the Least; he was the furthest from it in the World; & I question whether ever any man has produc'd more Experiments to Establish his Opinion, without Dogmatizing: He was a Corpuscularian without Epicurus,[14] Addicted to no particular sect; but as became a generous & free Philosopher, before all preferring Truth; In a Word, a Person of that singular Candor, Comity[c] and worth, that to draw a just Character of him, one must run thro all the Vertues, as well as all the Sciences; And tho' he tooke the greatest care imaginable to conceale the most Illustrious of 'em, his Charity, & the many Good Works he did;[15] it is well known how large his Bounty was upon all Occasions:[d] Witnesse the Irish, Indian, Lituanian Bibles, to the Translations, printing & publishing of which he layd out considerable summs, The Catechisme & Principles of the Christian Faith, which I think he caus'd to be put into Turkish, and dispers'd among those Infidels; And here you will take notice of the Lecture he has Indow'd and ‹so seasonably› provided for:[e]

As to his Relations, so far as I have heard, his Father (Richard Boyle) was *Faber Fortunae*, a person of wonderfull Sagacity in Affaires, & no lesse Probity; by which he compass'd a vast Estate, & greate Honors to his Posteritie, which was very numerous, & so prosperous, as has given to the publique both Divines, Philosophers, Souldiers, Politicians and States-men; and spread its branches among the most noble and Opulent of our Nobility: Mr. Robert Boyle, borne I think, in Ireland, was the Youngest; to whom yet he left a faire Estate; to which was added an

---

[a] Adapted to several uses (OED).

[b] Sic. On Boyle's interest in making saltwater drinkable and in medicines, see above, p. lxviii.

[c] Courtesy (OED).

[d] At this point there is a marginal note by Evelyn: *A greate Benefactor to Dr Sanderson Bishop of Lincoln, se his dedicatio De juramento &c.*[16] Evelyn evidently refers to Sanderson's *De Obligatione Conscientiae Praelectiones Decem*: see above, pp. lxxiii, 74–5.

[e] See above, pp. xxiv–v, 48–9.

honorary-pay of a Troope of Horse, if I be not mistaken:[a] And tho among all his Experiments, he never made that of Marriage Yet I have ben told, he Courted the beautifull & Ingenious Daughter of Carew, Earle of Monmouth, to which is owing the Birth of his *Seraphic Love*, and the first of his productions:[b] Descartes's, was not so innocent:[c] In the meane time, he was the most facetious, & agreable conversation /fol. 59/ in the World among the Ladys, when he happnd to be Ingag'd; and yet so very serious[17] & Contemplative at all other times, tho far from Morose; for indeede he was Affable, and Civile rather ‹to› excesse, yet without formality.

As to his Opinion in Religious Matters & Discipline, I could not[18] but discover in him the same free thoughts, which he had of Philosophy; not in Notion; but strictly as to Practise an excellent Christian, and the greate duties of that Profession, without Noise, Disputes or Determining; owning no Master but the Author of it, no Religion, but Primitive, no Rule but Scripture; no Law but Right Reason: for the rest, allways Conformable to the present setlement, without any sort of singularity. The Mornings after his devotions,[19] he constantly spent in Philosophic Studys, & his Laboratorie, sometimes extending them to Night; but he told me, he had quite given-over reading by Candle, which Impair'd his sight: This was supply'd by his Amanuensis, who sometimes Read to him, and wrote-out such passages as he noted, and that so often in Loose Papers, pack'd-up without Method, as made him sometimes to seeke upon occasion, as himselfe confesses in divers of his Works; And indeede, his very Bed-Chamber, was so extreamely crowded with Boxes, Glasses, Potts, Chymical & Mathematical Instruments; Bookes & Bundles of Papers; that there was but just roome for a few Chaires; so as his whole furniture was very Philosophical, without formality: There were yet other Roomes, and a small Library (and so you know had *Descartes*[d]) as learning more from Men, Real Experiments, &[20] his Laboratory (which was ample and well furnish'd), than from Books.

In his diet, (as in Habite) he was extreamely[21] Temperate & plaine; nor could I ever discerne in him the least Passion, transport or censoriousnesse, whatever Discourse, or the Times suggested; all was tranquill, easy,

[a] For Cork's bequest to Boyle see his will, printed in Townsend, *Great Earl of Cork*, Appendix III; however, no troop of horse is there mentioned. Cork's Septipartite Agreement, ibid., p. 469, gave Boyle feudal responsibility for ten horsemen.
[b] Elizabeth Carey (1631–76), daughter of the 2nd Earl of Monmouth. Evelyn's suggested link with *Seraphic Love* cannot otherwise be substantiated. See above, p. lxxvii.
[c] The Letterbook version has the following marginal note at this point: *who confesses he had a Bastard daughter. Se M:* Baillet in Vita *Des[cartes]*. Cf. Baillet, *Vie de Descartes*, vol. ii, pp. 89–90.
[d] The Letterbook version here has the following marginal note: *One at Egmond desiring to see his Library, he brought him into a roome, where he was dissecting a Calfe.* Cf. Baillet, *Vie de Descartes*, vol. ii, p. 273.

serious, discreete & profitable; so as besides Mr. Hobbs, whose hand was against every body, & admir'd nothing but his owne; & Francis Linus, (who yet with much Civility, writ against him) I do not remember he had the least Antagonist:[a] – In the Afternoones he was seldom without Company, which was sometimes so incommodious, that he now & then repaird to a private Lodging, in another quarter of the Towne,[b] and[22] at other times (as the season invited) diverted himself in the Country, amongst his noble Relations.

He was rather talle & slender of stature than otherwise; for most part, Valetudinary, pale & much Emaciated; nor un-like his picture in the R: Society;[23,c] which, with an almost impudent Importunity, was hardly extorted, ‹or rather stolln› from this modest gentleman, by Sir Edmond King; after he had refus'd it, to his neerest Relations.[24]

I have say'd nothing of his style, (improperly mention'd here[25]) which those who are Judges, thinke he was not altogether so happy in, as in ‹his› Experiments: I do not call it Affected, but doubtlesse not answerable to the rest of his greate[26] parts; and yet, to do him right, it was much Improv'd in his *Theodora* & latter writings.

In his first Addresses, being to speake, or Answer; he did sometimes a little hesitate, rather than stammer, or repeate the same word; Imputable to an Infirmity, which since my remembrance, he had exceedingly overcome; This, as it made him somewhat slow & deliberat; so after the first effort, he proceeded without the least Interruption, in his discourse: And I attribute the Imperfection much to those frequent Attacqus of Palsey, contracted I feare not a little by his frequent attendance on Chymical Operations. It has astonish'd me to [have] seene him so often Recover, when he has not ben able to moove an hand to His mouth, and indeede, the contexture of his body, [in][27] the best of his Health appeard to me so delicate; that I have often compar'd him to a Chrystal or Venice-Glasse, which tho wrought never so thinn & finely, carefully[28] set up, would outlast the harder Metal of daily use: and he was withall, as cleare, and candid; not a bleb, or spot to tarnish his Reputation; and he lasted

[a] The version of the letter published in Evelyn *Memoirs* (1818), vol. ii, p. 307, has the following footnote at this point: *Viz. Tract. de Corporum Inseparabilitate, &c. 8vo London 1661 J.E.* This is presented in a manner which matches that in which marginal notes from the Letterbook version are presented earlier in this printed text; however, this note does not appear in the Letterbook. One must therefore conclude either that Bray was working from yet another copy of the letter which is now lost, or that he accidentally attributed to Evelyn a note which was in fact his own. Since the text is otherwise virtually identical to that in the Letterbook, and since the texts of the other letters from Evelyn which appear in this edition all derive thence, I consider the latter of these two possibilities the likelier.
[b] On Boyle's house at St Michael, Crooked Lane, see Maddison, *Life*, p. 275.
[c] On the Kerseboom portrait to which Evelyn evidently here refers see Maddison, 'Portraiture', pp. 163–4; on the links of Sir Edmund King (1629–1709) with this painting, ibid., pp. 164, 169.

accordingly, tho' not to a greate, yet a competent Age; Threescore Yeares I thinke, & to many more he might, I am perswaded,[29] have arriv'd; had not his beloved Sister, the Lady Ranalagh (with whom he lived, a person of extraordinary Talents, & suitable to his Religious & Philosophical Temper, dyed before him: But it was then that he began[30] to droope apace, nor did he (I think) survive her above a fortnight: But of this last scene, I / fol. 59v/can say little, [being] unfortunately out of Towne, & not hearing of his danger, 'til 'twas past recale [sic].

His Funeral[31] was decent, and tho' without the least of pomp; yet accompanied with a greate appearance of Persons of the best & noblest quality, besides his owne numerous Relations:

He Lies Buried (neere his Sister) in the Chancel of St. Martins-Church; The Lord Bishop of Salisbery preaching his Funeral Sermon, with that Eloquence natural to him on such & all other Occasions: The Sermon you know is Printed, with the Panegyric, so justly due to his Memory: Whether he has yet[32] any other Monument erected on him, I do not know, nor is it material, His Name (like that of Joseph Scaliger[a]) were alone a Glorious Epitaph.

_____[33]

And now, Sir, I am againe to Implore your Pardon, for giving[34] this Interruption, with things so Confus'dly huddl'd-up, this very Afternoone, as they crowded into my thoughts: The subject you see is fruitfull, & almost inexhaustible; Argument fit for no mans pen, but Mr. Wottons, Oblige then ‹all› the World, and with it,

> Your most humble servant
> J. Evelyn

I have got so very greate a Cold & hoarsenesse, by stepping out when I was hot, two days since; that I feare I must suspend my journey to London for a day or two longer.

Mr Bently writes me word, that he is made one of the Kings Chaplains, & with seeming Regret, requiring so much ‹of› his attendance at Kensington, & Interrupting his Sweete Retirement. I tell him, while men seeke Honor, they loose Liberty.[b]

‹this blurr'd Rhapsody, & other defects,[35] I have not patience to write it out againe [?] & I believe you'l have lesse to Reade it once alone.

[a] J. J. Scaliger (1540–1609), the well-known classical scholar.
[b] Richard Bentley (1662–1742), the great textual scholar, became chaplain in ordinary to the king in 1695.
[c] Preceded by three or four words obliterated by repair to manuscript.

DOCUMENT 6B

John Evelyn's letter to William Wotton
12 September 1703

Add. Ms 4229, fols 56–7. Holograph. This letter was first published in William Bray's edition of Evelyn, *Memoirs* (1827), vol. iv, pp. 403–16, from the version of it kept by Evelyn which is now Add. MS 28104, fols 21–2. In John Forster's 1850–2 edition of Evelyn's *Diary and Correspondence*, vol. iii, pp. 390–8, extra footnotes were introduced recording places where the Add. MS 28104 text published by Bray differed from that in Add. MS 4229 from which the text published here is taken. Trivial differences between the two versions have not here been tabulated, and neither have passages which appear in this version but not in the copy in Add. MS 28104 (referred to in the notes as 'Evelyn's copy'): these can be readily identified by comparing this text with Forster's edition of 1850–2. However, for the sake of completeness, substantial passages which appear only in the other version and not this one have here been noted in endnotes. For the background to this letter, see above, pp. xli–iv.

Wotton X September, 1703[36]

Worthy Sir,

I had long ere this given you an Account of yours of the 13th past (which yet came not to me 'til the 20th) If a Copy of the Inscription you mention, & which I had long since among my Papers, could have ben found, upon diligent search; but lost or mislaid, I believe (with other loose notes) upon my remove *Cum pannis*[a] from Deptford, whence they were brought: To supply *This*, it is now above 12 days,[37] that I sent to Dr Stanhop (Vicar of Deptford[b]) to send me a Fresh Transcript: But hearing nothing from him 'til this day (& that dated 4 days since) I believed my Letter might not come to his hands, & then a Servant of mine (who looks after my small concerns in that place) told me, that the Doctor was at Tunbridg drinking the Waters, &, as I conjectur'd,[38] my Letter might lie-dormant at his house (as it seemes it did) expecting his returne: Now altho I have bin almost all

---

[a] *Bag and baggage.*
[b] George Stanhope (1660–1728), Boyle Lecturer, 1701; Dean of Canterbury from 1704.

this summer, & especialy at present, very much Indispos'd, fev'rish & febricitant, & consequently very feeble: Yet unwilling you should remaine any longer in suspense, or thinke me Negligent or Indifferent in promoting so desirable a Work: I send you this.

To the first of your Queries: Mr Hartlib[a] was I think, a *Lituanian*, who coming for Refuge hither to avoyd the Persecution in his Country; with much Industry and addresse, recommended himself to many Charitable Persons, & amongst the rest, to Mr. Boyle, by Communicating to them, many seacrets in Chymistry, & Improvements in Agriculture, & other usefull Noveltys, by his generall Correspondents [sic] abroad, of which ‹he› has publisht Severall Treatises: Besids this, he was not Unlearned, Zealous & Religious with so much Latitude, as easily recommended him to the Godly Party, then Governing, among whom (as well as Mr Boyle & others, who usd to pity him, & cherish strangers) he found no small subsistence during his Exile: I had very many Letters from him, & often Reliev'd him. *Claudius*,[b] whom next you Inquire after, was his Son-in-Law, a profest Adeptus, who by the same *Methodus Mendicandi* (& pretence of Extraordinary *Arcana*) Insinuated himselfe into the Acquaintance of his Father-in-Law: But when, or where they either of them died, or what became of them (tho' I think the poore man died of the stone) I cannot readily tell; no more than I can, who it was Innitiated Mr. Boyle among the Spagyrits, before I had the honor to know him; tho' I believe it was whilst he resided so frequently at Oxford (after his Travels abroad) where there was an Assembly of Virtuosi:[c] Dd. Bathurst of Trinity, Dikinson of Merton; Wren, now Sir Chr) [sic], Sharrock, Scarbrough, Seth Ward, afterwards Bishop of Sarum; and Especialy Dr Wilkins (since Bishop of Chester) head of Wadham College; where These, & other Ingenious Persons usd regularly to meete, to promote the study of the new philosophy, which has since obtaind:- It was in that College, where there was, I think, an Elaboratory for tryals, and other Instrument[s,] Mathematical, Mechanical &c.; which might probably be that you speake of as a Schole, and so lasted 'til the Revolution following, when (every body seeking after preferment) this society was dispers'd:- This Sir being the best Light I can at present give you having since lost so many of my Worthy Friends, who might possibly have ‹Inform'd› me[39] better.

---

[a] Samuel Hartlib (d. 1662), social reformer and intelligencer; his correspondence with Boyle survives from 1647 to 1659.

[b] Frederick Clodius, chemist and projector. Corresponds with Boyle in the 1650s and 1660s.

[c] On the Oxford experimental philosophy club see Webster, *Great Instauration*, pp. 153f., and Frank, *Harvey and the Oxford Physiologists*, esp. ch. 3. The members noted here by Evelyn are: Ralph Bathurst (1620–1704), Edmund Dickinson (1624–1707), Sir Christopher Wren (1632–1723), Robert Sharrock (1630–84), Sir Charles Scarburgh (1616–94) and Seth Ward (1617–89).

As to the Date of my first Acquaintance with this honourable Gentle-man, It sprung from a ‹courtious›[40] Visite he made me at my house at Saye-Court; Which as I constantly repayd; so it grew, reciprocal & familiar: Divers Letters passing between us at first in Civilitys; and the style peculiar to him: upon the least sense of obligation: but those Compliments lasted no longer, than ‹til› [?][41] we becum perfectly Intimate, and had discover'd our Inclinations of Cultivating, the same studys & designes, especialy in search of natural, & Use-full things: My-selfe then Intent on Collections of Notes in Order to an History of Trade, & other Mechanical furniture, which he Earnestly incited me to proceede in: So that our Inter-courses of Letters, were now mostly upon that subject, & were rather so many Receipts & processes, than matter of letters: What ‹I› gatherd of this nature (especialy for the Improvement of Gardning & Silva, Kalendarium, Acetaria &c., being but part of that Work (a plan whereof is publish'd):[a] would astonish you to see the[42] Bundles & Packetts, among other thinges in my Chartophylatium[b] here, promiscuously ranged among multituds of papers, Letters, Divine, Politicks, Law, Poetry &c. Some as old as Henry VIII,[43] left by my Father-in-Law, Sir Richard Brown,[c] whose Ancestors ‹were› in great & publiq Imployment, as well as him-selfe for 19 yeare Resident in the Court of France & whose Dispatches, of what past abroad, & we know little of here make up ‹two›[44] considerable folio Volum, beside a number of othe[r] Correspondencys — But to Returne from this digres-sion. These designes & Apparatus's,[45] growing beyond my forces, were left Imperfect, upon the Restauration also of the banish[ed] King, when every-body expected a New World, and had other prospects, than what the Melancholy, & almost despaire,[46] suggested to passe-away anxious Thoughts, by those innocent Imployments I have mentiond;[47] The Estab-lishment of the Royal Society, taking in all these Subjects, which made our personal Meeting, unless at Gresham College, where we assiduously met & convers'd) made our usual Conversation, at one another houses & lodgings less necessary. I have by me Severall Letters; but such as can be of no Importance to your noble Work, unlesse you have a mind to see, his Answer to my Letter of thanks, for his presenting me with his Seraphic-Love, which may rather passe for a Compliment.[48] That which exceedingly Afflictd me, was my Absenc from London: & in the Country, which before his decease, I came not to know of, [sic] by which I lost many Opportunitys.

---

[a] The work in question is Evelyn's *Elysium Britannicum*, which has never been published; a list of its contents was published in his *Acetaria* (1699), as Evelyn states at this point in his copy of the letter. (See endnotes for details of other significantly different passages in Evelyn's copy.) He also refers to his *Sylva* (1664), which included his *Kalendarium Hortense*. On the background in Evelyn's work on the 'history of trades', and on his correspondence with Boyle, see Hunter, 'John Evelyn in the 1650s'.

[b] A Latinate construction of Evelyn's own, evidently meaning an archive.

[c] Sir Richard Browne (1605–83), English Resident in France during the Interregnum.

I can never give you so accurate an Account of Sir William Petty, (which is another of your Inquirys), as you'l find in his owne Will, (that famous & Extraordinary piece which, I am sure cannot have escap'd you), in which he has Ommited nothing concerning his owne Original, Birth, Life & wonderful progresse he made, in so prodigious /fol. 56v/ a fortune as he has left his Relations: Or if I could say more of it, I would not deprive you of the Pleasure, you must needs receive in Reading it.[a]

The onely particular he has [?] (I find) taken little[49] notice of, is the misadventure of his double Botome; which yet perishing, in the Tempestuous Bay of Biscay (where 15 other Vessals were lost, in the same storme) ought not at all Reproch perhaps, the best & most Usefull Mechanition in the World:[b] For such was this *Faber Fortunae*: – I need not acquaint you with his Restoring a certain Criminal Wench, who having ben hanged at Oxon, & taken down for the Usual dissection in the Lent-Assizes,[50] obtaining a pardon by the Intercession of the Scholars, & Professors of Medicine (of which Mr Petty was one) was afterwards marryd, & had many Children, surviving ‹her Condemnation›[51] 15 yeares.[c] – These, amongst many other Things extraordinary made him deserv'dly famous,[52] not forgetting the easy and Expeditious Method, by which, getting the Office of generall Surveyor of the Kingdome of Ireland,[53] Instructing Ignorant Souldiers to Assist in the Admesurement, & reserving to his own share, the Acers assign'd him for his Reward, & the dispatch, which gained him the favour of the Souldiers, whose pay & Arrears,[54] was to be out of the pretended forfeited estates (‹giving›[55] him opportunity): He purchas'd their Lots & Debentures for a little ready-money: & which he got Confirm'd, after the Restauration, tho' probably not without Acknowledgements to the Greate men, who were as greedy of Mony as others.[56,d]

I need not tell you of his Comparative & Political Arithmetic, & the Computations going under the name of Mr. Graunt,[57] who (tho' but a Bodice-maker) was an Excellent Accountant, & knew much of the Bills of Mortality, of which it seemes he had a great Collection.[e]

Sir William, was with all this[58] politely learned, a Wit, & a Poet (se his

[a] For the will of Sir William Petty (1623–87), adventurer, inventor and pioneer of political economy, see Fitzmaurice, *Life of Petty*, pp. 318–24. The whole of this passage closely parallels Evelyn's account of Petty in his *Diary*, vol. iv, pp. 56–61.

[b] On Petty's catamaran-style ship, see Fitzmaurice, *Life*, pp. 109f., Lansdowne, *The Double Bottom*.

[c] On the case of Ann Green, see Lansdowne, *The Petty Papers*, vol. ii, pp. 159–67.

[d] On Petty's comprehensive and effective land survey of Ireland for the Cromwellian regime see Fitzmaurice, *Life*, ch. 2, Larcom, *The Down Survey*.

[e] Petty wrote a number of books on what he called political arithmetic. These are listed in Keynes, *Bibliography of Petty*, who also examines the issue of the extent to which Petty might have assisted his friend, John Graunt (1620–74), in his *Natural and Political Observations upon the Bills of Mortality* (1662) (ibid., pp. 75–7).

Paraphrase on the 104 Psalm,[a] & severall other Loose pieces) in a word, nothing was too hard for ‹him› which renderd him the most Charming, & Instructing Conversation in the World.

But all these Excellent Talents of his, rather hindred, than Advancd his Reputation at Court; where some wretched favourits (whom for their worth, one would not have set with the dogs of the flock)[59] afraid of his Abillities & wit, stopt his progresse there; nor indeed ‹did› he neede or affect it, being by my long observation & acquaintance with him, a man of sincerity &c.[60]

I mentiond his Poetry; but said nothing of his Preaching; which tho' rarely, & when he was in perfect humor, to diver[t] his Friends: he (as the terme is) ‹would› hould forth, in Tone, Action & unintellible [sic] Nonsense; Passe from the Court Pulpit[,] ‹Learned› distinctions, ‹various›[61] readings & Inference to the Presbyters doctrines, reason & uses; & fall into ‹the› Independent, Familist, ‹Quaker› ‹Anabaptist›[62] Enthusiastic way; & thence slide into the Jesuite, Friars, Capuchins (whom I hold to be as errant, Enthusiasts) transports; Putting-on the Person of those Sectarys, in such Variety of Postures,[63] Countenances & affected Zeale, as entertained the Company to their admiration.

This coming to be told the King, they prevaild (with no small Importunity) with him, to shew his faculty one Evening at Court, when declaiming upon the[64] Vices of it, & miscarriages of the Greate ons & favorits, so warmly, & almost personaly; Particularly in the Miss-government of Ireland, &c., as (tho' it exceedingly diverted the King, who bore Raillery the best in the World) so touch'd the Duke of Ormond,[b] then present, as making him very un-easy, our Preacher, observing it, alterd[65] his style, into a Calmnesse, and Composure as charm'd the King exceedingly: – One instance more (which possibly you may not have heard of) was his Answring & accepting of a Challenge, Sir Allin Brodrick (a greate favorite of the Lord Chancellor Hyde)[c] sent him: The Appelant you know, by the Law of Armes yeilding the Choice of the List, field of Battel, & Weapon: Sir William, for the first Chose a Dark Cellar, & for the second Hatchets: being himselfe pur-blind, and not so skillful at the Rapier, having never learnd at the Fencing schole: And so the Challeng was turn'd into Raillery, & Sir Allin was Laught-at

But after all this, This[66] Wonderfull man, an excellent Physician also (& professor of it) was sudenly taken away, by a Gangrene in one of his legs (too long neglected, a few Days after we had din'd together in cherefull Company.

---

[a] See Fitzmaurice, *Life*, p. 172.

[b] James Butler, 1st Duke of Ormonde (1610–88), Lord Lieutenant of Ireland, 1662–9, 1677–85.

[c] The Earl of Clarendon. Sir Alan Broderick (c. 1623–80) was Surveyor General of Ireland. On this quarrel see Evelyn, *Diary*, vol. iv, p. 59n.

He Chose for his Coat-Armour (& which he caus'd to be depicted on his Coach) a Mariners Compass, the Needle pointing to the Polar-star, and for his Crest, a Bee-hive, the *Lemma*[a] (if I remember[67] well) *Operosa et sedula*:[b] than which nothing could be more apposite.

And now Sir I am extremely sensible of my detaining you so long, in giving you rather the History of Sir William Pettys Life, instead of Answering a short Query concerning his lady: She was Widow of Sir *Maurice Felton* (not *Fenton*), a Norfolk family, of which was *Felton* who assacinated the famous Duke of Buckingham)[c] Daughter of that Arch-Rebell, Sir Hardresse Waller:[d] whom for her Beauty, Wit &, I suppose Mony, Sir William Maried, & had 3 or 4 /fol. 57/ Children, to whom he left ample fortunes, & to the Eldest the Title of Baron: this Lady is still living in Dublin, a very stately Dame, in one of the statliest Palaces of that city. This I have from my Daughter in Law, who was well aquainted with her, whilst ‹my› unhapy Son (one of the Commissioners of the Revenue of the Kingdome) lived there:[e]

Asking you pardon againe, for these Impertinent Abberrations: I Returne to Mr. Boyle: who had (besides all those we have enumerated, that were his Acquaintance & Admirers) The Lord Viscount Brounker, first President of the Royal Society[,] That worthy person & honest Scot, Sir Robert Morray Secretary of that Kingdome; the famous Sir Kenhelm Digby, Dr. Goddard, Milington, Needam, Sharrock, Lower, Lister, Wallis, severall of them yet Living;[f] & of later date, Dr. Burnet, the present Bishop of Sarum: Indeed generally all Strangers & Learned Persons, Pretending to Chymistry,[68] the Corpuscularian Philosophy, & whatsoever was Curious, Usefull & Uncommon: Besides his Correspondent [sic] abroad in Holland, Paris, Florence, Rome: Nor did any Ambassador[69] Person of ‹Quality› or Learning; Travellers come over to see this Kingdom, thinke he had seen any-thing, 'til they had Visited Mr. Boyle.

As to the Affinity & Relation of my Wifes Family to Mr. Boyle Take the following Account, as she Receiv'd it from that most Religious & Excellent

---

[a] *Motto.*

[b] *Active and diligent.* On Petty's coat of arms see Keynes, *Bibliography of Petty*, p. 85 (where the motto is given as *Ut apes Geometria*).

[c] John Felton (1595?-1628).

[d] (1604?-1666?), Regicide. Elizabeth, Lady Petty, née Waller, was created Baroness Shelburne in 1688; she died c. 1708–10. In fact, her first husband was Sir Maurice *Fenton*, not *Felton*.

[e] John Evelyn junior (1655–99) married Martha Spencer (c. 1661–1726) in 1680: Evelyn, *Diary*, vol. iv, pp. 194–5.

[f] In addition to Brouncker (c. 1620–84), Moray (1608–73), and Digby (1603–65) Evelyn refers to the following doctors, many of them early Fellows of the Royal Society: Jonathan Goddard (1616–75), Sir Thomas Millington (1628–1704), probably Walter Needham (1631?-91), Richard Sharrock (1630–84), Richard Lower (1631–91), Martin Lister 1639–1712) and John Wallis (1616–1703: a doctor of divinity rather than medicine).

Lady, his Neipce, late Countesse of Clancarty:[a] Who coming downe one day, to Visite my Father-in-Law Sir Richard Brown, lying Incommoded with the Gout, & sitting by his Bed-side: Upon some accidental Discourse of her Family, & how they allways esteem'd & lookd upon them as Kindred: Related this Pretty Passage of a[70] kindsman of Sir Richards mothers first first [sic] Husband, whose name was [Sir][71] Geofry Fenton;[b] that neglecting his study (design'd for a Lawyer) so exceedingly displeas'd his Unkle; that he sent him into Ireland, as an abandoned Youth, to seek his fortune there: The young Man, Considering his Condition, soone Recover'd his Unkle's favour; by so dilligently applying himselfe to that study: that in short time he became one of the most eminent of that Profession. Now the first Earle of Cork, being then but Mr. Boyle (a Kentish-man, & perhaps as I may have told you, a School-Master at Maidstone[,] but this particular being nothing of the Countesses narrative, & a seacret 'twixt you & my selfe onely, & perhaps Uncertaine) coming to Advise with Sir Geoffrey Fenton, now knighted) finding him Ingag'd with another Client & seeing a pretty Child in the Nurses Armes Entertain'd himselfe with them, 'til Sir Geoffrey came to him, making his Excuse for causing him to waite[72] so long: Mr. Boyle pleasantly told him, he had ben Courting a Young Lady for his Wife: And so it fortun'd, that sixteene-yeares after'd, Mr. Boyle made his Addresses in good Earnest, and Married the Young Lady, from whom is sprung all this Numerous Family, of Earles & Lords, branching now into the noblest & most Illustrious of the King-dom: – How many sons & daughters, he left, I do not remember; Onely that Roger Boyle was Eldest Son, whom his Father sent young into Eng-land, to be under the Care of his Relations Grand-mother at Deptford, where was then a famous Schole for young Gentlemen of the best quality, it being neere the Court at Greenewich, where both King Edward VI & Queen Elizabeth were borne & delighted often to be:[c] – Thus, Sir, have you the Original of the Relation, & Compellation of that you Inquir'd after;[73]

This Roger Boyle is the Young Gentleman, who Dying at Says-Court (then Sir Richards & since my house in Deptford) was buried in the Parish-Church, with the Inscriptions I here Inclose, For I find, there was a *Fenton* interrd in the same place: –[d] I will now then Indeavor to commute for your Patience, with a pleasant story, Current among the Boyles. When King

[a] See above, p. 85.
[b] Sir Geoffrey Fenton (d. 1608), Irish adventurer and officeholder.
[c] Sir Richard Browne's mother, Thomasine Gonson, was formerly married to Edward Fenton, (d. 1603), sea captain and explorer. Nothing further appears to be known about the Elizabethan school at Deptford mentioned here: see Dews, *History of Deptford*, p. 177. In the next sentence, a *compellation* is a style of address (OED).
[d] See previous note and below, pp. 100–2.

Charles IId, newly come to his Crowne, ‹then›[74] among other diversions, frequently, usd to saile downe the River in his *Yachts*, accompanyd by all the Greate-men & Courtiers, waiting upon him: It was observ'd, that when ever the Vessel, pass'd by a certaine place, opposite to the Church, at Deptford, My late Lord Burlington, constantly pull-off his hat, with some kind of Reverence. This being remark'd, by some of the Lords, standing by him on the Deck: They one day desir'd he would tell them, what he Meant by it? To which he Replyd, Do you see that steeple there? Have I not reason to pay Respect to that Place, where my Elder-Brother Lies Buried, by which I Injoy the Earledome of Corke? /fol. 57v[75]/

When I speake of the Family, perhaps, it were not amisse to see what Sir W. Dugdale says of it, in his Baronage:[a] Tho' what the Heraulds write, who prepare the *Diploma*,[76] are sufficiently Mercenary, & able to bring any Upstart, as far as Sir Thomas Urchart, dos his:[77,b]

You will Wonder at the Blurrs, Mistakes & untoward Character, in the midle part ‹& all ensuing› of this hasty scrible, 'til you shall understand, that a fit of a quartan surprizing me (of which I have had two within a moneth) causd my hand so to tremble,[78] that I could no longer hold my pen: & my Grand-son[c] being not with me, I had no Amanuensis to write what I could onely dictate. So as begining my Letter not before Friday noone, when I receivd the paper from Dr Stanhop, I was not able to finish it 'til this Afternoone (Saturday) my Indisposition ending in such a sweat, as detaind me in Bed 'til past 2 a Clock.

Me-thinks you speake, of your not being at London, 'til next spring: A long Day for *Octogenarius* to hope for that hapynesse, who of late have seene so few moments I can call so. A greate part of the yeare past, my health has much declind, nor do I murmur, considering that ‹I› hardly had occasion to keepe my Bed in above 60 yeares: But that which brought upon me my present Suffering, was from so slight an Accident, as nothing could be lesse apprehended: Walking one day to take the aire in Mr Wises Garden at Brompton Parke neere Chelsy,[d] my foote sliping, I fell upon a stump, which scraz'd one of my Shin bones; onely to make it bleede, I got into the house & bathed it with a little Hungary-water, 'til I came home; where by the perswasion of an eminent Chirurgion, (who being my God-Son, I bound Apprentise to an Excellent Artist, now 40 yeeres since)[79,e] I sufferd him to lay a plaister, which he assurd, would heale it in a day or

---

[a] Sir William Dugdale (1605–86), *The Baronage of England* (1675–6).

[b] Sir Thomas Urquhart (1611–60) claimed his ancestry from Adam: see his Παντοχρονοχανον; *or a Peculiar Promptuary of Time* (1652) and *The Jewel*, pp. 55–6.

[c] John Evelyn, first baronet (1682–1763).

[d] Henry Wise (1653–1738) was partner with George London in the nursery at Brompton Park.

[e] Nothing is known about this figure, who does not appear in Evelyn's *Diary*.

two (as a piece of paper & faire water would have don, for my flesh never rankles). But this plaster instead of healing brought down a sharp humor, which for all the fomentation prescribed, kept me within neere 3 Moneths. But deliverd from this defluxion...[a] its Effects tormented me with the Haemoroids, if so I may call, blind piles which do not bleede, & are often troublesom to me. I have at this age certwise [?] let blood this yeare, & now, as I tell you, have intermitting Parosysms which make me very weake in Body, & Minde.

My young Grandson Improves laudably in his studys, both Laws, History, Chronology, & practical Mathematics, & went as far as Algebra: 'Tis pitty he has not a Correspondent that might provoke him, to write Latin Epistles, in which, I am told by some able Judges (that have seene some of them) he has a grave & masterly style: Nor dos he forget his Greeke, having read from Herodotus, Thucydides, Zenophon, ‹Polybius›,[80] &c.: He having from an Infant ben with a French Master has that Tonge perfect, & has now conquerd Italian. I do not much Incourage his poetry, in which he has yet a prety Vaine: My desire being to render him an honest Usefull man, of which I have greate hope, being so grave and steady, & vertuously Inclind: He is now gon to see Chichester & Portsmouth having already Travelld most of the Inland Countys, & went the last summer as far as the Lands End in Cornwall. – Thus Sir I make you part of [my n]earest concerns, hardly abstaining from the Boasts of men of my Dotage.

I have yet adventurd to pay my Duty to my Lord Garnsey,[b] who did me the honor to visite me Recumbent in Dover-street, when I was not able to stir, & here late[ly] at Wotton, since his coming out of Kent to Albury.

I have also payd the Visite lately received from Mr. Hare,[c] & his Lady, very glad to find them both in so perfect good health. He longs to see Mr. Wotton, and so dos, Sir Your most humble obliged Servant,

J. Evelyn

[P.S. written smaller]
Roger Boyle Memorial, plac't over the Vestry-dore, is a small Basse-relieve in Alabaster, where the young Gent. is kneeling in a praying posture, & in a Tent, behind which is a map, or prospect of Ireland: Some little compartment of carving is about it. The Inscription under it.

My Wifes most respectfull service to your Lady; & your selfe.[81]

---

[a] The MS is damaged at this point.
[b] Heneage Finch (1647?–1719), later 1st Earl of Aylesford; created Baron Guernsey in 1703.
[c] Francis Hare (1671–1740), later Bishop of Chichester.

DOCUMENT 6c

George Stanhope to Evelyn, 6 September 1703, enclosing transcript of Roger Boyle's funerary inscription

Add. MS 4229. fols 64–5; in the MS, the transcript has been bound before the covering letter. Fols 64v and 65v are both blank, except for the endorsement on fol. 65v 'Dean Stanhope's Letter' in Wotton's hand. On Stanhope see above, p. 91.

Honoured Sir.

My absence at Tunbridge Wells for some time, till the end of last week, kept back the honour of yours from me several days. Immediately, upon finding it I ordered the Inscription to be taken, but would not adventure to send it you, till yesterday's duty at[82] Deptford had given me an opportunity of examining and comparing it. The Monument hath exactly the situation and figure you describe, and was not only preserved but refreshd at the rebuilding of our Church.[a] I have inclosed the whole belonging to it, as well that of Fenton as Boyle.[b] It is a satisfaction to hear that a good hand undertakes to do right to Mr Boyle, the best cannot honour the subject more than he will be honoured by it; and it is very happy that so excellent a person as your self do still survive to contribute to so worthy a Remembrance of that great man, whose intimacy with such Friends is one shining part of his Character. His elder Brother, as you will find by the Epigram, lyes in his uncle Fenton's Tomb, which I presume is the reason of another Escutcheon graven upon the Marble ‹(or Alabaster)› opposite to that of Boyle. Winter drawing on I promise my self shortly the honour of waiting upon you in Dover Street,

---

[a] The church of St Nicholas, Deptford [Green], was rebuilt in 1697; it was severely damaged in the Second World War, when various monuments were destroyed, though the Boyle one survives: Cherry and Pevsner, *London 2: South*, pp. 416–17.

[b] See above, p. 97.

where I shall be glad to receive any farther commands, which may give me fresh occasion of approving my self

> Honoured Sir
>> Your ‹most› obliged and most humble
>> Servant

September 6. 703

> George Stanhope

I beg leave to tender all due respects to your good Lady and Family.

/Fol. 64/

Close under the Device, is this Inscription, all in Capitals.

> M:S.

H.S.E.Rogerus Boyle Richardi Comitis Corcaciensis Filius Primogenitus; Qui in Hiberniâ natus, in Cantio, solo Patris natali, Denatus, Dum Hic Ingenii Cultum Capessit, Puer Eximiae Indolis, Praecocitatem Ingenii Funere Luit Immaturo. Sic Luculenti, sed Terreni, Patrimonii Factus Exhaeres, Coelestem Crevit Haereditatem. Decessit Anno CIↃ DC.XV. IV.EID.VIIIBris.

Beneath upon the Base you have the following Words in Capitals also.

Ricardus Praenobilis Comes Corcaciencis uxoris Suae Patruo. B M P.

Memoriae Perenni Edwardi Fenton, Reginae Elizabethae olim pro Corpore Armigeri, Jano O'Neal, ac post eum Comite Desmoniae in Hiberniâ turbantibus fortissimi Taxiarchi: Qui post lustratum improbo ausu Septentrionalis Plagae Apocryphum Mare, et excussas variis Peregrinationibus inertis naturae Latebras, Anno CIↃ IↃ LXXXVIII in Celebri contra Hispanos Naumachiâ meruit, navis praetoriae Navarchus. Obiit Anno CIↃDC III.

> Cognatos Cineres et Amicam Manibus Umbram
> O Fentone Tuis excipias Tumulo.

> Usuram Tumuli Victuro Marmore pensat,
> Et reddit Gratus pro Tumulo Titulum.

Translation:

Here is laid Roger Boyle, eldest son of Richard, Earl of Cork, who was born in Ireland, and died while pursuing his studies in Kent, his father's native place. He was a boy of unusual intelligence, and paying for his too rapid proficiency in study by an early death, lost a splendid inheritance on earth to obtain a still nobler one in heaven. He died October 12th, 1615.

Richard Earl of Cork erected this monument to the uncle of his wife.

This monument is placed to the perpetual memory of Edward Fenton, formerly Esquire of the body to Queen Elizabeth, and who afterwards served with great distinction as Brigadier in the civil commotions occasioned by Shane O'Neil, and afterwards by the Earl of Desmond, in Ireland. He subsequently undertook many bold and adventurous voyages in the then unknown regions of the North Seas, where he made many valuable additions to our geographical knowledge of countries in that portion of the globe. Finally he commanded the Admiral's flagship in the famous naval engagement against the Spanish Armada. He died A.D. 1603.

> Mid kindred dust, Fenton, we lay thee down,
> Where kindred shade shall greet thy high renown.
> Not that the living marble shall set forth
> To future times a sailor's, soldier's worth;
> Recorded but the line 'Here Fenton lies,'
> Shall living marble's self immortalize.[a]

[a] From Dews, *History of Deptford*, pp. 79–81.

DOCUMENT 7

Thomas Dent's letter to William Wotton, 20 May 1699

Add. MS 4229, fols. 50–1. Endorsed by Wotton, sideways at top of fol. 50: 'Dr Dents Letter concerning Mr Boyles Life'. Dent's punctuation and paragraphing are eccentric: the breaks marked here by paragraphing are sometimes so denoted in the MS, sometimes indicated by dashes (which have not been reproduced). In addition, it is sometimes unclear whether or not Dent intended a capital letter at the beginning of a word. See above, pp. xl–xli.

Reverend Sir,   Westminster Abbey the 20th may
                                                       (1699)

I am well pleased to see, by our public advertisments, that, the world is likely to be soe soon obliged to you for the history of mr Boyles life etc; This gave me occasion to reflect on my promise to you, which I had perform'd long since, had I not waited for what (I had promised you & myself) shou'd have been perform'd on the part of my Lord Shannon, but that is now at an end, by his unexpected death;[a] he having recovered of his severe fit of the gout, & gather'd strength to that degree, that we all hoped for a continuance of his life, at least five or seven years longer, he was taken of[f] in four daies at his mansion house (call'd Shanon parke) in the County of Corke Ireland; I had a letter of the 18th of march last, taken notice of ‹by› me, to have been the most steady, & well pen'd of any, he had writ to me for a twell-month past; & now, I cannot yet hear, whether he left any memorialls, of what I had requested among his papers – but supposing, nothing was done by reason of his constant Indisposition, since he left this kingdome – As to my poor recollections, which I have here Transcribed, I know not whether they will be worth your perusall; but I have thought of all, I judged to be (in any respect) material; tho' they lie in confused order, yet you may insert them digested, as suitable to your severall paragraphs

---

[a] Francis Boyle, 1st Viscount Shannon, was buried at Youghal, 19 April 1699: Maddison, *Life*, p. 296.

My Lord Shannon & mr Boyle were both sent to travell together under the conduct of one mr Markham,[a] whom I have heard both commend, for an Ingenious, pious & Learn'd person; they lived together two or or [sic] three years at Geneva; where mr Boyle was Greatly admired, & address'd to by the greatest & most learn'd /fol. 50v/ upon the account of his great sobriety, & that nice observance which was ‹there› made of his Extraordinary piety & diligence in his studies

I have often heard my Lord Shanon say – That he was so intent upon his studies, that he wou'd seldom loose any vacant Time, but often, ‹if› on the road, as[1] they happen'd to walk down a hill, or be in rough way, he wou'd pull out a Book or two, (his pocketts being always well stored) & continue reading, as he had time & opportunity, & do the same constantly in the evening before supper; my Lord also Told me that he wou'd frequently sollicit his Governor to confer with him, & satisfy him of the meaning of any particular passage; he did not well understand in any author;

my Lord said, his constant Example in the office houre of morning & evening Devotion of which ‹when they were abroad› he never once failed[2] to his remembrance; was[3] a good memento to him at the time of rising & going to bed & that he wou'd often affectionally admonish him of his ‹Christian› duty; & was a check to all his lighter words or follies. he wou'd severely reprimand his servants, when he heard them use any rash oaths, or Indecent Expressions;

my Lord Shannon being married to mrs Killigrew (then maid of honour in king Charles the 1st court) return'd from his travells sooner, then mr Boyle, & left mr Boyle with mr Markham at Geneva; when the warrs of Ireland broke out, & my Lord of Corke was confined to Youghall, the town being closely besieged by the Irish; when there was no opportunity to remit money – which did not a little mortify mr Boyle, (as I often heard him mention) yet mr Markham was so Indeared in kindnesse to his pupil, seeing such forward Indications of his great pregnancy, & Improvement in his studies; That, notwithstanding, he had heard of the Earles death (viz. mr Boyles father) & that Ireland was in confusion, & my Lord of Corkes estate in the hands of the Rebells, so that he had nothing left in possession, but his ‹one› mansion house at Youghall; where the Earle ‹soon after› died, & was ‹there› privately Interr'd during the siege; & where, (by the by) the Earle /fol. 51/ of Orrery his ‹3d› son, & the late Lord Shannon, now lie both Interr'd, (with the countesse their mother) who soon died after mr Boyle was born; & used to mind the Earle of making provision for her poor child Robin, (as she often Express'd) & the Earle gave her assurance, (which he perform'd to a little [sic]), sweet heart, fear not I will make

[a] Sic: i.e., Isaac Marcombes.

‹our›[4] child ‹Robins› estate (if he lives) as good as any of his younger ‹brothers›,[5] – but to stop in this digression, which casually dropt in remembrance – The worthy mr Markham having been often ‹before› Governour to ‹noblemen› abroad, & gain'd good credit & esteem in all places – supplied mr Boyles expences during his Travells in expending by Bills of Credit £500 if not £600 for mr Boyle; for which he Ingaged himself by Bonds – of which mr Boyle was so Tender, & sensible ‹in honour› at his return, that he took care to satisfy that debt, out of the first product or rents of his estate; & often mention'd this with great honour & respect to the memory of his Good Governor.

he lived with great affection & Endearment to all his relations, & particularly, to the Lady Ranelagh & Lord Shannon, & had a sort of decent regard to his Elder Brother the late Earle of Burlington; whom he knew[6] to be his heir; & therefore seldom Transacted any thing in his Temporalls, without his advice & concurrence

I never mention'd any person in distresse or want, but his hand was liberall suitable to the person or occasion.[7] It was his custom every year at Christmas to order his Bayliffe to distribute five pounds in money among the poor at Stalbridge, & he wou'd often say in my hearing, do not oppresse my poor tenants by any hard rents or usage

He often told me, The main scope & design of his studies lay in reading & understanding the scriptures, which he had in the originalls so perfect, that he never failed of repeating the Hebrew & Greek with great readinesse, & suitable to the occasion or subject of discourse[.] I have known him[8] severall mornings (when I had the honour to wait upon him) entertain persons of severall nations viz – french – spaniards – Germans & English & that in different dialects, most readily with a most pleasing aire & Genious; ‹&› which was in discourse alwaies /fol. 51v/ agreable & Instructive to the whole company – what was remarkable in Experiment or occurrence, he note[d] down ‹Every day› when the company parted

he always kept a Laboratory, not only for Experiments, but [to] prepare[9] medecines for his friends, & the poor, of which he was most freely communicative; & used to attend the present, with money, if he heard any person was necessitous

I alwaies heard him Expresse his Judgement & Inclination to the Church of England – but he was for moderation to those, who dissented from us, & not to force Tender consciences – ‹for›[10] which he seem'd to expresse great aversen[ess]. He had frequent conferences on this subject with the present Archbishop[,][a] Bishop [of] Sarum – but particularly, the ‹late› Bishop of Worcester ‹the Learn'd› Dr Stillingf[leet][b] – for whose depth of

---

[a] Thomas Tenison.
[b] Edward Stillingfleet (1635–99).

Learnin[g] & solid Judgement he had alwaies the greatest Value & Esteem as he often Express'd himself to me – & I remember about a fortnight or ten daies before his death, desired me, to send that Learn'd friend to him; It was (I Guesse) upon some nice & criticall case, which did affect the Good man's mind

But the only person, with whom (both at Oxon & London [)] he chiefly confer'd, & corresponded in things of this nature was Bishop Barlow, (whom he call'd his confessor,) & had been so near twenty years, or more to his death; & as I take it, he lived at Queens colledge with him.[a]

I wish, you cou'd have a free recourse to Bishop Barlows[11] papers – in which no doubt you wou'd find many letters ‹with› Casuisticall Enquiries, very proper materialls for your historicall account; for, I took mr Boyle to have been most excellent & accurate in all points of that nature, & therefore a stroke of this will not be amisse

I wish, what I have noted may be of any use to your other Ingenious & accurat ‹adversaria or› remarks, of which I am ever Impatient to see the birth; & believe, It wou'd not be to your disservice,[12] if you honour'd me with an Inspection, before it be transmitted to the presse; for by the Advertisement I perceive you are in great forwardnesse[.] I heartily pray for your health; & wish you a good ‹Event›[13] of successe, suitable to the merit of your excellent Genius, & that Elaborate study whereby you will oblige the ‹learnd›[14] world; & in it, the microcosm of

Dear W[otton]

Yours with service
& affection
T.D.

I must needs say, you have been unkind in your correspondence of late – but believe you to be a person of that sincerity which you professe

[a] On Boyle's place of abode at Oxford, see Maddison, *Life*, p. 88.

DOCUMENT 8

James Kirkwood's letter to William Wotton, 22 June 1702

Royal Society Boyle Letters 3, fols 108ᵛ-9. Ibid., 3, fol. 120ᵛ, is endorsed by Wotton: 'Mr Kirkwood's Papers concerning Mr Boyle', and it is clear that this letter accompanied transcripts of various documents, now ibid, 3, fols 108 and 118–20, and 5, fols 119–20, together with others which no longer survive, presumably because they were returned to Kirkwood. Kirkwood's copies, which were published by Birch in an appendix to his *Life*, *Works*, vol. i, pp. clxxxviii–cciv, will be included in the Pickering Masters edition of Boyle's correspondence, in conjunction with the original letters concerning the affair to which they relate. The letter published here has been endorsed at its head by Birch: 'No. 1. Mr Kirkwood to Dr Wotton'; it was printed by Birch in *Works*, vol. i, pp. clxxxviii–ix, without the list at the end. It is written in a scribal hand, like the material that it accompanied; the first insertion (only) is in Kirkwood's own hand. See also above, pp. xli–ii.

Reverend Sir, June the 22, 1702

I am glad you are going to publish the Life of the Excellent Mr Boyl who was so great an Ornament to his Country & to our Holy Religion: I Reckon it one of the[1] Blessings of my Life to have been aquainted with so extraordinary a Person whose company I found always ‹very› Delightfull & Edifying: It was soon after our acquaintaince was begun, That I had the Opertunaty of talking with him of the sad State of Religion in the Highlands of Scotland where they had Neither Bibles nor Catechism in their own Language.[2]

This gave him an Occasion to tell me of his having caused Five hundred Bibles in Irish In a Quarto Volume in the Irish Character to be printed for the use of those in Ireland who understood not the English. He then offerd a few of those Bibles to be sent Into Scotland to see what Reception they might meet[3] with there[:] a Dozen of them was first sent and Afterwards two Hundred which made one for Each Parish. In such places where these Books were Distributed They had a very good Effect as you may perceive more fully By the papers Relating to that Affair which I send you: after some time I heard from some Ministers in those Parts that It was the Earnest Desire of many who wished well to our Religion to have A new Impression of the Bible in the Latine Character In a Small Volume for the use of such persons in the Highland Parishes as had been taught to Read English tho they did not Understand it and for the Advantage of such Children as should be sent to school, Especially those of the Poorer sort who could not purchase Books for them selves: To answer The pious Desires of such persons, Indeavours were used in Scotland to procure another Impression but in this attempt we mett not with success: the first Encouragement that was given me to goe on[4] with it in this Kingdom was by the worthy Mr Boyl, who told me he would subscribe for printing One Hundred Bibles. This Example Disposed others whom I acquainted with the Design, to subscribe very freely & Largly; most of the subscriptions you will see in the Printed Paper I send you Relating to that affair:[a] I need not mention Other Particulars, only in short after some few years this work was happily finished the Books was Printed, transmitted into Scotland, and Long agoe the greatest part of them was Dispersed In the severall Countys of the Highlands: wee have had many Accounts of the happy Effects which have attended our Endeavours In behalf of thos poor People /fol. 109/ Who[5] have been so Long Neglected & suffered to remaine In a state of Ignorance, and Barbarity. By the printed paper you will Likewise see that there ware six thousand[b] Catechisms and prayerbooks in Irish printed at the only Charge of Mr Boyle for the use of the Highlanders which Were Accordingly sent Down into Scotland many of which have been[6] likewise Dispersed among the Highlanders.

The Catechism and prayers was composed by Mr Charteris,[c] & were translated into Irish by Sir Hugh Campbell of Caddell[d] and afterwards Revised and Corrected by Mr Kirk;[7,e] There were added to the Catechism some Passages of Scripture Containing the principal heads of the Christian Religion; to serve As a short and playin [sic] Tract of Devotion & Christian Morality: It was this Catechism and Prayers that are mentioned in the

---

[a] Kirkwood refers to this as printed, and a unique copy of the printed version survives at Christ Church, Oxford (classmark 133.Z.252): *An Account of the Design of Printing about 3000 Bibles in Irish, with the Psalms of David in Metre, for the Use of the Highlanders* (n.d.). In addition, however, a manuscript text of the *Account* accompanied Kirkwood's letter (see No. 1 below), which differs from the printed version, not least in that the names of the subscribers other than Boyle, to whom he here refers, are omitted. Possibly Kirkwood sent both printed and manuscript texts.

[b] Though Kirkwood here states that 6000 were printed, the *Account* has 3000 in both its MS and printed versions (*Works*, vol. i, p. cxc; above, n. a).

[c] Lawrence Charteris (1625–1700), Professor of Divinity at Edinburgh.

[d] Or Cawdor, of which Campbell was Laird. Campbell also saw the volume through the press in London. See Johnston, 'Notices of a Collection of MSS', pp. 5, 9.

[e] Robert Kirk (1644-92) was a Gaelic scholar and minister of Aberfoyle, Perthshire.

Bishop of Rosses Letter.[a] As for the papers I send you which have some Relation to that great man, I Leave it to you to make use of them, or any Part of them in such sort as you Judge may best serve your Design: When you have done with them bee Pleased to return them to me again, To be Left for me att Mr Milbourns a Watchmaker at the blew boar In the Old Baylie Near Ludgate London. As for the Letter from[8] Mr Charteris, I send it you, that you may see the High Opinion that Pious, and Primitive Person had of the famous Mr Boyle. As for Mr Boyls own Letter the Cheif reason why I send it is upon the Account of some few seasonable Expressions in it about[9] Education, which may be very well Improv'd in some part, or Other of his Life.[b]

As for the Letter from Mr Kirk a Highland Minister (who was Corrector afterward of the Press, when the Irish Bible was Reprinted) Perhaps it will furnish you with some Hints not unusefull to your Design. I need say no more as to this matter[.] I heartily wish you may have good success In what[10] you are about, that you may be able to set forth so great a Pattern, & Example in such a Light as ‹may› move others to have a higher Regard for solid Piety, & usefull Learning. Before I conclude this Letter I must acquaint you with another Design which has been[11] set on foot In Behalf of the Ministers, schoolmasters, and Probationers In the Highlands: the Reasons of this Design you will see In the Printed paper I send you: they who are now In the Government there have soe far Espoused it, as to Recommend it to the severall Presbyteries to promote it. I know your good affection to Religion & Learning will Dispose you to Incourage a Work of this Tendency, which Is Likely to prove of ‹very› great Use & advantage:/ fol. 109v/ Your Neighbour Mr Frank[c] can Acquaint you with Some[12] farther Particulars Relating to this Affair.

I am,
    Reverend Sir,
        your afectionate Brother,[d]
          & Humble Servant,

                              Ja: Kirkwood

The Papers I send you are as Follows. Viz
No 1   Proposals for printing the Irish Bibles[e]

[a] James Ramsay (1624?-96). For his letter see *Works*, vol. i, pp. cciii–iv.
[b] On Boyle's letter, see above, p. xlii.
[c] Thomas Frank, Rector of Cranfield, Beds.; Archdeacon of Bedford in 1704; d. 1731. An active correspondent of the Society for Promoting Christian Knowledge: Bahlman, *Moral Revolution*, p. 77.
[d] Kirkwood had until January 1702 been a fellow minister of Wotton's as Rector of Astwick, Beds., not far from Wotton's living at Milton Keynes.
[e] BL 5, fols 119–20, printed in *Works*, vol. i, pp. cxc–i. See above, p. 108 n. a.

2　An Answer to the Objections against Printing them[a]
3　Mr Kirks Letter[b]
4　My Letter to Bishop of Ross ‹Decemb. 15 1687›[c]
5　A Letter from the Bishop of Ross[13,d]
6　A Letter from the Clergy of the Diocese of Ross ‹Oct. 15 1688›[e]
7　A Letter from Mr Boyl ‹Octob. 18 1690›[f]
8　A Letter from Mr Spalding Clerk of the General Assembly to Mr Boyl ‹June 28 1690›[g]
9　A Letter from Mr Charteris[h]

Be pleased to Let me know by a Letter Your Receit of these Papers.

[a] BL 3, fols 118–19; printed in *Works*, vol. i, pp. cxci–iii. The printed version of this, a broadside entitled *An Answer to the Objection against Printing the Bible in Irish*, is reproduced in facsimile in Johnston, 'Notices of a Collection of MSS': on the differences between this and the text from Kirkwood's MS printed by Birch, see ibid., pp. 13–14.
[b] No longer extant. This item is marked with a cross in the MS.
[c] BL 3, fols 108–9, printed in *Works*, vol. i, pp. cxcv–vi. The date is added in Birch's hand.
[d] To Kirkwood, September 1690; BL 3, fols 119ᵛ–20, printed in *Works*, vol. i, pp. cciii–iv.
[e] BL 5, fols 119–20, printed in *Works*, vol. i, p. cxcix. The date is again added in Birch's hand; it should be 16 October.
[f] BL 5, fol. 119, printed in *Works*, vol. i, p. cci. The date is again in Birch's hand.
[g] BL 5, fol. 118, printed in *Works*, vol. i, p. cc. The date is again added by Birch.
[h] This entry is again marked with a cross; the item in question is no longer extant.

DOCUMENT 9

Chapter from William Wotton's life of Boyle

Add. MS 4229, fols 1–32. Holograph text by William Wotton, including his characteristic (ab)use of apostrophes. Endorsed on fol. 32^v: 'An Account of Mr Boyles 1st Experiments' and 'No 140', written sideways on the right hand side of top half of page in a hand which is not Wotton's but could be that of William Clarke (see Introduction, sects 6–7). On fol. 1 is the note '4229 111E', relating to the classification of the volume as part of the British Museum collection. The versos are blank throughout, except for fol. 32^v. The text is apparently incomplete: the material on fol. 32^v seems to be an addition keyed to a text that formerly faced it: this presumably dealt with one of the experiments following no. xxxv, Wotton's description of which ends the recto of fol. 32 (this is probably complete, since it lacks a catch-word: cf. fol. 24). The original foliation appears in the inner margin at the top of each leaf. See further above, pp. li–iv.

An Account of Mr Boile's first Physico-Mechanical Experiments concerning the weight & Spring of the Air: of other Experiments concerning the Relation between Air & Flame: of other Experiments concerning Bodies shining in the dark; of several Experiments of the Academy del Cimento & Mr Hauksbee upon these subjects; of Otto Guerick's Magdeburgic Instrument[1]; & of Mr Pascals Tract of the[2] the Weight of the Air.[a]

Having now resolved to form his Notions of natural Bodies, according as upon Trial he should discover their Quality's, He considerd that Air was one of the first which he ought to examine; it being in itself one of the noblest of all material substances, & to us certainly the most necessary. For there is scarce any other known Body, without which we cannot for some time live: tho the absence of some things which we daily stand in

---

[a] Wotton here announces his intention to intersperse his account of Boyle's *New Experiments Physico-Mechanical, Touching the Spring of the Air and its Effects* (1660) with details both of related material published by Boyle in *Philosophical Transactions* in 1668 and in his *Tracts, Containing New Experiments, Touching the Relation betwixt Flame and Air* (1672), and of the experiments of various contemporaries and successors. The items by Boyle will be found in *Works*, vol. i, pp. 1f., and vol. iii, pp. 157f., 562f.; since Wotton's account follows them closely, it has not seemed necessary to tabulate this in detail. The works by the other authors will be referred to piecemeal below.

need of, may if it be long, be very grievous. We can live without Light, &
Water for some Time: Cold must be almost beyond imagination intense, if
it destroy's us in a few Minutes: It is but at some Times, that we call for
Meat & Drink; we can subsist many Hour's without Sleep: & other
Things are rather Conveniency's than necessary's of Life. But few even the
strongest Men can live without Air seven Minutes: & this too is not
peculiar to us, for all Brutes, Birds, Fishes, Vegetables & Fishes [sic] share
proportionably in the same Condition. Nothing that lives at all in this
Terraqueous Globe can live without Air, so that that truly ought to be
esteemed as the perpetual supporter of Animal Life. Its good or bad
Temperature usually gives & destroy's Health: Its constant presence, &
powerful pressure influences all living Bodies, & even very many times
when its Agency is not suspected. So that a thorough Knowlege of the
nature & Quality's of the Air, goes a great way[3] in all the Branches of
natural History.

By *Air* here is meant[4] that thin, fluid, transparent, compressible &
dilatable Body, in which we breath, & wherin we move: which encloses
this Globe on all sides to a great Height above the highest Mountains, tho
different from that thinner Body which is in the intermundan or inter-
planetary Spaces, in that it refracts the Ray's of Light, which are constantly
omitted by the Sun & Stars, & Planetary Bodies:

The Air seem's to contain three sorts of Bodies. I. An /fol. 2/ innumerable Variety of Particulars that perpetually exhale from the Earth, the
waters, & the Bodies in them contained. II. More subtil Exhalations from
the heavenly Bodies, to which may be added magnetical Effluvia from the
Earth. III. Elastical or Springy Particles, which are properly aëreal; &
which distinguish the Air from any other known fluid. The other Particles
it has sometimes in greater, & sometimes in lesser Quantity's, so that they
rather swim in it, than are its constituent parts. But these it never is
without, & the Effects proper to this their Elastical Nature, they[5] alway's
exert in whatsoever Position or Place they can be put. So that these may
truly be called the constant & permanent Ingredients of the Air, & the
other its temporary & transient ones.[a]

The two most remarkable Quality's of the *Air* which this last Age has
discoverd, are its *Weight*, & its *Elasticity*. By *Weight* is meant its actual
Gravitation towards the Center of the Earth, such indeed as all material
Bodies[6] whatsoever within the Atmosphere of this Earth actually have. By
virtue of this a Bladder filled with Air, is heavier in a Ballance, than the
same Bladder when 'tis empty. This property of the Air, to which the
Ancients were in a manner wholly strangers, was discovered in our Fathers
memory by the great *Galileo*, who perceiving that Water would not rise by

---

[a] For the sources of this whole passage see above, pp. li–ii.

pumping in any Tube above 34 or 35 Foot. Thence he concluded that the *Fuga Vacui* of the Ancients did not cause those Effects, which had bin ascribed to it. For if Water rose 34 Foot for Fear of a *Vacuum*, why would it not rise 36 Foot, & why not more? Something then (as he judged) must be a counterballance in this Case, which could be nothing but the Weight of the Air, that kept the water suspended, at one height, when it would not in another.[a] But Tubes of 34 Foot, being extreamly difficult to manage, & without something of that sort it not being easy to compute the mutual weight of Air & Water, his Scholar *Evangelista Torricellius* try'd whether Mercury would stand at any determinate height in a Tube, whose lower End was immersed in the same Metal. Upon Trial he found that without much regard to the greatness or smallness of the Diameter of the Tube, Mercury in a Tube would stand at about 29 Inches from the stagnant Mercury in which it was immersed.[b] Here then was a manageable Ballance, by which the weight of a cylindrical Column /fol. 3/ of Air, whose Base was a Circle of any given Diameter, might be certainly found.

By *Elasticity* or The *Power of the Spring* is meant such a Power as causes Bodies which by compression are moved out of the Place they formerly filled, to return into it agen. Such a Power, Mr Boyle having upon making several Experiments judged to be in the Air, he resolved upon contriving an Engine by which the Air might be Exhausted out of a given Space, which if it could be done, would certainly decide both Questions of the positive Gravity, & Spring of the Air at once; & so thereby at least two of its most remarkable Properties would be discovered.

Something had already bin done this way, which gave him some Light how to proceed farther. For some few Years before one *Otto Guerick* a Consul of Magdebourg, a Man not of much Learning, but who had a Head excellently well turned to Mechanical Disquisitions, had made an Engine, by which the Air might be in some measure exhausted out of a Receiver that before was full.[c] The contrivance of Otto Guerick's Engine lay in plunging a Glass-Vessel under Water in such a manner that the Air might be sucked out of its Mouth. For whenever the Suction was intermitted, the Air would rush with great force into the Orifice of the emty'd Vessel, or violently force the Water up in its stead. But this Engine was useless as to what Mr Boyle designed it upon two Accounts. For ‹first›[7] It continually required the labor of two strong Men for several Hours to empty it. And then[8] the Receiver or Glass-Vessel that was to be empty'd was not so

---

[a] Wotton's wording is here similar to that of Perier in his preface to Pascal, *Physical Treatises*, pp. xv–xvi, except that Perier gives the height as 32–3 feet. Cf. Galileo, *Dialogues concerning Two New Sciences*, pp. 16–17, where the height is given in cubits.

[b] On this famous experiment by Torricelli (1608–47) see DSB, vol. xiii, pp. 438–9. See also Pascal, *Physical Treatises*, pp. 163f.

[c] This experiment by Otto von Guericke (1602–86) was first described by Kaspar Schott in his *Mechanica hydraulico-pneumatica* (1657).

contrived as to put in, & let out any thing into ‹or out of› it with any Ease. And yet without that, the influence of Air upon other Bodies could never be discovered. These Defects made Guericks Engine serviceable for but one single purpose, which was to demonstrate the positive weight of the Air, to prove which alone it was at first contrived.

This did by no means answer Mr Boyle's purpose. He desired to find out all the properties of the Air that could be discoverable; to search into its Nature in itself, with its several Effects upon terrestrial Bodies: to observe how long, or how little while, any Kinds of living Creatures could subsist either absolutely without Air, or with any given Quantity: what different Phænomena it would exhibit in different seasons of the Weather, or Times of the Year: whether it could be generated /fol. 4/ or destroyed, or whether the same Quantity remained (as far as we can judg) immutably the same from the first Creation till now. To make such Trials as would be satisfactory in Enquiry's of this sort, it was absolutely necessary to find out way's whereby Bodies might be kept without Air either wholly or in part for long Periods of Time, whereby certain Portions of Air might be entirely at command, & in short ‹whereby› it might be let into, or let out of Vessels, in which other Body's living or inanimate were inclosed at pleasure: Nothing of this could be done in the Magdeburgic Engine, & all this & a great deal more was done by Mr Boyle, tho not at first by Engines which he procured to be made for such different purposes as he proposed to execute.

The contrivance of Mr B's first *Air-Pump* (as it has since bin commonly called) with which his first *Physico-Mechanical Experiments* were made, & by which the world was first taught what improvements in natural Knowlege a right Theory of the Air would produce, was owing to Dr Robert Hook. The first Essay towards framing such an engine as would suck Air out of a Receiver was made by Mr *Gratrix* but what he did was to[o] gross to perform any great matter, & for that reason laid aside.[a] Afterwards about the year 1658, Dr Hook who was then at Christ-Church, whilest Mr B. lived in Oxford, contrived & perfected this Air-Pump, & gave thereby a noble Specimen of that wonderful skill in Mechanics, for which he was afterwards so deservedly famous. This Engine I shall now describe, because the Experiments which Mr B. made with it, will by that means be better understood, & the force of those Conclusions which were drawn from those Experiments more plainly perceived.

The Engine itself consisted of two parts, a *Receiver*, & a *Pump* to draw the Air out of that Receiver.[b]

[a] On the instrument-maker, Ralph Greatorex (1625–1712), and his efforts see Taylor, *Mathematical Practitioners*, p. 229, and Maddison, *Life*, pp. 92–3.
[b] See facing plate: the illustration of Boyle's first air-pump which appeared in his *New Experiments Physico-Mechanical, Touching the Spring of the Air and its Effects* (1660), whence Wotton's description of it derived.

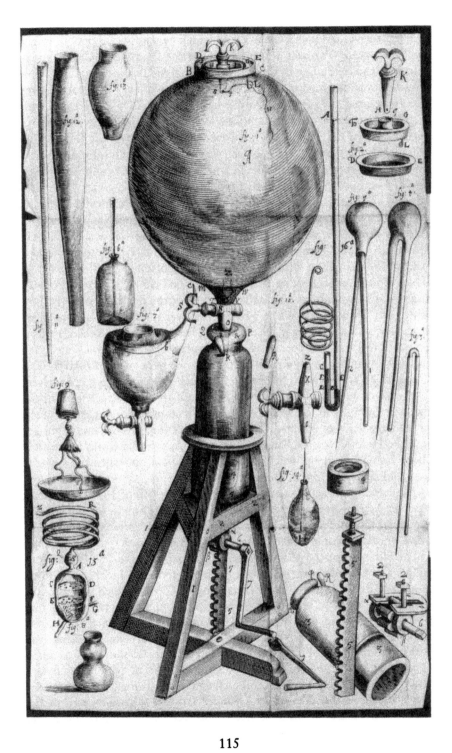

115

The Receiver (A) was a Glass Vessel which held about 60 pound weight of Water; at the Top was a round Hole, whose Diameter (BC) was about 4 Inches; round the Edg of this Hole was a Lip of Glass near an Inch high. This Lip was designed to hold a Cover in this manner.[9] A Brazen Cover (DE) was fitted as exactly as possible upon the Lip (BC) & cemented with a strong Cement: it was made hollow & tapering with a Cavity about 3 Inches over at the Top. Into this Cover was fitted a stopple (FG) which stopple & the Cavity of the Cover (DE) were carefully ground to one another, that they might touch each other in as many parts as possibly they could,[10] within the Cover. In the middle of the stopple was a Hole (HI) about ¼ an Inch over, bound about with a Ring or Socket of Brass, into which a Handle made like the Key of /fol. 5/ a Stopcock (K) was exactly fitted by grinding it with Care into the Hole (HI) which it was to fill. This Handle (K) would turn round with Ease in the Hole (HI) & neither let the Air into or out of the Receiver. It had at the bottom a small Hole (8) to which if need were a String (8,9,10,) might be fastned, which String was run thro a Ring (L) that was fastned to the bottom of the Stopple (FG) by which any Body that was put into the Receiver might be moved at plea-sure, without unstopping it. The last thing belonging to the Receiver was a Stopcock (N) ‹to› the upper ‹shank›[11] of which (X) was soderd a Tin-Cover (MTUW) which was long enough to cover the neck of the Receiver. The Use of this Cover was to fasten the Stopcock by cement to the Receiver. That was done thus. The Cavity of the Tin Plate was filled with Cement made of Pitch, Rosin & Wood-Ashes, well incorporated, & then the neck of the Receiver well warmed was put into it: & that the Cement might get into the Hole (Z) of the Shank (X) it was stopt with a Cork, that had a string fastned to it, by which the Cork was taken out when the Cement was cold at the Bottom of the Receiver.

The *Pump* itself was fixed in a woodden Frame (III) of 3 Legs, so placed that one Leg might stand perpendicular upon the Ground for the Ease of the Operator. In the middle of the Frame was placed a cross Board (222) upon which the Pump rested, & to which it was fastned.

The *Pump* consisted of four parts, a *Cylinder*, a *Sucker*, a *Handle*, & a *Valve*. The *Cylinder* was made of cast Brass; about 14 Inches long, thick enough to be very strong, notwithstanding its Cavity, which was about 3 Inches Diameter, & as exactly Cylindrical as could be bored. To this *Cylinder* was fitted a *Sucker* (4455) consisting of two parts. One (44) somewhat less in Diameter than the Cavity of the Cylinder. This was covered with a strong piece of tanned Neats-Leather, so that if it was taken at any time out *of the Cylinder*, it could not be put in agen unless it was rammed in with great force. This prevented the insinuating of any Air between that & the sides of the *Cylinder*. To this cylindrical part of the *Sucker* was fastned a thick & narrow ‹plate›[12] of iron (55) something longer than the *Cylinder*, smooth on one Edg, & indented on the other. To fit these Teeth, an Iron Nut (αβ) was fastned by two staples, to the underside of the Cross-Board (22) on which the Cylinder rests, which was[13] turned to & fro by a *Handle* (77) like that with which Jacks or Clocks are usually wound up.

The last part of the Pump was the *Valve*, (K) which consisted of a Hole bored tapering thro the Cylinder, & a brass Plug ground to that Hole, that it might be thrust in & taken out at pleasure.

The Engine thus framed & fitted together, was set to work in the following manner. Some Sallad Oil was put into the Top of the Receiver to fill up the Cavity's between the inner surfaces of the Stopcock, as likewise /fol. 6/ to help the Key to move every way with Ease. The Cylinder likewise was[14] well moistned with Oil in the inside to assist the Sucker in its Motion, & to prevent the Air from getting in between that & the Cylinder; for the same reason the Valve was likewise oiled. If Oil would not sufficiently lubricate the Cylinder (as sometimes it would not) Oil & Water mixed together would do it effectually. Then the Cover of the Receiver & its Stopple were put into the Ring at Top, & carefully plaisterd with the Cement which was spread round the Edges with a hot Iron, by which means all the little cavity's & cranny's were so filled up that no Air could get in between them.

Then the lower shank (O) of the Stopcock was put into the Cylinder at the Hole (&c) to which it was exactly ground. The Operator then forced the Sucker[15] into the Cylinder by turning the Handle of the Nut upwards, & so forced the Air contained in it out thro the Valve. When the Valve was shut with its Plug agen, he turned the Sucker the other way. By this Operation the Cylinder was empty'd of its Air, since what was in before was driven out thro the Valve, & there was no other passage for fresh Air to come in. If then the Key of the Stopcock was so turned, as that thro its Cavity the Air could pass between the Receiver & the Cylinder, part of the Air enclosed within the Receiver would pass with great swiftness into the Cylinder. And then if the Key of the Stopcock were turned agen, the Air in the Cylinder could not get up into the Receiver, but must by the Sucker be thrust out thro the Valve. Thus by repetition of the motion of the Sucker upwards & downwards, & by turning the Key, & stopping the Valve as occasion required, more or less Air might be drawn out of the Receiver, according to the design of the Experiment, & the Intention of him that made it.

This is the famous Engine, which from Mr Boyle who was the first Director, at whose Instance it was first made, has bin usually called Mr *Boile's Air-Pump*. By it he demonstrably proved the Spring as well as weight of the Air, by which Discovery more Light has bin given to the understanding the greatest & noblest Operations of Nature in this lower world, ‹then› by any single Discovery whatsoever that has bin made by any Philosopher in our Times. What the Experiments were which he first communicated to the world that were made by this Engine, & what the Inferences were which he drew from those Experiments we are now to enquire.

Experiment I.

Mr Boile begins with those Phænomena which the working of the Pump constantly produce. I. Upon drawing down the Sucker, if the Valve be shut, the Cylinder is void of Air. If then the Key be turned so, that[16] the Air can get out of the Receiver, it rushes with a violent Noise /fol. 7/ if the Receiver be full into the Cylinder, till the Air in Receiver & Cylinder be brought to an equal measure of dilatation. II. Upon shutting the Receiver by turning back the Key of the Stopcock, after the first Exsuction,[17] one may perceive by turning up the Sucker, that the Cylinder was in a manner full of Air; Afterwards ‹upon› every Exsuction less Air & less will go every Time out of the Receiver into the Cylinder, & consequently the remaining particles of Air will not press so much upon one another. So that when the Receiver is almost exhausted, the Sucker may be forced almost to the Top of the Cylinder, without unstopping the Valve to let in the Air: & then, if whilest the Valve is shut the Handle of the Pump is let go, the Sucker will be forcibly carried up to the Top of the Cylinder by the Protrusion of the outward Air: which will be a certain mark to know how far the Receiver is at any Time empty'd.

These Phænomena convinced Mr Boyle, that there was ‹an›[18] actual Spring in the particles of the Air: For here was no external force to drive the Air out of the Receiver into the Cylinder, & yet as soon as the Cylinder was exhausted & the Key turned, the Air rushed into it, & was driven with a manifest force thro the Valve into the external Air; & this still from time to time, till at last so little remained in the Receiver & Cylinder, that the Sucker would (spontaneously to outward appe‹a›rance) run up to the Top, & there stay.

To explain this he supposed that the Air consisted of Heaps of[19] very little Bodies, which might be compared to Flocks of Wooll, lying one upon another, & piled up, with different degree's of condensation, to the Top of the Atmosphere. Now, as Wooll consists of small flexible Hairs, which when prest by a Hand, & much more by an Engine be squeezed into a very narrow compass, & at the same endeavour to dilate itself to its former expansion, which Endeavor may sensibly be perceived: So the lower Air in which we breath being prest by what is above it quite round the Globe, will when freed (as in the Receiver) from the pressure expand itself into as large a room as the Nature of its innate spring will allow.

This indeed supposes the actual Weight of the Air, since where there is pressure, there[20] must be weight. To put that therefore beyond dispute, Mr Boile weighed a dry Lambs-Bladder which held near ⅔ parts of a Pint, when full of Air, & found it to lose 1⅛ gr[ains] of its former weight when it was empty. And the Magdeburghers found that their Glass Vessel which held 32 measures, was lighter when the Air was exhausted by 1¹⁄₁₀ oz. then it was before. And Mr Pascals Experiment /fol. 8/ which was made in the Year 1648 with the Torricellian Tube at the Foot, middle & Top of the *Puy de Domme*[21] (a high Mountain in Auvergne) in which the Mercury subsided 3 Inches lower at the Top of the Hill, than at the bottom, demonstrated that the shorter Column of Air at the Top of a very high Mountain could not counterpoize so high a Column of Mercury, as the larger one which was at the bottom of the same Mountain.[a]

Experiment II. III.

III. Another Phænomenon which this Pump constantly exhibits is this: When the Receiver is well exhausted it is a difficult thing to lift up the stopple of the Brass Cover; & when but a moderate Quantity of Air be drawn out, after it is lifted up a little, so as to be somewhat loose from the sides of the Socket, it will be so difficultly taken quite out, that tho there were onely a Bladder ty'd to it, one would imagine there were some great weight ty'd to the bottom of it. The reason of which Phænomenon is this, That the Air in the Receiver being almost exhausted, the Spring of what remains which is exceedingly dilated is so far weakned, that it can but faintly press the lower End of the Stopple, & consequently the Hand of him that lifts the Stopple, must support a weight equal to that Disproportion by which the external Air presses more than that within. That pressure is in this Case ‹much›[22] greater than one unused to these Things would be apt to imagine.

That this is the true Reason of this Phænomenon is plain, because the Pressure upon the Stopple lessens when Air is let into the Receiver, & when the Receiver is full, it may be lifted up without difficulty. Now wheras this supposes that the little Quantity of Air enclosed within the Receiver, makes an Æquilibrium[23] with the whole Incumbent Atmosphere without, which seems unlikely: Yet it necessarily follow's from the Nature of Elasticity. For if (for example) a Lock of Wooll be comprest in a Mans Hand, & then put into a Box, where it cannot reexpand itself, all those little springs that compose that Lock, will continue as strongly bent as they

---

[a] This famous experiment by Blaise Pascal (1623–62) is noted by Boyle in *Works*, vol. i, p. 14. See DSB, vol. x, pp. 333–4, and Pascal, *Physical Treatises*, p. xviii and pp. 103f.

were in the Hand. So in like manner, the Air inclosed in a Receiver was before ‹its inclosure› bent by the whole weight of the Atmosphere, &[24] is now shut up in a Glass where it cannot dilate ‹itself› any more than it could ‹in its seeming free Estate›[25] while the Atmosphere ‹onely› prest upon it: Consequently then it must now press as violently against the lower part of the Stopple by virtue of its Spring or[26] Restituent force, as that does upon the upper, when it is prest down /fol. 9/ by the incumbent Atmosphere. IV. From the same Cause it alway's happens that when the Sucker is drawn to the bottom of the Cylinder, & the Valve & Stopcock are shut, it will readily ascend of itself when the handle is let go since the Spring of the outward Air finds nothing to resist it, when it forces up the Sucker. This will equally happen if the Communication between the Receiver & the Cylinder is free, as if it is not, provided the Receiver be exhausted. And in drawing down the Sucker, there will be no more Difficulty to draw it down when it is near the Bottom, of the Cylinder, then when it was at the Top. These Phenomena constantly occurr during the Operation of this Pneumatic Engine, & when well considered will go a great way to determin in general both the weight & Spring of the Air.

Experiment IV.

Having now examind the Appearances which his Engine by itself exhibited, Mr Boile went to try various Experiments relating to the Operation of Air upon Bodies of all sorts in these Receivers. These he continued making at Times from the Year 1658, in which his Air-Pump was first brought to perfection, till some Years before his Death. The first Volume[a] of his *Physico-Mechanical Experiments* came out in 1660. Three years after that Mr Pascals Discourses *of the Æquilibrium of Liquors & the Weight of the Air* were published in French by his Brother in Law Mr Perier, which, as the Editor ‹says›, were written ten years before.[a] Mr Pascals Doctrine shall be carefully considered, & compared with Mr Boile's, when I shall have described this first set of Experiments, & given some Account of the Inferences which he drew from those Experiments, which were all made at Oxford before the Restoration of King Charles II.

Hereby we shall see how much is originally owing to Mr Boile in this matter; & then let the progresses which other Men have made be never so great, as indeed they have bin very surprizing, yet if Mr Boile first led the

[a] See Pascal, *Physical Treatises*, pp. x, 1f. Although in the next sentence Wotton promises a careful comparison of Pascal's doctrine with Boyle's, this does not appear in the extant MS.

way, part of that Glory which they are justly entitled to, will of right belong to him. This I chuse to observe here, before I begin to give an Account of these Experiments, because the Excellent Mr Hauksbee whose Discoverys of this sort are the latest we have, do's in the Preface to his *Physico-Mechanical Experiments* printed in 1709, own that *the principal subject of his Book, is an Account of great & further Improvements of that noble Machine the Air-Pump, & of many new Experiments made thereby, the Invention of which most usefull Instrument he ascribes to Mr Boile.*[a]

The first Experiment made upon any Body besides the Pump itself, which /fol. 10/ Mr Boile try'd, was upon a large, weldryed limber[b] Lambs Bladder, not above half full of Air, & ty'd very closely at the neck. This Bladder being put into the Receiver began to swell after 2 or 3 Exsuctions, & when the Receiver was quite empty, it ‹was› perfectly distended, as if it had bin blown up by a Quill. When Air was let in, it grew flaccid as it was at first. As often as this was repeated, so often did the same Phænomena occurr. Now that there might be no doubt whence these Phænomena proceeded, two other Bladders, one not ty'd at the neck, the other squeezed together, that no Air might lurk in its wrinkles & then ty'd at the neck, were put into the Receiver along with the first: & wheras the Bladder which had Air in it, swelled proportionably as the Receiver was emptyd, the other two which had none, lay wrinkled as they were put into the Receiver without any remarkable Alteration. Another Bladder also was ty'd at the middle, & the Neck left open, & then as the Receiver empty'd the lower part in which the Air was included swelled, the upper continuing as it was before.

Experiment V.

A dry Bladder well ty'd, & moderately full of Air, was hung up in the Receiver by the string above-described. Upon exhausting the Receiver before it was near empty, the Bladder broke with a fierce Noise, & with so crooked a Rent, that one would judg it was made with Hands. The same Effect will be produced, if a Bladder so orderd be held for some time against a Fire, but not near enough to burn the Bladder. (*N.B.* To make these Experiments succeed, Mr Boile advises, that the Bladders may be ty'd when they are brought from the Butchers, before the wrinkles dry.)

[a] Cf. Hauksbee, *Physico-Mechanical Experiments*, sig. a1ᵛ. On Francis Hauksbee the elder (c. 1666–1713), see above, p. xlv.
[b] Supple (OED).

Experiment VI.

The first Trials which Mr Boile made to calculate the Dilatation of a given parcel of Air were these. First he took a Lambs Bladder thoroughly wetted to prevent the lurking of any Air within its Folds, & ty'd it about the neck of a small Glass which[28] would hold 5 full Drachms of Water, & put them both into a Receiver. Upon exhausting the Receiver, the Air within the Glass filled the Bladder 'till it was blown up to its full extent. Then the outward Air being let in, it grew flaccid as before. It was then taken out of the Receiver, & a Hole made at the Top, but without severing it from the Glass, at which Water enough was poured in to fill both Glass & Bladder. That Water weighed 5 oz. & 5½ Drachms. It is plain therefore by this Experiment, that Air is dilatable to at least nine Times its former Dimensions.

He then took a Cylindrical Glass of ¼ Inch Diameter in the Bore /fol. 11/ & about 7 Inches long, open at one end, & sealed hermetically at the other, at which End was a Glass-Bubble which was to contain the Air whose Dilatation was to be measured. The Tube was then covered on one side ‹for a good way› with a long piece of Parchment which was pasted on, & marked with ‹26› Divisions in such a manner, that the Air ‹which was then› inclosed in the Bubble would not in its natural state, fill more of the Tube than was marked by one Division. This Tube so marked was filled up to the Bubble with water, & the whole including the Bubble, might be divided into 31 parts ‹such as the 26 were before›[29]. When it was filled, it was put into a Receiver, & after some Exsuctions the Air in the Bubble drove all the Water out of the Tube into the Vessel in which it was immersed, & consequently it was dilated full 31 Times more than it was in its former State. He then ‹tryd the Experiment with another›[30] Bubble not above half so big, & the Air included within that drove down all the Water that was within the Tube, & so was dilated to above 60 Times its former Room. And by another Trial he found that Air was capable of taking up 152 Times the room it took in its Atmospherical, & to outward appearance uncompressed State.

Experiment VII. VIII.

It may sometimes be useful to know what Figure will best resist the Airs pressure. To resolve that Mr Boile contrived the following Experiments. He had a round Glass Bubble blown at a Lamp, holding about 5 oz: of Water, with a slender Neck about the bigness of a Swans-Quill. A Receiver was then moderately emptyd, & taken out of the Pump, & the neck of the

Glass was applied to the Hole of the bottom-shank of the stop-cock, & they were carefully cemented together. When the Key was afterwards turned, the Air in the Bubble was let into the Receiver & then both were alike empty. Now[31] tho the Bubble was as thin as Paper, it did not break, but resisted the whole weight of the ambient Atmosphere. That this Resistance was owing to tbe Figure appears from hence, because he afterwards applied such a *Glass-Head* as Chymists use in Distillations instead of a Receiver to the Pump. For then, having sealed the Nose hermetically, & made a Hole at the Top, he fitted a stop-cock to that, & then turning the *Head* upsidedown, he fitted to the wide Orifice a Cover of Lead cast exactly[32] which was carefully cemented to it. The other shank of the Stop-cock being then fitted to the Pump, the Exsuction of the Air was endeavourd. In a little time there was a Crack almost half round the Glass that bent inwards, which Crack was made with a great Noise; & as the pumping was continued the Crack ran farther /fol. 12/ tho the Glass was in that place judged to be near twenty times as thick as the Bubble.

Experiment IX.

But since the sealing Glasses hermetically rarify's the contained Air, & consequently weakens its Spring, Mr Boile was willing to prosecute this Enquiry farther other ways. He took then a small Glass Viol, & filled it almost full of Water, into which he put a Glass Pipe open at both Ends, & put it into a Receiver in such a manner, that the Top of the Pipe came thro the Cover of the Receiver. Upon the first Exsuction, before the Sucker was drawn down to the bottom of the Cylinder, the Viol burst with a violent Noise, & with so great a force that it crackt the Receiver. And when the Experiment was try'd a second Time, with a Viol enclosed in a Bladder, to save the Receiver, the leaden cover not being exactly fitted to the Ring of the Receiver, the external Air prest in the Cover so forcibly that it split the Receiver tho new. And trying it a 3d time with a Cover exactly fitted, the Sucker being drawn down quick, the Viol was broken into above a hundred pieces, & the Bladder was torn in several places.

Lest this should be attributed to Natures abhorrence of a Vacuum, the Experiment was tryed with stronger Viols, & tho the weight & pressure of the Atmosphere was the same than that it had bin before, yet no Exhaustion of the Air out of the Receiver would break the Viols.

Here Mr Boile subjoins a Receipt of ‹the›[33] Cement by which he mended his crackt Receivers, & by which Glasses that may be crackt in distilling the most subtle Chymical Liquors may for that Time be made useful. It is onely Quick Lime finely powderd, mixt with scrapings of Cheese & Water into a Past, which should be spread upon a piece of Linnen Cloth, & put upon the crack of the Vessel before it hardens, & then covered with a Diachylon-Plaister.

Experiment X.

Mr Boile try'd afterwards several Experiments upon Fire & Flame in an exhausted Receiver, that he might find what Influence the Air had upon their Duration. First he put a Tallow-Candle of eight to the pound into a Receiver & hung so that the Flame might burn about the middle of the Vessel. The Flame went out upon pumping in half a minute. Afterwards before he pumpt he freed the Receiver of Smoke by blowing with Bellow's, & thereby clearing the Receiver & the Flame went out with pumping in two Minutes. At first the Flame contracted itself in all its Dimensions. Then it appeared exceeding blew, except at Top, & still went from the Tallow, till it got all up to the Top before it went out. It lasted much longer when the Receiver was full of Air, & as it expired it went from the Tallow, tho slower then it did before. When it went out of it self, & the Receiver was unexhausted /fol. 13/ the Wick would remain kindled, & would emit a Smoke which ascended to the Top in a slender Cylinder. Wheras when it was extinguished by exsuction, it sometimes emitted no Smoke, & at other Times very little, which[34] instead of ascending upwards, unless perhaps a very little way, usually fell down. If Wax Candles were used & the Flame was large, tho the lower Stopcock & Valve were left open, yet for want of Air sufficient to cherish the Flame, it could go out before every thing could be got ready for pumping. If a small Taper of white wax were put in, tho that emits much less Smoke than Yellow, yet the Flame would not last upon pumping above[35] a minute; wheras if the Receiver was kept full of Air, & carefully cemented, the same Taper would keep its Flame above five Minutes.

Experiment XI. XII.[36]

Live Wood-Coals enclosed in a Wire Screw contrived on purpose[37] to hold them safe, & yet leave them every way accessible to the Air, were then put into the Receiver. The Coals lay in a heap 5 Inches high, & they were thoroughly lighted: the Wire hung pendulously in the Receiver, that by the agitation the Fire might be longer preserved. At the first Exsuction the Fire grew dimmer, & in two minutes it wholly disappeared. When the Coals were immediately taken out it appeared agen; & when they were thoroughly lighted they were put into the Receiver & the Pump was not used [?] The Fire was then visible four Minutes. In the open Air, the Coals retained the Fire, when they were hung up by a String in the Wire above half an Hour.

A piece of Iron of the bigness of a middle sized Charcoal, red hot, was put into the Receiver within the Wire. But that sending out few steams the change upon pumping was not so manifest. The redness continued visible 4 minutes, & when the Air was let in agen no change appeared. The Receiver grew very hot at last, tho the Air had bin considerably exhausted.

A lighted Match was put in afterwards. Its steams filled the Receiver so much, that by the Time they could get ready to pump, nothing was visible. Upon pumping Smoak & Air went out together, & the Fire in the Match seemed very languid, & in a little Time neither Light nor Smoke could be discerned. When the outward Air was let in, that Fire tho to view extinguisht some time before revived, & emitted Smoak as formerly.

Experiment xiii.

Mr Boile was willing afterwards to see whether Smoak in the Receiver would /fol. 14/ hinder the Dilatation of Air inclosed in a Bladder, which was in the Receiver at the same Time; as also whether want of Air, or the pressure of its own Fumes would extinguish the Fire of a lighted Match. He put[38] then together with the Match, a Bladder which would hold about 12 Pints of Air into the Receiver; The Bladder was well ty'd at the Neck, & contained not much above a Pint of Air within it. Upon exsuction the Match burnt more dimly, tho the Fumes actually went out with the Air upon nimble pumping. When upon darkning the Room the Fire seemed to be quite out, if the Air were readmitted the Fire revived, & the steams quickly filled the Receiver, & then a 2d Time if by nimble pumping the steam was carried away, in less than 8 minutes the Fire went quite out, & could not by letting in fresh be again kindled.

These remarkable Phænomena were also then exhibited. When the Receiver was full of Smoke, if the Cylinder were empty'd, immediately upon turning the Stopcock the Receiver appeared manifestly darker to any one that looked into it; & yet afterwards when there were much fewer fumes in the Receiver the darkness seemed less.This alteration was instantaneous, & seemed to proceed from the change of situation in the Exhalations by the vent suddenly afforded them. The Receiver also when full of Steams exhibited a kind of *Halo* that appeared a good while about the Fire. It was[39] observable that if the Receiver into which the lighted Match ‹was›[40] put, was very small, the Smoak would choak the Fire in a Minutes Time; but when Air was let in thro the Key, after once or twice pumping, it would blow the Fire, & rekindle it, & upon pumping away the Air, the ‹Match›[41] would (as formerly) leave smoking. But these Fumes never sensibly hindred the Air in the Bladder from dilating itself; for before the Match was quite out, the Bladder appeared to be swelled to six or seven times its former Capacity.

Experiment xiv.

He try'd next to produce Fire in this new Vacuum. A Pistol might be let off in it without sensibly weakning the Spring of its Lock, & upon the Collision of the Flint & Steel, there would be ‹to appearance› the same sort, &[42] the same Number of Sparks in this highly rarify'd Medium as in the open Air. It was difficult to make the Gunpowder take Fire, because the Pistol hung almost perpendicular in the Receiver, & by that posture the Powder was shaken out of the Pan before the Piece went off. Once however it succeeded, & the kindled Gunpowder seemed to make a more expanded Flame, than it would have made in the open Air, & mounted upwards as usually. When it went out, the Receiver was full of Smoak, which if Fresh Air were let in, would circulate much faster than before. In a *small* Receiver he set combustible matter on Fire, with a burning-Glass in a bright Day, but not in a *large* one, where the Glass was less transparent than the other;[43] /Fol. 15/

This Experiment of producing Sparks by the collision of Flint & Steel in *Vacuo*, has bin much farther carried by Mr Hauksbee ( ) in whose Hands every Thing of this Nature receives Improvement.[a] He therefore having invented a way to produce a brisk attrition between those two Bodies in a Receiver found, That if no Air were exhausted Sparks might be produced in great plenty: That if some[44] Air were withdrawn, the Sparks were neither so numerous, nor so bright & lively as they were before: That still as more & more Air was drawn out of the Receiver the change in the Sparks so produced would be more manifest: That when the Receiver was well empty'd, tho the attrition were more violent than it was at first, yet the Sparks wholly disappeared, & in their stead succeeded a faint, continued, little streak of Light, visible on the Edg of the Flint that was rubbed with the Steel: & That as the Air was let in the Sparks agen appeared numerous & vivid according to the Quantity of admitted Air.

This show's as Mr Hauksbee rightly concludes, *that the Airs presence is absolutely necessary to that vigorous expansive Motion of the Air wherin the Nature of Culinary Fire consists.*[b]

The *Academy del Cimento* founded at Florence by Leopold Prince of Tuscany, made several Experiments about Fire and Flame in the *Torricellian*

---

[a] See Hauksbee, *Physico-Mechanical Experiments*, pp. 21f.
[b] Ibid., p. 23.

Vacuum, & they found that inflammable Bodies would take Fire in it, by Ray's reflected from a concave speculum, but then the Smoak in stead of ascended [sic] as it commonly does, descended like the spout [?] of a Fountain in a *Parabola*, ‹&› when the Air was let in it rose immediately to the Top of the Ball in which it rolled. ( ).ᵃ

But here it may be justly doubted whether the *Mercurial Vacuum* which those *Florentine Academians* used, be so exactly free from Air as Mr Boile's Receiver. They say indeed ( ) that a Vessel cannot be so perfectly empty'd that way as by Mercury: But as Mr Boile observes ( ) the Torricellian Vacuum as 'tis made all at once, so 'tis made once for all, & therefore if there be any aërial particles lurking in the ‹Mercury›,[45] (as there will be a good many, if Mercury enough be used to make a large Vacuum) they ‹will› ascend into the deserted space, & there by their Expansion spoil the Experiment. Besides Experiments made there can be made onely upon dry Bodies, of which scarce any will stick to Mercury except Metals, & then great numbers of aërial particles will be intercepted between the surfaces of those Bodies & the Mercury to which they will not adhere, so that these will join to those that will emerge from the Mercury. Wheras in the Air-Pump the Air is empty'd by degree's; & the Suctions may be repeated as often as the Experimenter pleaseth, so that if any Air emergeth out of the included Body, as many times a great deal will, it may afterwards be drawn out. So that tho these excellent Florentine Academians (as Mr Boile very truly calls them) may perhaps have /fol. 16/ prosecuted their Trials in their *Torricellian Vacuum*, (which for their Countrymans sake they seem to have bin more than ordinarily fond of) further than any other Virtuosi ever did; yet some of them owned to Mr Boile that they could never make a Glass-Bubble sealed up with Air in it burst in their Vacuum, tho he had shewn them that it might be easily broken in his Receiver.ᵇ

About 12 Years after these *Physico-Mechanical Experiments* were published, Mr Boile put out an account of some new Experiments concerning the *Relation between Flame & Air*, most of which were made in his Pneumatic Engine, for which reason I shall mention them here, since they will naturally give Light to one another. He sent them[46] to Mr Oldenburgh, with a large Letter prefixed to them, in which Mr Boile takes notice of the great difficulty there is in making Experiments of that kind succeed; tho (as he frequently observes) unsuccessful Trials may sometimes afford as useful hints as successful ones, & for that reason do almost equally deserve to be remembered.

ᵃ Wotton's information is derived from *Saggi di Naturali Esperienze fatte nell'Academia del Cimento* (1667): see Middleton, *The Experimenters*, pp. 146–7. On their vacuum experiments generally, see ibid., pp. 263f., and Shapin and Schaffer, *Leviathan and the Air-Pump*, pp. 276–8.
ᵇ Wotton here quotes Boyle's *Tracts* (1672), the work which he then goes on to discuss: see *Works*, vol. iii, pp. 565–6.

The Experiments he then try'd are ranged under three Titles.
I. *The Difficulty of producing Flame without Air.*
II. *The Difficulty of preserving Flame without Air.*
III. *The extream ‹Difficulty› of propagating Flame in an exhausted Receiver.*

I. Under the first Head *of the ‹Difficulty of› producing Flame without Air* he sets down the following Experiments. I. Into an earthen Vessel nearly cylindrical was put a Cylinder of Iron, an Inch thick, & a little hollow at top, made red hot; these were put into a Receiver & when the Air was hastily pumpt out, a piece of Paper with Flowers of Brimstone in it, were let down by the turning-Key upon the Iron, which burnt the Paper, & made both it & the Sulphur smoke, but produced no Flame that could be perceived from either of them. II. Flowers of Sulphur were put into a Glass-Bubble which when the Air was exhausted out of it was laid upon burning Coals: The Sulphur instead of taking Fire rose to the opposite part of the Glass, in the Form of a fine powder: If that side were turned downwards, the Sulphur rose agen in a transparent Substance, something like yellow Vernish. III. However since by other methods ( ) he found way's to make Sulphur flame in his Receivers, he was willing to see what Influence the Air would have to revive the Flame when once it was out. And he found that when the Iron was not hot enough to make the melted Sulphur burn without Air, yet if some little while after the Flame was extinguished the Receiver were removed, the Sulphur would take Fire agen & flame as vigorously as before. But here it may be questiond whether this was wholly owing to the readmission of fresh Air, or to the removal of those Sulphureous Steams, which when the Receiver was taken off had Liberty to fly away. IV. Another Time he found that if /fol. 17/ a very little Air was let into the Receiver, that had bin very well exhausted, & the Flame quite out, the sulphur would flash with faint blew Flames, which upon two Exsuctions would disappear. This was done three times successively. V. Three or four Grains of Gunpowder were put into a Receiver, & when that was well exhausted, he try'd to kindle them with a good burning Glass, & after the collected Sun-Beams had shone upon them a good while, instead of Flaming they melted & smoaked. VI. Some very strong Gunpowder was put into a Receiver,[47] which would hold 16 [pounds] of Water, upon a red-hot Iron. There appeared onely a blew sulphureous Flame, which lasted a good while: The Paper next the Iron was burnt; most of the Grains of the Gunpowder were untouched, but seemed disposed to take Fire tho so much of the Brimstone was burnt away. Another

Time after long waiting Air was let in, when the Iron was much cooled, &
when all success from the Experiment was despaired of, the powder went
off with so much force, that tho the Receiver would hold 2 Gallons of
Liquor, & there was but one Grain of Gunpowder in it, yet the explosion
was so violent, that it endangerd the throwing the Receiver down. VII.
Some Corns of Gun powder were put into a Glass-Bubble. When it was
well exhausted & secured against the return of Air, it was set upon live-
Coals covered slightly with Ashes. The Sulphur was kindled with the Heat,
& burned blew with a large Flame, a good while in proportion to the
powder, & still the powder never took Fire, but sent up a considerable
quantity of Flowers to the Top of the Glass, which being held to a Candle,
exhibited several vivid Colors like those of the Rainbow. VIII. Having
formerly try'd to fire Gunpowder in an exhausted Receiver by discharging
a Pistol ( ) the success of which Experiment, might (as he feared) be owing
to some unheeded Air left within; he try'd it agen, & took care that his
Receiver should be empty'd with great exactness; the Event was that the
powder would not take Fire: But upon letting in a little Air, he found that
upon a 2d pulling of the Trigger, tho no new Powder was ‹put into›[48] the
Pan, save only what remained after the late Trial, that it readily took Fire
& flashed in the Pan, & yet in both Trials Fire had bin struck out by the
Collision of the Steel & Flint. IX. Some ‹Grains of› *Aurum Fulminans* was
put into a Receiver, & when it was exhausted, Mr Boile try'd to fire them
with a burning-Glass: the Ray's in some Time made them go off with
violence, & a yellowish Dust was scatterd about the sides of the Receiver.
He try'd this afterwards in the dark with a hot but not candent[a] piece of
Iron upon which he let down ⅛ gr[ain] of good *Aurum Fulminans*, which,
when the Paper in which it was wrapt was thoroughly heated, went off all
together with a luminous Flash.

II. *The Difficulty of preserving Flame without Air.*

How long Fire in Wood-Coals, lighted Matches, Candles & heated Irons
may be preserved *in Vacuo* Mr Boile had seen already. He /fol. 18/ was
willing now to try Some Mineral Flames, who ‹tho'› not easily produced
without Air, yet as he judged might be well preserved without it, the one
being certainly easier than the other. His success was as follow's. I. Flowers of Sulphur kindled in the Air, were put into the Receiver upon a thick
Plate of Metal: The pumping was performed with all possible speed, & yet
at every Exsuction the Flame visibly decayed, & before the Air was quite
withdrawn it expired. II. Next time a thicker Cylinder[49] of Iron was used,

[a] At a white heat (OED).

which was placed upon a pedestal of Tobacco pipe-clay, both to save the Receiver & to shew the Flame. The Iron Cylinder was put red-hot upon the[50] Pedestal; when the Brimstone was let down upon the Cylinder, it was covered with a convex piece of Iron like a flattish Button, which being layd red hot upon the Sulphur encreased the Heat, & by lessening the surface of the Flame, kept it from spreading too fast. The Event was, that the Heat being very intense, tho upon pumping the Flame seemed gradually to lessen, yet it continued to burn much longer than from former Trials one would have expected. III. An inflammable Solution of *Mars* was made by dissolving some filings of bright steel, perfectly free from Rust, in a sharp piercing saline Spirit. The Solution was put into a Viol which would hold 3 or 4 oz. of Water. Whilst the Menstruum was working it emitted copious stinking Fumes, which, as a Candle was brought near, would take Fire, & burn with a blewish-Green Flame at the top of the Viol. This flaming Viol was put into a Receiver, that might be empty'd at six Exsuctions: The Receiver being well cemented, at the first Exsuction the Flame flashed out, so as to seem to be 4 or 5 times as great as it was before; thus it blazed 3 times & then went out, & could not be rekindled tho Air was let in agen hastily. These Flashes plainly proceeded from the numerous Bubbles that lay in the agitated Menstruum, & which broke out with violence when once the weight of the ambient Air was diminished. IV. Spirit of Wine strongly impregnated with a prepared Metal, was poured into a Glass-Lamp & being set on Fire, was put into a Receiver that would hold about 16 lb. of Water. In that capacious Glass the Flame would not last after Exsuction much more than ½ minute. And tho Air was let in by several contrivances, on purpose to preserve it, yet the Flame would dy in very few Minutes. V. A mixture of Lead & Tin fell by chance upon a red hot Iron in an exhausted Receiver which when melted sent forth a pale blewish Flame, which lasted longer than any other Flame could ever be made to continue in that rarified Medium. This mixture not being purposely made for that End, the proportions were unheeded: So that a 2d Trial to produce Flame by a mixture of those two Metals in the open Air proved unsuccessful. VI. How easily Flame may be preserved without Air these Experiments shew. It was thought then worth enquiry whether it could be preserved in Water. That *Naphtha* /fol. 19/ and Camphire will burn in Water has bin commonly asserted. Mr Boile could never upon frequent Trials make them do so. He therefore made the following composition which would burn tho Water touched it, & that with great vehemence. He took Gunpowder [three ounces], good Charcoal [one drachm], Flowers of Brimstone neare [half a drachm], Saltpeter [near a drachm and a half]. These were mixed dry, & very fine. With this mixture he filled a goosequill cleared of the Feather, & stopt at the End with the same matter wrought into a Past with some Chymical Oil that would bring it to a consistence. Being kindled in the Air it was fastned to a weight, & let a good depth under Water, where since the Water could get in onely at the Top, the kindled matter came out with such vehemence that it beat off the water till the whole was consumed.

Here the Suspicion which Mr Boile starts seems manifest, which is that the Air inclosed in the Nitre being let loose by the Fire, had more force to preserve the Flame by beating off the Water, than the Water had by its weight to get into the Quill.

III. The extream[51] difficulty of propagating Flame in Vacuo.

I. A red hot of Cylinder [sic] was put into a Receiver[52] with some Sulfur kindled upon it; when the Receiver was well exhausted, a piece of *Spunck*, (which is a sort of *Fungus* that will when it is thoroughly dry take Fire instantly in the open Air upon the Touch of a single spark,) was let down upon the Sulphur. In most places the Flame would not touch it, & even where it was most exposed it became onely black & brittle like Tinder. II. Camphire would not take Fire, in the same place, tho the Sulphur let in along with it did when it touched the red hot Iron. One piece of Sulphur inflamed would not set another on Fire, tho put close to it in the Receiver: Nor would every sort of Sulphur flame at all, tho the Iron would melt it, & make it ready to boil. A match made with a piece of Card was once let down into the Receiver; the sulphur flamed, & the Card was hardly any where discoloured, tho carefully dipt beforehand on purpose. III. A Train of Gunpowder was laid upon the Plate on which the Receiver was placed; when the Receiver was well exhausted a Trial was made to kindle it with a burning-Glass.[53] The Receiver was chosen of the clearest Glass on purpose, because if the Receiver be thick & coloured the Experiment will certainly miscarry. Some Grains onely melted; others took Fire, but those that took Fire never communicated the Flame to any of the rest, upon which the Ray's united in the Focus of the Glass did not actually fall. IV. Some Gunpowder was put into one of those Instruments they try powder with. What was in the Touchhole took Fire with the Ray's of the burning-Glass: but no Trial could communicate the Flame to the Gunpowder in the Box; tho when the Experiment was try'd in an unexhausted Receiver, the priming in the Touch hole was readily inflamed in the Sunbeams, & when inflamed would as readily kindle the Grains in the Box. V. Gunpowder in a Glass Bubble /fol. 20/ free'd from Air rose in Flowers to the Top of the Glass, when laid near live-coals coverd with Ashes, & neither flamed nor burnt. Tho another Time the Gunpowder in two other Bubbles, from one of which the Air was in part, & from the other very carefully emptyd, when laid near the Fire, went off with a Noise like a[54] Musket, but without any Flame.

131

This last Experiment may seem to contradict the former about the propagation of Fire in the Receiver, but if we consider that the Grains ‹were›[55] probably heated all at once, & so there ‹was›[56] but contemporary explosion of the whole Mass heated together, as if so many Granado's had burst[57] in the same instance, the wonder will cease.

Besides these Mr Boile communicated other Experiments to Mr Oldenburgh made in October & December 1667 ( ) relating to this matter, & therefore since Fire & Flame are properly condensed Light, put into a violent & expansive Motion, I shall subjoin an Account of those Experiments in this place.[a] I. A piece of shining Wood was put into a middle sized Receiver, & after a very few Exsuctions the Light grew sensibly more dim, & at the 10th went quite out. II. When the outward Air was let in tho onely by Degree's, the extinguished Light revived so fast that it was like a Flash of Lightning, so that it seemed to be greater than it was at first. In a small Receiver the Light grew faint at one Exsuction, & quite disappeared by the 7th And all this was performed so quickly, that the Room could be darkned, the Pump exhausted, the Wood quenched & revived by the readmission of Air, & Candles brought in agen in Six Minutes Time. III. If the wood were left extinguisht in the exhausted Receiver, for a quarter of an Hour, it would rekindle if Air were let in agen tho less vividly than before. IV. Another luminous Substance was put into the Receiver; upon Exsuction the Light manifestly lessened; upon readmission of the Air it revived; in a short Time the Light wholly disappeared. V. The Exsuction of Air put out the Light, yet Mr Boile could not find that putting in a greater Quantity than was before in the Vessel in which the wood was enclosed made any Alteration, in the Quantity of Light. VI. Shining wood enclosed hermetically in a Glass-Pipe out of which the Air was not exhausted, would shine for several Day's (perhaps for a much longer Time) without any sensible Diminution of its Light. VII. A Red hot Iron put into a Receiver did sensibly preserve its Light notwithstanding it was pumpt with the utmost care & diligence, as long as it would have preserved it in the open Air. VIII. Shining wood hermetically inclosed in a Glass Pipe full of Air was put into a Receiver. The Light which it gave appeared as visibly when the Air was pumped out, as it did before. It is plain therefore from thence that Air is not necessary to transmit Light. IX. That the Light of shining wood will go out in rarify'd Air is evident. Mr Boile was willing to see whether the same Air condensed agen, would reproduce the Light which before went out. To effect that he used this contrivance. He put a piece of shining Wood into a Cylindrical Glass of a pretty large Bore, sealed[58] at one End, & ‹the Wood wedged in›[59] with a Cork that it might

---

[a] These had appeared in *Philosophical Transactions*, 2 (1668), pp. 581–600, reprinted in *Works*, vol. iii, pp. 157–69.

not stirr. The Cylinder was then /fol. 21/ inverted, & put into another Glass, in which was a good deal of Mercury, & then the whole was put into a Receiver. Upon exhaustion the Air in the Cylinder expanded itself very much, lower than the Surface of the Mercury, & the Light of the wood went out. When the outward Air was let into the Receiver agen, the Mercury rose, & when the Air had recoverd its former Density, the wood shone agen as formerly. X. Fish when rotten shines much more vividly than Wood. A whole Fish almost all over luminous was put into a Receiver. Its Light was very vivid, & many of its parts were at some distance from the Air. When the Air was well exhausted, what was at first the weaker Light considerably lessened, & upon the readmission of Air it manifestly gathered Strength. And when the Experiment was try'd upon another Fish whose Light was much weaker[60] at first, the dimmer Light vanished, & the rest was greatly eclipsed, but upon the letting in of fresh Air the whole recoverd. XI. Light in shining wood, tho in as large Quantity's as could be procured, could never be preserved without Air in a Receiver, it being never so vivid as that of putrify'd Fish. XII. Small pieces of shining Fish[61] in an empty'd Receiver lost their Light, & after it had bin kept out 24 Hour's, upon letting in Fresh Air it revived agen. *N.B.* Here Mr Boile observes that the Light of shining Fish will not continue in the open Air many Day's. That therefore ought to be heeded, least the Loss of Light proceeding from one Cause may not be ascribed to another. He observes also that of Fish commonly to be procured, Whitings grow luminous the most easily. XIII. When shining Fish was inclosed in Water, & so put into a Receiver the absence & return of the Air had no great Influence upon the immersed Body. XIV. ‹Several pieces of› Shining Fish was put into Receivers which were kept Exhausted for several Day's, & upon readmission of Air recovered ‹their›[62] Light. One piece after 48 Hours absence of Air recover'd its Light when new Air was let in upon it so vividly, that tho Fire & Candle were in the Room, it would appear to shine, if a Hat onely screen'd it from their Beams.

From these Experiments Mr Boile drew several Corollaries from which both the Resemblance, & Difference between Light emitted from these sort of Bodies, & what comes from Bodies actually hot, may be plainly understood.[a]

They agree in the following particulars. I. Their Light actually resides in them both, & is not reflected as in Looking-Glasses from any external Beams falling upon them. II. Both require the presence of Air, & that too of a determinate Density to make them shine. III. Both will recover their Light upon the readmission of fresh Air, even when the Air has bin

---

[a] Wotton here summarises Boyle's further communication in *Philosophical Transactions*, 2 (1668), pp. 605–12, reprinted in *Works*, vol. iii, pp. 170–4.

withdrawn[63] for some Time. IV. Both may be[64] quenched by Water &
many other Liquors. This was try'd upon shining wood with Spirit of
Salt,[65] Spirit of Sal Armoniac, Spirit of Turpentine, & Spirit of Wine
highly rectify'd. Neither Live-Coals nor /fol. 22/ shining Wood will be
extinguisht by the coldness of dry Air, tho that Cold be greater than
ordinary. But if shining Fish be kept in places extraordinarily cold, its
Light will go out.

They differ in the following Cases. I. Live-Coals may be extinguisht by
compression, but no compression that can be used without spoiling the
Operation will much injure the Light even of small pieces of shining
Wood. II. Live-Coals will be quite quenched by Exsuction of Air, wheras
shining Wood when quenched in a Receiver will recover its [sic] upon
admitting fresh Air, within half an Hour after it was quite withdrawn.
III. Live-Coals inclosed in a small Glass will not burn many Minutes,
wheras shining Wood so inclosed will emit its Light ‹some whole›[66] Days.
IV. Live-Coals send forth steams, which shining Wood do's not. V. Live-
Coals will wast with burning, which shining Wood will not. VI. Live-
Coals are actually & vehemently hot, wheras shining Wood is not so
much as lukewarm. Nay so much otherwise, that when a piece of this
luminous matter was applied to a Thermometer the Liquor would sensibly
subside.

These Experiments were all made upon Bodies which exhibit Heat or
Light or both in the open Air, & where a rarify'd Medium lessens it if not
put it out. There are other Cases in which the more rarify'd the Medium is,
the more Light is produced. If Mercury in a well cleansed Barometer be
shaken, it will exhibit a brisk Light. Mr Hauksbee try'd several curious
Experiments upon this Mercurial Phosphorus.[a] He found that in a well
exhausted Receiver if Air could be let thro a small Tube into a Glass filled
pretty full with Mercury,[67] the Motion which the Air caused as it past thro
the Body of the Mercury, would make the whole appear like one great
flaming Mass, composed of numberless glowing Balls, which kept their
Light till the Receiver was half filled with Air: That if Mercury was let fall
into an empty'd Receiver thro a narrow Funnel, in which was a long Glass
included with a Convex Crown, the Mercury as it broke upon the Crown
of the included Glass, would appear like a shower of Fire breaking on its
Top, & trickling down its sides: That if the Quantity of the Mercury was
great, it would exhibit in falling great & unusual flashes of a very pale
Flame like Lightning: That this Light was never produced without Motion:
That tho it might be produced by shaking the Mercury in ‹the› open Air,
yet then it appeared onely in small distinct sparks, wheras in a highly
rarify'd Medium it appeared continued & vivid, & still the more rarify'd

---

[a] Hauksbee, *Physico-Mechanical Experiments*, pp. 5f.

the Medium the greater the Light. These Trials were made upon the Mercurial Phosphorus.

But Mercury was not the onely Body out of which Mr Hauksbee by Attrition & violent Motion produced[68] Light.[a] Glass rubbed against Glass, if the Motion was swift would produce copious Light *in Vacuo*, & a very sensible Light in ‹common Air›[69] & under Water. The Light was[70] heatless, tho to view ‹the Body was› red hot. Amber /fol. 23/ rubbed *in Vacuo* upon Woollen produced a sensible Light & great Heat. In ‹common›[71] Air the Light was very small & faint. Glass rubd upon Woollen *in Vacuo* produced a fine vivid purple Light, which lessend as Air was let in: But then if the same Glass were used often, the purple Light would disappear, & a pale Light would appear in its stead.

But I shall recite no more of Mr Hauksbee's Experiments upon this Subject. The curious Reader will be more pleased & instructed by seing how he prosecutes his Enquiry in the Book itself. The Trials already made will go a great way to instruct us yet farther in the nature & difference of Light & Heat. The clearing of that Question is probably left to Posterity. My Business is onely to shew how far Mr Boile opend the way, & what has bin since raised upon his Foundation. I shall now go on with those *Physico-Mechanical Experiments* which Mr Boile first published in the Year 1660.

Experiment xv[72]

The next Experiment was a successless Trial to kindle Bodies easily inflammable in his Vacuum. The Event was that if the Receiver was very thin & clear, & stood in a right position, he could produce Smoak, but never Fire. This answers to what Mr Hauksbee try'd before.[b]

Experiment xvi.

The presence or absence of Air in a Receiver seemed to have no Influence upon the operation of a Loadstone upon Iron. A needle touched before was put into a Receiver, & when the Air was pumped out, upon applied [sic] a Loadstone to the sides of the Receiver, the turned irregularly,[sic] according to the Directions of the Magnet, & when it was removed, the Needle stopd pointing North & South.

---

[a] Cf. Hauksbee, *Physico-Mechanical Experiments*, pp. 23f.
[b] See above, p. 126.

Experiment xvii

The next Experiment which Mr Boile registerd in this part of his *Physico-Mechanical Experiments,* was the Phænomenon (now so well known) of the subsiding of Mercury in the *Torricellian* Tube[73] when put into his Vacuum. This alone demonstrated the Spring as well as the weight of the Air, & shewd them both to have their proper shares in keeping the Mercury suspended in that Tube. And first he found that the Mercury would stand at 29 Inches in the Receiver when full of Air, as freely as in the open Atmosphere. so that the included Air operated as strongly by its Spring alone when inclosed in the Receiver as by its weight when at Liberty in the open Atmosphere. Then upon the first Exhaustion he found that the Mercury sunk in his large Receiver 1⅜ Inch, at the 2d as much: & so on still sinking every Time tho not so much as at first, till it at last it [sic] would not sink above the breadth of a Barlycorn at a time, & never wholly so as to be exactly levell with the stagnat [sic] Quicksilver, because the Air could never be totally exhausted. In less Receivers the Mercury would sink 1½ Inch at one Exhaustion; & yet the Mercury would stand as high in the smallest Receivers if full of Air, as in the greatest. Afterwards, to put the Cause of this Phenomenon out of dispute, the Quicksilver would rise agen upon letting fresh Air into the Receiver, & proportionably according to the Quantity of Air that was admitted. Tho it would never rise so high as it was at first, when /fol. 24/ the Tube immersed in the stagnant Mercury was put into the Receiver. The Reason of that was evidently this, that the Tube was never quite cleared of Air; & ‹so› those formerly inconspicuous Bubbles were now gathered into one great Bubble at the Top of the Tube, which upon the exhaustion of the Air in the Receiver expanded itself, & so kept the Quicksilver from rising so high as it did before.

This Experiment first led Mr Boile to calculate the different Gravity's of the Air at different Times, by means of this Torricellian Tube, & the Trials he made now were afterwards improved by him to ‹those purposes to› which this *Barometer* (as from its[74] Use in calculating the weight of the Air at several Times it is commonly called) is now constantly applied. The first step which he took to make this noble Discovery is related in the following Experiment.

Experiment xviii.

Mercury was put into a Tube 3 foot long, & being tolerably well cleansed of Air was put into a Cylinder of stagnant Quicksilver, &[75] fastned to a woodden Frame, & so set against a wall for several weeks. Mr Boile did not yet know what the success would be. Heat & Cold would (as he found) make some small Alterations. This he rightly attributed to the greater or lesser Expansion of the Air included in the Top of the Tube. But then agen he perceived that many Times without regard to Heat or Cold, the Mercury would very remarkably rise & fall in the Tube yea many contrary to the Indications which those seasons might seem to require. It was plain therefore that the encrease & diminution of the Air's weight & spring (at least as to its Operation upon us) depended upon other Causes, which Causes & the Phænomena thence arising he saw plainly were to be left to other Trials before they could be sufficiently found out. And thus far Mr Boile went with the *Barometer* at Oxford before the Year 1660 when King Charles II was restored; whose happy return to his Kingdoms was soon followed by the Institution of the *Roial Society.*[a]

Experiment xix.

He was willing in the next place to see what Effects the Exsuction of Air would produce upon an inverted Tube of Water about 4 Foot long, when immersed in a Vessel within this Vacuum. It is well known that a Cylinder of Water of 32 Foot will answer to one of 29 Inches of Mercury. Accordingly therefore when this Tube of 4 Foot ‹which› was filled with Water was put into the Receiver, it was necessary that a good part of its Air should be drawn out before any Air would appear at the top of the Tube; it being requisite that much of the Air[76] in the Receiver should be exhausted before it could come to an *Æquilibrium* with that short Cylinder of Water which was but 4 Foot long. But when once it had attained that *Æquilibrium*, as the Air was still more & more exhausted out of the Receiver, the water subsided still lower & lower; tho no Exsuctions could draw it lower than[77] within a Foot of the stagnant Water in the Vessel, seldom so low, that being scarce equivalent to the Cylindrical Inch of Mercury, to which that in the Tube would fall upon Exsuction in the Receiver./fol. 25/

Experiment xx.xxi.xxii.

This led Mr Boile to enquire whether Water were an Elastical Body as well as Air. He found that Water put into a Glass Egg, with ‹the Water in› its neck reaching about a span above the Globe, would, if put into a[78] Receiver, upon repeated Exsuctions manifestly ascend in the stem, &

[a] Boyle's work played a crucial part in the development of the barometer: see Middleton, *History of the Barometer*, pp. 65f.

diverse Bubbles of Air would get loose, & break thro the Body of the Water into the Receiver; The usual method of its Ascension was, that ‹when›[79] once ‹it› appeared to swell, it would upon every Exsuction rise the breadth of a Barlycorn in the neck of[80] the Glass, & when fresh Air was let in, it would immediately subside to its first Mark. Afterwards he filled a peuter Globe with water, which when full was beaten on the outside with a woodden Hammer,'till it was considerably comprest. Then a small Hole was made in the Globe with a Needle, out of which when the Needle was taken out, the Water would spurt several Feet high. This seemed to evince the reciprocal compression & expansion of Water, proceeding from an inward Spring analogous to that of Air. But then since this Expansion might possibly arise from the Dilatation of the particles of the Air lurking before in the Water, he found it necessary to enquire, Whether the Atmosphere truly gravitates upon Bodies under Water, & if so, Whether the Expansion of Water upon the removal of the Atmosphere to some Matter more subtile than Water residing within itself.

To satisfy himself therefore in that matter, he put a Glass Viol filled with water of which it would hold above a pound into a Receiver into which Viol was put a Glass Pipe open at both Ends, reaching several Inches above the Viol, filled also about half way up to the Top with Water. These were carefully cemented together, so that no Water could get out any where but thro the Pipe; & this whole *Apparatus* was let down into the Receiver. Upon pumping great Numbers of Bubbles came thro the Pipe & broke in the Receiver, & at last the Bubbles which ascended very slowly, bore up the whole ‹column of› water in the Pipe with them for near an Inch together to the Top of the Pipe; all which Bubbles immediately vanished when Fresh Air was let into the Receiver.

Now because it might be questiond, whether these Bubbles were parcels of Air inclosed in thin Films of water; or onely the finer parts of the Water itself, He took a Tube of three foot long, & filling it with Water he inverted it into a Viol of Water, & put it into a small Receiver, & he observed, that when he had drawn the Water by Exsuction as low as he could, the Bubbles which appeared plentifully gathered at the Top into such a Body as hindred the Water from filling the whole Cavity when fresh Air was let in, tho had it not been for such a hindrance it would have filled a Tube of many Times that length. Then he drew the Air out a second Time, & the Water in the Pipe fell a great way beneath the Level /fol. 26/ of the Water in the Viol, which it would never do before. The surface also of the Water in the Pipe was much hollower than ordinary; nay to put the Thing out of debate, when the Water in the Pipe & viol were about a level, the Warmth of a Hand applied to the Pipe sunk the Water speedily & remarkably below the Mark, & when a live-Coal was applied to the Tube, the Water was driven to the very bottom. He found also that after the first Trials,

Bubbles would not arise in future Exhaustions till the Receiver was very much empty'd, & the Water sunk very low. All these things put together shew, that upon the removal of the Atmosphere Air actually included before in Water will expand itself, & depress the Water more or less, according as it is more or less rarified. Even Mercury which by reason of its weight discovers itself to be one of the closest Bodies known, can never be so perfectly drained of Air, but that repeated Exsuctions of Air out of a Receiver, in which a Mercurial Tube is inclosed, will discover latent Bubbles which will immediately vanish as soon as fresh Air is let in.

But here a Question may naturally arise, whether Air be an original, primitive Body, or whether it may be generated from, or changed into Water or any other Body. That Vapors which arise in Distillations are onely particles of Water sublimed by Heat, so as to become invisible, which Particles afterwards fall down in the same form of Water they had before, is manifest. The same will hold in Water evaporated by Heat out of an Æolipile; there indeed as Mr Boile found by experience, so much Air will be contained in that Instrument, (especially if it be a large one) that it will be difficult to distinguish Vapors from rarify'd Air, tho even there if a smooth, hard Body be applied to the Mouth of a heated Æolipile, that is filled with Water, the Water that will slide off, will shew what those Vapors consist of. In short all the Experiments that have[81] yet bin made, have not sufficiently proved that Water will turn into true Elastical Air, or that Air by any condensation can be alterable into Water.

The *Academy del Cimento* found ( ) that a weight of 80lb of Mercury would not compress 6lb of Water inclosed in a Viol, one Hairs Breadth below its proper Bulk, when it had bin thoroughly cooled in Ice. And when Water so cooled was put into a thin large Vessel of Vessel [sic], so as to fill it, & the cover was screw'd on with a very close screw, if that Vessel was by hammering contracted into a less capacity than it had before, the water would sweat thro the little pores of the Metal, as Mercury will thro Leather, rather than admit any compression.[a]

These things seem to shew that Air & Water are Bodies no way convertible into each other: Tho that true Elastical Air may be generated Mr Boile discoverd even then, as appears by the following Experiments. Into about an equal mixture of Oil of Vitriol & ‹Water›[82] he put 6 small Iron Nails. This mixture was put into a long necked Glass-Globe, which was quite filled with it. Then covering it with some Diapalma-Plaister to keep the Air out, he inverted it /fol. 27/ & put it into a wide mouthed Glass in which there was some of the same Liquor. In some Time Bubbles arose at the Top of the Globe, which by degree's filled it, & some part of the Neck, & continued so to do for several Day's, & if warmed, tho onely by a Hand,

---

[a] See Middleton, *The Experimenters*, pp. 219–20.

it dilated itself farther & broke out at the small Orifice like common Air. The same thing was try'd with Nails dissolved in *Aqua Fortis*. What was then attempted with corrosive Liquors, was try'd afterwards several way's in the Air-Pump, as we shall see in the following sheets.

Experiment xxviii. (   )[a]

A new Glass Viol holding about 6oz. of Water, in which there 2 or 3 spoonfuls, [sic] was put into a Receiver wellstopt with a Cork. The Air imprisond within the Viol did not stirr the Cork. Another Time when it was more loosely stopt, the inclosed Air made its way thro the Pores of the Cork. For in this second Trial, Bubbles appeared at the Bottom of the Viol, which when fresh Air was let into the Receiver immediately shrunk up, wheras in the first Case no Bubbles at all appeared. From hence it was evident, that as long as no Air got out of the Viol, so long the Air was equally prest as it would have been by the weight of the whole Atmosphere, wheras when once any of the inclosed Air got out, the Bubbles (i.e. the Air enclosed within the Viol) being freed from the Air that was over them began to rise. And that this could be ascribed to no other Cause appeared farther from a Glass-Egg that had some water in it, which was sealed hermetically & so placed, that the Top of the Neck might be broke off at pleasure. For when, after the Receiver was empty'd, the Neck was broke off, tho the Water was still before, yet the Bubbles rose then so quick[83] & so numerous, that they appeared like a violent shour of Rain; & when Red Wine was used instead of Water, the Bubbles rose like a curious white Froth, like that of Bottled Ale unwarily opend, which soon disappeared.

Experiment xxiii.

The last Experiment which Mr Boile sets down of the Influence of the Air upon Water, in his *Vacuum*, was made upon Water carefully distilled. And then, tho the water was put into a Glass Globe whose neck was not above ⅛ of an Inch in Diameter, yet after a more than ordinary Exsuction of Air, no swelling of the Water nor any ascending Bubbles appeared. Tho another Time[84] when a Rod of solid Glass was put into the Vessel, so as almost quite to ‹close›[85] the Neck, after many pumpings numbers of large

---

[a] Boyle stated that this experiment was intended to accompany the twentieth (*Works*, vol. i, p. 64), so Wotton has accordingly moved it forward in his exposition. His brackets were perhaps intended for a note indicating this.

Bubbles appeared fastned to the bottom of the Glass-Rod, & between that & the sides of the Vessel, so that the water was raised above a Fingers breadth[86] higher than the Mark. These upon the admission of fresh Air immediately vanished. Still, here was a manifest [sic] between ‹distilled›[87] & common Water, the Bubbles being much less, & the swelling less conspicuous in the one than in the other.

Experiment xxiv.

Sallad-Oil upon Exsuction in a Receiver afforded more Bubbles, & sooner than Water, & they continued still to rise in great Numbers, even when the Operators were tired with pumping. Oil of Turpentine likewise afforded them copiously; Solution of Salt of Tartar in fair Water, which is one of the heaviest Liquors that is known, was long before it afforded any /fol. 28/ Bubbles, & when it did, they were the smallest & the fewest of any Liquor that Mr Boile made use of. Spirit of Vinegar afforded a moderate number, & nothing else worth mentioning. In Red Wine the Bubbles appeared at Top like a shallow Froth. In Milk which is a more unctuous substance, the Bubbles did not easily break at Top, but thrusting one another forward, the swelling seemed greater than in common Water.[88] Hens Eggs came out of the Receiver whole as they went in. Spirit of Urine carelessly dephlegmed swelled near 1½ Inch above mark, with a Froth at Top, & over that 8 or 10 Bubbles upon each other, which made a Cylinder near ½ Inch high. When Fresh Air was admitted, that Cylinder lessened by degree's & subsided. Spirit of Wine & Water mixt afforded Store of Bubbles & nothing more. Spirit of Wine rectified shew'd many Bubbles, which rose very swiftly, & bore up the Top of the Liquor till they broke, & ascended streight up, seeming to come as it were from determinate[89] places in the Viol, ascending in nearly equal Distances. This was peculiar in the Spirit of Wine this was peculiar, [sic] that it retained its new Expansion, & kept it for very many[90] Hour's after Fresh Air had bin let in.

Experiment xxv.

Mr Boile then was willing to see whether Air would expand in like manner under Water, if the weight of the Atmosphere were removed as it would in its natural State. Accordingly he put two small Essence Glasses into a Crystal Jarr, in which was near ½ Pint of common Water; into one was put enough of a heavy Mercurial substance to make it sink, & it was stopt with soft Wax. The other had Water in it more than enough to make it sink; & they both sunk with their Mouths downwards. Upon pumping for a good while onely Bubbles rose, & broke upon the Top of the Water, but the Bottles did not stirr. At last the open Glass rose to the Top, swimming with the mouth downwards, & after some Time when most of the Water was thrust out, a great Bubble appeared, which broke, & then it sunk agen. This it did nine times successively, the Bubbles being about the bigness of Pease; the stopt Glass never rose. When Air was let in, the open Glass sunk to the bottom. The Experiment was then renew'd, & after much Labor from the stopt Glass about fourty Bubbles arose, & then the open Glass began agen to float. At last the heavy Viol floated too, being buoyed up by a Bubble at the Top, which when it broke, the Viol sunk. After the pumping was quite over, new Bubbles rose nine or ten Times at the Mouth of the Glass, upon the rising & breaking of every one of which, it alternately rose & sunk. So that the grand Maxime in *Hydrostatic's*, that a Body will swim in water, when it is lighter than so much water as equals it in Bulk, holds as well when the presence of the Atmosphere is removed, as when it is greatest; since these Bubbles raised the Wax, & so encreased the Bulk, tho they had not strength enough to trust it quite out.

Experiment xxvi.

It is known that Pendulums vibrate slowest in thickest Mediums. To try /fol. 29/ the difference between the Vibrations in common Air, & in exhausted Receivers, two Pendulums of steel of equal weight & length were set agoing in both; but whether the Receiver was too small, or what other Reason is uncertain, the Event did not answer expectation.

Experiment xxvii.

Mr Boile was then desirous to ‹try›[91] whether Sounds could be conveyd as easily in an exhausted Receiver as in the open Air. For that purpose a Watch whose Case was open was put into the Receiver, & tho before the Air was pumpt out the noise of its Ballance might be very easily heard, yet upon every Exhaustion it grew less & less, & at last could not be distinguished tho ones Ear was applied close to the Receiver. The Motion however continued undisturbed all the while, & when fresh Air was let in the noise was heard as audibly as before. When a Bell of two Inches Diameter was put in, the Noise which it made upon shaking the Receiver did not seem to be diminished. In the communication of the Sound of the Watch it was remarkable, that tho it hung by a Packthread fastned to the Top of the Receiver, yet the Sound seemed to come in a right Line thro its sides,[92] & not obliquely thro the Stopple. The noise of the Flint when the

Gunpowder was fired in the xivth Experiment, was considerably weaker than it would have been in the open Air.

The Experiment of propagating Sounds in *Vacuo* was try'd by the *Florentines* (   ) with small success.[a] They used no *Vacuum* but the *Torricellian*, & when they sounded which was [sic] fastned by a string to the top of the Globe that was at the upper End of the Tube, they observed no discernible Alteration in the Sound. But, as they rightly suspected, that Trial was by no means sufficient to decide the matter. For the Bell being thus fastned to the Tube, it was uncertain whether the ‹Motion into which› Glass was necessarily put upon the sounding of the Bell, (which was fastned by a string to the Top of the Glass) might not strike the outward Air with a vehemence nearly equall to that with which it would have bin stricken, if the Bell had sounded in the open Air. Upon which occasion they lament their not being able to try this Experiment in Mr Boile's Engine for want of Workmen that could make the Apparatus; & they do him there the Justice to own *that he had with admirable success in his curious & noble Experiments shewn how Air inclosed in a Vessel may be exhausted by Attraction.*[b]

But, what neither Mr *Boile* nor the *Florentines* were able to do for want of proper Engines, Mr *Hauksbee* has done by his Instruments, & thereby both shewn the constant *Relation between Sound & Air*, & opened a Way for future Discovery's which may prove of singular Use.[c] He (   ) inclosed a Bell in a Receiver which was placed at one end of a Room about 50 Yards long. If no Air was injected into it by his condensing Instrument, the Sound might just be heard at that Distance if diligently attended to. If one Atmosphere was injected (*i.e.* if as much more Air was put in as was in at first, when the Receiver had its full natural content of Atmospherical Air) the Sound was considerably augmented, & so it /fol. 30/ increased on, upon the Intrusion of a 3d Atmosphere; a 4th, & a 5th, after which the Sound did not increase proportionably as it did upon the injection of the first & second. He tried also the same thing in the open Field, & then, if the sound might be heard at 30 Yards distance when onely one Atmosphere was in the Receiver, upon the injection of a 2d it might be heard 60 Yards, & of a 3d ninety; & after that no Injections could carry the Sound above 20 Yards farther; the Reasons of which difference of encrease he gives at large. On the contrary in a rarifyed Medium, as the ‹Air was› Gradually exhausted, & the several stops made, if the Bell was shaken at the several different degree's of Rarefaction, the Sound lessend manifestly at every stop. When the Receiver was quite exhausted the Sound was so

[a] See Middleton, *The Experimenters*, pp. 151–4.
[b] Ibid., p. 151: this is Wotton's paraphrase.
[c] Hauksbee, *Physico-Mechanical Experiments*, pp. 97f.

little that the best Ears could just distinguish it, it being like a small shrill note heard at a mighty distance; & still as the Air was let in the Sound proportionably encreased.

From hence he draw's these Observations. I. That Sounds are augmented in condensed Air; & diminished in that which is rarify'd. II. That Distances at which equall Percussions of the same sonorous Body may be equally heard by the same Person, are in Proportion to the Density of the several Mediums thro which those Sounds are propagated: Consequently were the Proportions exactly known, from given Density's we might inferr Distances, & from given Distances we might determin the proper Density's to make[93] Sounds of a given degree equally audible at those Distances. III. That Distances by which Percussions unequally strong of the same sonorous Body may be equally audible by the same Ear, are in a *Ratio* compounded of the strengths of the Percussions, & the Density's of the Mediums. IV. That the same Sounds at[94] equal Distances from the Surface of the Earth, are not alway's equally audible, by reason of the different Rarefactions & Condensations of the Atmosphere. And V. That Diminutions of Sounds in Ascents from the Surface of the Earth will bear some Proportion to the[95] Descents of the Mercury in the Barometer, at those Elevations.

These Conclusions we may reasonably imagine Mr Boile would have made, had he had Instruments to have try'd these Experiments with, & since they are immediate Deductions from Trials to which what he did first gave birth, they will not be thought to be brought in improperly in this Place.

Experiments xxix.xxx.

Fumes which ascend in open Air, descend like Liquors in exhausted Receivers. Mr Boile try'd that with a metalline Liquor, which when exposed to the Air alway's emitted copious white Fumes, so thick that they seemed like Dust of Alabaster thrown into the Air; & yet when kept close it was clear & transparent. For a Viol[96] with some of that Liquor in it, being put stopt into a Receiver was unstopt ‹when›[97] the Air was pumped out. The few Fumes which it then emitted, did not[98] rise up, but trickled down in a stream about the bigness of a Swans Quill along the sides of the Viol, & the Lead which was fastned to it, to keep it from falling down in the Receiver. The smoak of a Match did the same in the same Place, it kept at the bottom, & there rowled about /fol. 31/ keeping a horizontal situation when unstirred, sometimes undulated & curled at top if gently moved, just as any other Liquor would have done. These Experiments shew that is [sic] the Motion & Weight of the Air which impell Vapors upwards, & keep

them there rolling about in the Atmosphere, till being sufficiently condensed by meeting together, whilest they are floating about, they sink
when urged [by][99] their own Weight.

Experiment xxxi.

That Plates of Marble or of any other hard Body's made exquisite[ly]
smooth will when joined cohere so closely, that a very great strength is
s[ome]times requisite to pluck them asunder is very well known. The
Cause o[f] this Mr Boile attributed to the mighty pressure of the Air upon
the [low]er Surface of the undermost Marble, in comparison of the very
[strong] Pressure of Air upon its upper Surface. Now according to those
Principles if two Plates of Marble so joined were put into a Receiver &
there kept suspended, the lowermost must of necessity fall down, when
once the Receiver was exhausted; & yet the contrary happend. For having
moistend the Plate with nothing but pure Spirit of Wine, that the undermost might not too quickly drop off, the Exhaustion of the Receiver did
not part them. Mr Boile ascribed this to the force of the Spring of the
remaining Air which could not with all their Pains be perfectly drawn out.
How great this force was, he try'd to find out by the following [Ex]peri-
ment.

But before I give an Account of ‹the next›[100] Experiment it will not amiss
[sic] here to o[b]serve that Mr Boile having assigned the Airs pressure as
the sole ‹Cause of the› cohesio[n] of the smooth Marbles, in which he
‹was›[101] very probably mistaken, it was no wonder if the Experiment did
not answer his exspectation. That the pressure of the Air do's a great deal
in[102] producing that Phenomenon, is I think out of all Question. But there
ought to be another Cause taken in & that is *Attraction*, by which Power
every particle of Matter attracts another, & that as Mr Keill has demonstrated with a force infinitely greater at the *point of contact* or extreamly
near it, than at any determinate Distance from it. Consequently then the
smoother the Marbles are, the stronger is the Attraction, because the more
numerous will be the Points of Contact, & the strong [sic] the Attraction
is, the closer & stronger will be the Cohesion. But the Discovery of that
general Law of Matter was not then known, that was reserved to the
incomparable Sir Isaac Newton to find out, to whom, what Mr Keil & Mr
Hawksbee ( ) have since built, is originally owing, which they have made
no Difficulty of acknowledging.[a]

[a] See Keill, 'Epistola', p. 102. This was a point that Newton himself had very briefly made
in *Optice* (1706): see Newton, *Opticks*, p. 390, and Shapin and Schaffer, *Leviathan and the
Air-Pump*, pp. 200–1 (I am grateful to Simon Schaffer for his advice on this point). Hauksbee, *Physico-Mechanical Experiments*, pp. 156–8.

Document 9

Experiments xxxii.xxxiii.

Former Experiments having sufficiently shewn that the Pression & Pulsion of the Air causes those Phenomena which have usually bin ascribed to a *Fuga Vacui*, Mr Boile was desirous to find out ‹some› way to measure precisely what this Pressure of the Air is. The Methods he used to find it out were these. First he empty'd a Receiver & took it off the Pump; He then fitted a Valve of Brass, to the lower branch of the Stopcock, & hung a Scale to the bottom of the Valve, in which Weights might be put as there should be Occasion. The Key of the Stopcock was then turned, upon which the outward Air forced the Valve up into the Stopcock so strongly[103] /fol. 32/ that it sustained a weight of about ten pounds, without any visible Cause to keep it up more than the Weight of this outward Air. He try'd then what weight the Sucker of the Cylinder would bear, when the Cylinder was exquisitely empty'd, which was much easier to do, than to draw out all the Air out of a large Receiver. Now tho the Diameter of the Cylinder was not more than 3 Inches, & consequently a Column of Air greater than that could not impell the[104] Sucker into the Cylinder, yet when the Sucker was drawn down, & the Cylinder was exactly empty'd, it would fly up of itself, & bear at Top a weight which added to its own was above 100 pounds. Yet it is certain that there was nothing in the cavity of the Cylinder that could attract so great a weight, or sustain it when once it was driven up.

Mr Hauksbee's Trials go a great way farther. He took ( ) two brass Hemispheres, & fastned them together by a piece of wet Leather put between them; when they were so fastned he put them into a Receiver & found, That if the Hemispheres were full of common Air , & an Atmosphere of Air was injected into the Receiver, (which was also full of common Air before) it required 140lb weight fastned to the lowermost, to part the two Hemispheres: That if these Hemispheres were exhausted of Air & joined together as before, it required the same Weight to part them as in the former Case; tho here no more Air than the uncondensed Atmosphere prest the superficies of these Hemispheres: But if another Atmosphere was then injected upon these exhausted[105] Hemispheres, 280lb weight hanged upon the lowermost would not part them. This Experiment (as Mr Hauksbee truly observes) quite puts an End to all the Hypotheses of those who would solve these Phænomena by other Principles than the Pressure & Gravitation of the ‹Air›.[a]

---

[a] See Hauksbee, *Physico-Mechanical Experiments*, pp. 69f.

Experiment xxxiv.

It being certain in Hydrostatic's, that when two Bodies of equal Weight & unequal Bulk are weighed together in Air, as Gold for example & Iron, that if these be afterwards weighed in Water, they will lose their Æquilibrium, & Gold will sink lower than Iron; because the Iron by reason of its greater Bulk, hath more Water to displace than ‹the› Gold has before it can sink. Experiments of this kind he was willing to try in his Air-Pump. A Bladder closely ty'd about the neck, & about half filled with Air was hung to a Ballance that would turn with $\frac{1}{32}$ part of a Grain, & a metalline counterpoise at the other End. As the Receiver empty'd the Bladder sunk, & as it swelled the more was that sinking visible; & when the Bladder flagged, by the letting in of the outward Air, the Æquilibrium returned agen nearly as it was at first. When it was try'd with Lead & Cork, the Cork manifestly outweighed the Lead, when the Receiver was empty. So that this Experiment did not answer its End.

Experiment xxxv.

He try'd next what Effect Exsuction of Air would have in stopping the course of Water that was set a running thro a Syphon. After long exsuction he found that Bubbles would rise in the Water into which the shorter Leg of the Syphon was immersed & then gather at the Top; & so stop the running of the Water. This he try'd in several sorts of siphons, & still found that the same thing constantly & regularly happend.

*　　*　　*　　*　　*

[a]Future Trials have varied something, being perhaps made with nicer Instruments. Tho if we consider the different Density's of the Atm[o]sphere at different seasons, we have Reason to admire how such, ho[w] such [sic] early Trials as these of Mr Boile's were so exactly made. T[he] last Experiment of this sort that has bin[106] communicated (at least that I know of) is Mr Hauksbee's (　). He found upon a nice Trial made in *May*,

---

[a] The following passage appears on the verso of fol. 32. It is preceded by a mark apparently keyed to a passage on a facing page which is now missing, evidently dealing with experiments after no. xxxv in Boyle's work. Alternatively, it could be an annexe to an earlier experiment relating to seasonal variation of the atmosphere.

when the weather was warm, & the[107] Mercury in the Barometer stood at $29\frac{1}{2}$ Inches, that the weight of Air was to that of Wate[r] as 1 to $885\frac{1}{122}$:[108] The difference between $885\frac{1}{122}$ & $938\frac{2}{11}$ is so very inco[n]siderable, that the several Density's of the respective Atmosphere's when both Experiments were made,[109] may for ought we know easily reconcile them, so that it will be hard to judg which proportion is the exactest.

( ) *Physico-Mechanical Experiment. p.74 &c.*[a]

---

[a] A further reference to Hauksbee's book, pp. 74f.

NOTES

DOCUMENT 1

An Account of Philaretus during his Minority

[1] The manuscript is damaged at this point, and the word which formerly existed between *Lady* and *Fenton* is partly conjectural: there is insufficient space in the original for the full name *Catherine*. After *Fenton, &* deleted. Miles was also unable to read the passage, and Birch's edition has '*Catherine* daughter of Sir *Geoffrey Fenton*' in square brackets: *Life*, p. xii.

[2] Replacing *from* deleted. *Industry* has an *e* deleted at the end.

[3] Replacing *things* deleted. In the previous line, *built* is altered from *build*.

[4] Followed by *our s* deleted.

[5] Followed by *Ex* deleted.

[6] *-tangled* replacing *-gaged* deleted.

[7] Followed by *the* deleted.

[8] Replacing *Family* deleted.

[9] In margin.

[10] Altered from *secrecy*. Five words later, *&* followed by *da* deleted.

[11] Followed by *at* deleted.

[12] Followed by *an Inti* deleted.

[13] Altered from *Fondlynesse*.

[14] Replacing *Day* deleted.

[15] *of* in MS.

[16] Followed by *& who* deleted.

[17] Followed by *but* deleted.

[18] Duplicated by *humor*.

[19] Followed by *so* deleted.

[20] Followed by *co* deleted.

[21] Duplicated by *subject*. Three words later, *Sport* followed by *Many* deleted.

[22] Duplicated by *familiar*; followed by *of* deleted.

[23] The words *on his Father* are accidentally repeated.

[24] Followed by *im* deleted.

[25] Followed by *acci* deleted.

[26] Followed by *his* deleted.

[27] Followed by *many very narrowly scap't Drowning* deleted. Before *that* brackets are opened, but they are nowhere closed.

[28] Replacing *Empty* deleted.

[29] The *g* of *give* is altered from *l*.

[30] Followed by *them almost all up* deleted. Five words later, *them* followed by *up* deleted.

[31] Followed by *wh* deleted.

[32] Followed by *ass* deleted.

[33] Followed by *Publick* inserted above *thi*, all deleted.

[34] Followed by *Publicke,* deleted.

[35] Followed by *mor* deleted. Before *much*, *b* inserted above line but deleted.

[36] Followed by *desi* deleted; three words later, *improve* replaces *cultivate* deleted. Earlier in the sentence, *hi* deleted after *Schollership* (at end of line).

[37] Altered from *consideration;* later in the sentence, *e* deleted at end of *breeding.*

[38] Followed by *his Elder b* deleted.

[39] *put* accidentally repeated in MS.

[40] Replacing *happy* deleted. In the previous line, *It* deleted after *Dayes.*

[41] Duplicated by *landed.*

[42] Restored after deletion and replacement by *safelly.*

[43] Altered from *them*. Earlier in the sentence, *Eaton* is altered from something else.

[44] Duplicated by *by*. Three words earlier, the *g* of *goes* is altered from *is*. Five words later, brackets are closed after *culpable*, but they had never been opened.

[45] Duplicated by *of.*

[46] Replacing *not only* deleted.

[47] Duplicated by *Benefitts.*

[48] Followed by (*when the* deleted.

[49] Altered from *their*. The previous word, *seduce* has had a final *s* deleted. Two words later, *Traveller* replaces *Followers;* deleted.

[50] Altered from *them;* the next word, *to*, is followed by *like* deleted.

[51] In margin.

[52] Followed by *Unrip* deleted; the next word but one, *of*, is followed by *the* deleted.

[53] Followed by *whilst P* deleted.

[54] Replacing *let* deleted. Later in the sentence, *Dog* and *Child* are marked, probably to indicate an intention to reverse their order.

[55] Followed by *bef* deleted. Two words later, *its* is altered from *his.*

[56] Followed by *a* deleted.

[57] Followed by *of them* deleted. Later in the sentence, *may* is followed by *ta* deleted, *the* before *value* is duplicated by *his*, and *&* is followed by *his* deleted.

[58] Followed by *made* deleted. Seven words later, *Study* has been altered from *Studying* by bracketing the last three letters of the word.

[59] Replacing *attaine* deleted.

[60] Followed by *dr* deleted.

[61] Followed by *& con* deleted.

[62] Duplicated by *Appetite.*

[63] In margin.

[64] Followed by at *unawares &* deleted. Two words later, *giving* duplicated by *affording.*

[65] Followed by *&* deleted.

[66] Replacing *oft* (?) deleted; subsequently, *out* is followed by *of* deleted; *the* is followed by *rubbish* deleted; and *Philaretus had* is followed by *been* deleted.

[67] Followed by *an D* deleted.

[68] At this point, brackets are opened but never closed. Earlier in the sentence, the *r* in *run* is altered from something else.

[69] Followed by *he had* deleted.

[70] Replacing *so scap't* deleted.

[71] In margin. Replacing *This* deleted.

[72] Replacing *some* deleted. Later in the sentence, *vomit* is followed by *sc* deleted.

[73] Followed by *together with his br* deleted.

[74] Followed by *acc* deleted.

[75] Followed by *violen* deleted.

[76] Followed by *hid* deleted. Four words later, *into* followed by *greevous* deleted.

[77] Altered from *Fisitians*.

[78] Replacing *out* [?] deleted; duplicating *owe*.

[79] Followed by *st* deleted.

[80] Replacing *discern* itself replacing *Discover*, deleted.

[81] Altered from *having*.

[82] Followed by *some* deleted.

[83] Replacing *Recovery* deleted.

[84] Followed by *of th* deleted, Five words later, *other* followed by *Raving Bookes*; deleted.

[85] Followed by *did him more ha* deleted.

[86] Followed by *taught his fan* deleted.

[87] Duplicated by *steale*.

[88] Followed by *have* deleted.

[89] *reason-* duplicated by *consider-*.

[90] Followed by *but yet not so as perfectly to bridle their roving wildness* deleted. Later in the sentence, *thought* is followed by a deleted character.

[91] Followed by *his* deleted.

[92] Followed by *th* deleted.

[93] Duplicated by *a small*.

[94] Followed by an insertion mark but no insertion.

[95] Duplicated by *Servants*. Two words later, *the* followed by *Place* deleted. Earlier in the sentence, *aversion* is altered from *aversions*.

[96] Followed by *to* deleted. Four words later, *to* followed by *fling* deleted.

[97] Followed by *swallow'd it wi* deleted.

[98] Followed by *had* deleted.

[99] Duplicated by *tooke*.

[100] Replacing *waits upon* deleted.

[101] Replacing *Men, w* deleted. Three words later, *doted* followed by *on* deleted.

[102] Replacing *in* deleted. Two words later, *resemblance* duplicated by *likenesse*.

[103] Birch in his edition deleted *or to... Com[m]ands* (*Life*, p. xviii). In the catchword at the page turn, *, as* is deleted after *Courses*.

[104] Followed by *the* deleted.

[105] Followed by *to* deleted.

[106] Followed by *of* deleted.

[107] Followed by *neere 4 yeares* deleted.

[108] Followed by *mont* deleted.

[109] Duplicated by *solid*.

[110] Replacing *rare noble* deleted.

[111] Followed by *tis* deleted. Four words later, *much* followed by *fitter subjects no*

[sic] deleted, and *nobler* followed by *Ob* deleted; the gap in the text which then follows is Boyle's.

[112] Followed by *act* deleted. Later in the sentence, *onely* followed by *he* deleted.

[113] Followed by *w* deleted.

[114] In MS, *naturalarry*; followed by *much* deleted. Two words earlier, *he* followed by *much* deleted.

[115] Followed by *be h* deleted (i.e., two words begun and abandoned).

[116] Replacing *persuade himselfe* deleted. Three words earlier, *-wards* altered from *-would*, followed by *could* deleted.

[117] Followed by *Good were* deleted. Three words later, *be* followed by *s* deleted.

[118] Followed by *must p* deleted.

[119] Followed by *the* deleted.

[120] Followed by *in* deleted.

[121] Followed by *En* deleted. Four words later, *Copys* altered from *Copies*

[122] Replacing *let* deleted. Seven words later, *Fate* followed by *a* deleted.

[123] Followed by *Abo, A* deleted: i.e., prior to deciding to begin new paragraph.

[124] Replacing *learning to* deleted. Earlier in the sentence, *also* followed by *he* deleted.

[125] Duplicated by *soone.*

[126] Followed by *Philaretus* deleted.

[127] Followed by *Broghill & the Lord of* deleted.

[128] Followed by *his* deleted.

[129] Followed by *assig* deleted.

[130] Followed by *eit* (?) deleted.

[131] Altered from *knowing*; *well* written in margin.

[132] Followed by *Younger* deleted.

[133] Followed by *professt to* deleted.

[134] Followed by *not* deleted.

[135] Followed by *respect* deleted. Six words later // appears in the MS before *Eclipse.*

[136] Duplicated by *Cynically.* Four words later, *very* followed by *strict* deleted.

[137] Followed by *to exercise* marked for deletion.

[138] Followed by *an* deleted.

[139] Followed by *pro enclin'd* deleted.

[140] Duplicated by *Anger.*

[141] Replacing *came* deleted. Four words earlier, *that* followed by *at last* deleted; two words later, *governe* duplicated by *bridle.*

[142] Duplicated by *Enjoyment.*

[143] Followed by *God* deleted.

[144] Followed by *Manna* deleted.

[145] In MS, *all it.* Four words earlier, *so;* followed by *if* deleted. Two words later, *presently* followed by *corrupt, &* deleted.

[146] Followed by *the* deleted.

[147] Duplicated by *Promiscuous Discourse.*

[148] Replacing *upon* deleted. Earlier in the sentence, *his* is followed by *B* deleted.

[149] Followed by *the* deleted.

[150] Preceded by *the Lords of* deleted.

[151] Followed by *his* deleted; the next word, *the,* is followed by *Desparate lon* deleted.

[152] Followed by *w* deleted; earlier in the sentence, *freely* is added in the margin.

[153] Followed by *Orchar* deleted.

[154] *had no fancy to at all* duplicated by *was almost abstemious,* with *in* inserted before *the latter.*

[155] Followed by *time what* deleted.

[156] Followed by *& Rave,* deleted. Earlier in the sentence, *4* altered from *3.*

[157] Followed by *acti* deleted.

[158] Duplicated by *Fancy.*

[159] Replacing *ascrib'd* deleted.

[160] Replacing *-deed* deleted. Four words later, *Excursion* replaces *Effect* deleted.

[161] *yet untam'd* duplicated by *ramage,* both replacing *Ill countred* deleted.

[162] Followed by *so Ex* deleted.

[163] Replacing *the Kingdom affaires* deleted.

[164] Followed by (*Earle of Corke,*   ) deleted. The gaps are Boyle's. Brackets are also closed after *Queen,* but are nowhere opened. Later in the sentence, *with* is followed by *o* deleted.

[165] Followed by *th* deleted. Two words later, *to* followed by *th* deleted.

[166] Followed by *of about much cry'd up for her* deleted except for *of,* probably accidentally.

[167] Followed by *D* deleted.

[168] Altered from *their Fa.*

[169] Followed by *in* deleted.

[170] Followed by *Lodg* deleted.

[171] Replacing *lived* deleted.

[172] In the margin, an alternative version of this appears: *He was so suddenly deprived of his Joy; that he had scarcely found that it was reall, before he was forc't from it.* Within this, *found that it was reall* is duplicated by *leasure to consider it.*

[173] Replacing *sad* deleted.

[174] Followed by *ha* deleted.

[175] Replacing *find* (perhaps altered from *see*) deleted. Five words later, *to* followed by *see & learne new things* deleted.

[176] Followed by *t* deleted.

[177] Replacing *Gale,* deleted. Two words later, *safely* is followed by *blow them* deleted.

[178] Followed by *Röan* deleted.

[179] Followed by *u* deleted.

[180] Duplicating *did;* following *mus* deleted. The previous word, *he,* is followed by *wh* deleted.

[181] Followed by *did* deleted. Earlier in the sentence, *F* deleted before *Greate.*

[182] Followed by *wo* deleted. Four words later, *spirits* followed by *do* deleted; six words after that, *Ebbes* followed by *the fickle Fortune* deleted.

[183] Followed by *hi* deleted.

[184] Followed by *a* not deleted and by *better* deleted.

[185] Followed by *has* deleted. Four words later, *peculiar* followed by *w* deleted.

[186] Followed by *tying* deleted.

[187] *both* repeated; the second duplicated by *a.*

[188] Followed by *the* deleted.

[189] Followed by *sciences* deleted. Six words later, *handled* replaces *taught* deleted.

[190] Followed by *guilty* deleted.

[191] Duplicated by *attain'd a competent skill in*, which deleted.

[192] Duplicated by *Practicke*.

[193] Followed by *Sph* deleted. Previously, an insertion mark appears after *subordinades*, but there is no insertion.

[194] Replacing *Geography* deleted.

[195] Altered from *the* or *then*.

[196] Duplicated by *a greter*.

[197] Followed by *th* deleted.

[198] Replacing *diverts them*, deleted.

[199] Followed by *it* deleted.

[200] The bracket after *World* is missing in the MS.

[201] *yet un-* duplicated by *not yet*.

[202] Replacing *Diversions* deleted.

[203] Followed by */lang/taught* deleted. The bracket which follows is missing in the MS.

[204] Followed by *no* deleted.

[205] Followed by *of* not deleted (probably accidentally) and *im* deleted.

[206] Duplicated by *ever*.

[207] Replacing *voyd of* deleted.

[208] Altered from *diverting*.

[209] The bracket after *then* is missing in the MS.

[210] Duplicated by *by*.

[211] Altered from *one*.

[212] Followed by *mi* deleted.

[213] Duplicated by *dead*.

[214] Followed by *Eve* deleted.

[215] Followed by *scr* deleted.

[216] Duplicated by *carefully*.

[217] Replacing *the* deleted.

[218] Followed by *so* deleted.

[219] Duplicated by *fancy'd*.

[220] Followed by *was Christ borne in* deleted.

[221] Followed by *ea* deleted.

[222] Followed by *see* deleted.

[223] Duplicated by *so much*.

[224] Duplicated by *as*.

[225] Duplicated by *living vertuously*.

[226] Duplicated by *make it*.

[227] Altered from *himselves*.

[228] *-uss't* duplicated by *-ours't*.

[229] Duplicated by *at present*.

[230] Replacing *Motives* deleted.

[231] Duplicated by *recorded*.

[232] Duplicated with *paying* [*ser* deleted] *Dutys to our Maker*.

[233] Followed by *the* deleted.

234 Followed by *no* deleted; two words later, *the* is followed by *sole* inserted above the line but deleted. The whole passage, *& that true Gallant.... their Love* was omitted by Birch (*Life*, p. xxii).

235 Followed by *Som* deleted.

236 Altered from *no sooner*.

237 Followed by *Abbeys* deleted. Earlier in the sentence, *those* apparently altered from something else.

238 Two strokes appear between *so* and *sad*; later in the sentence brackets are opened before *tho his lookes* but not closed.

239 Followed by *many* deleted.

240 Followed by *shine on* deleted.

241 Duplicated by *satisfy'd*.

242 Followed by *Am* deleted. Two words later, *made* replaces *oblig'd* deleted.

243 Replacing *to compare* deleted. Four words later, *to the* deleted after *Nature*.

244 Duplicated by *However*.

245 Followed by *this Advanta* deleted. The gap before *Anxiety* is Boyle's.

246 Altered from *in*.

247 Duplicated by *consider* duplicated by *peruse*.

248 Followed by *m* deleted.

249 Replacing *neither* deleted.

250 Replacing *nor* deleted.

251 Followed by *not* deleted.

252 Followed by *ever* deleted.

253 Followed by *decei* deleted.

254 Followed by *//* in MS.

255 Duplicated by *in*.

256 Duplicated by *fortun'd*.

257 Followed by *to* deleted, and, before that, an undeleted opening bracket.

258 Duplicated by *by*; three words later, *that sits* is duplicated by *seated*.

259 *-oo-* duplicated by *-ee-*; *is* followed by *coi* or *ca* deleted.

260 Followed by *w* deleted.

261 Followed by *th* deleted.

262 Followed by *trav* deleted. Six words later, *beneath* is altered from *below*. The words *which... descent* are inserted in the margin.

263 Followed by *of whi* deleted. Two words later, *height* replaces *Bottome* deleted; three words after that, *they left* accidentally repeated.

264 Altered from *Territorys*.

265 Duplicated by *over*.

266 Followed by *g* deleted.

267 Duplicated by *skills*.

268 *passion* in MS.

269 Followed by *Resolu* deleted.

270 *-ing* duplicated by *-ant*.

271 In MS, *of*. After *Skill*, *Here Ph* deleted.

272 Followed by *Towne* deleted.

273 Followed by *ascri* deleted.

274 Replacing *like a Native* deleted.

155

[275] Followed by *readily* deleted.

[276] Followed by *Some Rabb* deleted. Earlier in the sentence, there is an insertion mark but no insertion after *readily*.

[277] Followed by *the* deleted.

[278] Duplicated by *Bookes.* After *Opinions*, brackets are opened but not closed; *perhaps* replaces *perhaps for* deleted.

[279] Replacing *be ruin'd* deleted; *be so* is inserted in the margin. The text is slightly garbled at this point.

[280] Duplicated by *his H.*

[281] The catchword on fol. 183v is *presuming his In*, thus suggesting an intention to continue slightly differently.

[282] Followed by (‹as› *unerringly*) deleted. Three words later, *Philosophy* followed by *as well* deleted.

[283] Followed by *This Famous Galileo,* deleted.

[284] Followed by *rich* deleted.

[285] Replacing *ever* deleted.

[286] Followed by *he still brou* deleted. Two words later, *retain'd* is altered from *return'd* and followed by *thence* deleted. Four words after that, *Chastity* replaces *honesty* deleted.

[287] In the catchword on fol. 184, this is followed by *Impudence of* deleted.

[288] Duplicated by *not above 15.*

[289] Duplicated by *faded.*

[290] Duplicated by *presst.*

[291] Followed by *Danger* deleted.

[292] Followed by *improve* [?] *heihten that Aversenesse* deleted.

[293] Followed by *mu* deleted.

[294] Followed by *by the Way* inserted above the line but deleted.

[295] Followed by *obs* deleted.

[296] Replacing *vast* deleted.

[297] Replacing *Miter being* deleted. Two words later there are two strokes in the MS above *bringing.*

[298] This represents the start of the earlier recension of the text, whereas fol. 184v, which represents the end of the fair copy, has *Ro* deleted after *& the* and *Barberine* altered from *Barberinian.*

[299] Followed by *& en* [?] inserted above the line but deleted.

[300] Replacing *both* deleted. Three words later, *nor* replaces *&* deleted. After *Language, w* deleted.

[301] Replacing *visited busily* deleted.

[302] Followed by *at the Chappell amongst all the Cardinalls,* deleted. Four words later, *one that find* deleted after *Pope.*

[303] Replacing *seeming* deleted. Five words later, *Assembly* followed by *appea* deleted.

[304] Replacing *much* deleted.

[305] Due to the damage at this point, Birch's version omits various short passages from here to the end of the text.

[306] Followed by *heard* deleted.

[307] Followed by *wo* deleted.

[308] Followed by *the* deleted.
[309] Followed by *seeme* inserted above *doe* [?] *often*, all deleted. The paragraph break at the start of this sentence has been inserted for the reader's convenience: it does not appear in the MS.
[310] Followed by *&* deleted.
[311] Followed by *to* deleted.
[312] Followed by *P.* deleted.
[313] Followed by *letting* deleted.
[314] Duplicated by *saw*.
[315] Replacing *calling in* deleted.
[316] Followed by *arriv'd* deleted.
[317] Altered from *Frances*.
[318] Followed by *Upon his Determi* deleted.
[319] Replacing *...ed the Way* deleted: the text is damaged at this point.
[320] Followed by *past* deleted.
[321] Followed by *bele* deleted.
[322] Replacing *shut up* deleted.

DOCUMENT 2

Biographical notes dictated by Boyle to his amanuensis, Robin Bacon

[1] Missing in the MS.
[2] Followed by *in the* deleted.
[3] *Munster in Ireland.* is written in the margin.
[4] Followed by *the*, deleted by Wotton.
[5] Followed by *and also the*, deleted by Wotton.
[6] Words written by Wotton fill gaps left in the original. After the place where Wotton has inserted *son*, Bacon's text continued: *child, and also the   son*; Wotton placed a line across the gap but failed to delete the words, presumably accidentally.
[7] Followed by *mention of the* deleted.
[8] Replacing *I* deleted.
[9] Replacing *For* [?] deleted.
[10] Followed by *the* deleted.
[11] Followed by comma deleted.
[12] Replacing *he* deleted.
[13] Replacing *of* deleted.
[14] Replacing *the* deleted.

DOCUMENT 3

The 'Burnet Memorandum': notes by Gilbert Burnet on his biographical interview(s) with Boyle

[1] Replacing *Controversies*, deleted; *Naturall questions* is in fainter ink.
[2] Followed by a deleted letter: *t*?

[3] *Rebellion* is partly obscured by an ink blot which also appears on fol. 59v.
[4] Followed by *being then more curl..* [sic] deleted.
[5] Replacing *nor* deleted.
[6] The first two letters of this word are altered from something else.
[7] Altered from something else.
[8] Replacing *glasse* deleted.
[9] Followed by an illegible deleted character.
[10] Replacing *shew* deleted.
[11] Altered from *them* [?].
[12] In MS, *ther[e]*, followed by *was an* deleted: clearly *ther[e]* (which appears at the end of a line) should also have been deleted and replaced by *this*.
[13] Followed by *he* deleted.
[14] In *script[u]re* what looks like *ou* has been heavily crossed out after the *t*.

DOCUMENT 5

Sir Peter Pett's notes on Boyle

[1] Replacing *most usefull* deleted. Three words later, *ruin* is followed by *his name is mentioned in pag. 21 of the Booke about the Milld Lead Invention* deleted.
[2] Replacing *he* deleted.
[3] Followed by *if their beliefe* deleted.
[4] Followed by *it, shall here set it downe as followeth*, deleted.
[5] Replacing *of* [?] deleted.
[6] Altered from *engraven*.
[7] From here to the end of this page, the text represents Pett's additions to the scribal text in his own hand.
[8] Fol. 35^v blank.
[9] Followed by *shame and* deleted.
[10] Followed by *one of* deleted. Three words later, *one* is altered from *ones*.
[11] This entire passage appears in the margin of the top left hand corner of the page, keyed for insertion into the text at this point (the word *as* is duplicated at the end as a kind of catchword). In the text, a letter or punctuation mark has been deleted at this point, as has an insertion above the line which is now illegible.
[12] Followed by *before* deleted.
[13] Altered from something else.
[14] Replacing *but 16* deleted.
[15] Followed by *with him* deleted.
[16] Followed by *enuf* deleted.
[17] The remainder of fol. 40 and the whole of fol. 41 are blank, except that the latter is endorsed on the verso in Miles's hand *Sir Peter Pett*. The following section, in Pett's hand, which begins on fol. 42, starts in mid-sentence: however, it clearly follows directly from the end of the text on fol. 40 with only one or two words missing: here *but* has been added to complete the sense.
[18] Written sideways in lefthand margin.
[19] Followed by *late* deleted.
[20] Followed by *mine to &* [?] deleted.
[21] Followed by *all* [?] deleted.

[22] Written sideways in lefthand margin, replacing *if there should be occasion for it* deleted in main text.

[23] Followed by *manuscript* deleted.

[24] Followed by *to* deleted.

[25] Followed by *But I which he* deleted.

[26] Replacing *it should be* deleted.

[27] Written sideways in lefthand margin. Six words later *of* followed by *the* deleted.

[28] Followed by *The Church of England Divines had the ball at their feet agai* deleted.

[29] Followed by *that* deleted. Three words later, *that* is followed by *Dr Bayly of Oxford his old true friend told him* deleted. This reference is to Richard Baylie (c.1586–1667), President of St John's College, Oxford, from 1633–48 and 1660–7 and Dean of Salisbury.

[30] Followed by *recan* deleted.

[31] Followed by *I should have mentiond it before & respecting the series temporum have inserted it after the* [gap] *paragraph* deleted.

[32] Followed by three short, illegible deleted words.

[33] Followed by *afor* deleted.

[34] Followed by *then* deleted.

[35] Followed by an illegible deletion: *sutle?*.

[36] Followed by *p* deleted. In the previous line, the opening bracket is repeated and the first deleted.

[37] Followed by *which* deleted.

[38] Preceded by *from* and followed by *82 to 88 hath* deleted.

[39] Written sideways in the lefthand margin.

[40] Followed by *cancel* deleted.

[41] Line begins with *And leavi* [?] deleted.

[42] Written in lefthand margin.

[43] Written sideways in lefthand margin, replacing *to which* deleted in text.

[44] Followed by *the Act* deleted.

[45] Altered from *Bishop's*.

[46] Followed by *the Bishop* deleted.

[47] Followed by *And Being* [?] deleted.

[48] Followed by *for* deleted.

[49] Replacing *against* deleted.

[50] This lengthy insertion is written sideways in the lefthand margin, most of which it fills.

[51] Followed by *trustees* deleted.

[52] Followed by *I* deleted.

[53] Followed by *f* deleted.

[54] Followed by *p* deleted.

[55] Replacing *the* deleted.

[56] Written sideways in lefthand margin.

[57] Written sideways in lefthand margin.

[58] An illegible letter has been deleted at the start of *my*.

[59] Followed by *&* deleted.

[60] Altered from *saying*.

[61] Followed by *acknow* deleted.

[62] Written sideways in lefthand margin, replacing *otiosa epitheta of Poets* deleted in text.

[63] Followed by *expe* deleted.

[64] Followed by *for* deleted.

[65] Preceded by *For* deleted.

[66] Followed by *boo* deleted.

[67] Written sideways in lefthand margin, keyed to a symbol in the text followed by *What Bishop Sprats* deleted.

[68] Followed by *in* deleted.

[69] Followed by *spent* deleted.

[70] Replacing *how* deleted.

[71] Followed by *his death* deleted.

[72] Replacing a deleted word which itself replaces a further deleted word, neither of which is legible.

[73] Followed by *old* deleted.

[74] Replacing *perhaps* deleted.

[75] Replacing *by* deleted.

[76] Replacing *having been* deleted.

[77] Followed by *repairing from* deleted.

[78] Followed by *him* deleted; a short illegible deletion precedes *pamphlet*.

[79] Followed by an illegible deletion.

[80] Followed by *the* deleted.

[81] Followed by *rem* [?] deleted.

[82] Written sideways in the lefthand margin.

[83] Replacing *mentions* deleted.

[84] Followed by *it mi* deleted.

[85] Followed by *And I* deleted.

[86] Followed by *quo* deleted.

[87] Followed by *w* deleted.

[88] Followed by *Doctor* deleted.

[89] Followed by a deleted letter.

[90] Followed by an illegible deleted word.

DOCUMENT 6A

John Evelyn's letter to William Wotton, 29 March 1696

[1] The Letterbook version is dated 30 March 1696. Despite this, however, the differences between the two versions suggest that the Letterbook copy was written first.

[2] Followed by *at parting* in the Letterbook version.

[3] Followed by *(as you desird)* in the Letterbook version.

[4] Following *from* deleted.

[5] Followed by *of their* deleted.

[6] Followed by *of* deleted.

[7] Followed by *of* deleted.

[8] Followed by *imaginable* in the Letterbook version.

[9] In fact *Incounted*; followed by *it* deleted. Earlier in the sentence, *Knowledge* replaces *Philosophy* deleted in the Letterbook version.

[10] Followed by *glorious &* in the version in Evelyn's Letterbook, as is *which* by *now & then* later in the sentence.

[11] Followed by *Air* deleted. Four words earlier, the Letterbook version has *made in* before *The Spring*.

[12] Followed by *and Anteluca* in the Letterbook version.

[13] Replaced by *insist on* in the Letterbook version.

[14] Followed by *a greate & happy Analyzer* in the Letterbook version, which also has rewording in the next sentence.

[15] Followed by ‹*could not [be] hid*› in the Letterbook version, which also has *the* before *many* and *continualy* before *did*.

[16] The Letterbook version has the slightly different marginal note: *Se Bishop Sanderson* De Juramento *&c 2nd edi: dedicated to him*.

[17] Followed by *compos'd* in the Letterbook version.

[18] Followed by *di* deleted. The Letterbook version has *& thought* deleted after *Opinion* earlier in the sentence; it also has the word *onely* after *Notion* and ‹*divine*› before *Author* later in it.

[19] The Letterbook version has *private* before *devotions*.

[20] Followed by *in* deleted.

[21] Followed by *pla* deleted.

[22] Followed by *Lo* deleted.

[23] *in Gressham Colledge* in the Letterbook version (altered from *at the Royal Society*), which also has *(at the request of the Society)* after *was* later in the next sentence, and *by Sir Edmond King* inserted.

[24] *Relatione* in the Letterbook version.

[25] In the Letterbook version, the bracketed phrase is omitted, and this paragraph appears before the previous three.

[26] Followed by *and shining* in the Letterbook version, which also has *better* before *Judges* earlier in the sentence.

[27] In MS, *is*, replacing *was* deleted. The Letterbook version has *during* instead; in addition, it has *plainly* after *It has* at the start of the sentence.

[28] Preceded by *being* in the Letterbook version.

[29] Followed by *he might* deleted.

[30] Followed by *evidently* in the Letterbook version.

[31] Followed by *(at which I was present)* in the Letterbook version.

[32] The Letterbook has *there have ben, since* instead of *he has yet*.

[33] At this point, Evelyn has put a pen stroke at a slight slant as if to mark the end of the formal part of the letter. The Letterbook has the final paragraph of the main text of the letter, but none of the postscripts.

[34] Followed by *you* in the Letterbook version.

[35] Followed by *'twere* deleted.

161

DOCUMENT 6B

John Evelyn's letter to William Wotton, 12 September 1703

[36] Dated 12 September in Evelyn's copy. However, the differences between the two copies suggest that the latter was written first. See esp. n. 37.

[37] From here to the end of the paragraph the two versions of the letter differ due to the fact that Evelyn's copy was written before Stanhope's letter had arrived.

[38] Part of word repeated and deleted.

[39] Replacing *be* deleted.

[40] Replacing *cort* [?] deleted; *a* has also been deleted, evidently accidentally.

[41] Replacing *when* deleted. Immediately after this, *we* has accidentally been repeated and the second deleted.

[42] Followed by *the* deleted. In the next line there is a 4 above *multituds*.

[43] Here, Evelyn's copy has the following, significantly different, passage (of which *My Wifes... Elizabeth* is inserted in the margin):

> My Wifes Ancestors having been Treasurers of the Navy thence to the Reign of Queen Elizabeth, & ‹exceedingly› increased by my ‹late› Father in [Law] Sir Richard Browne['s] Grandfather, who had the first Imployment under the Greate Earl of Lycester when Governor of the Low Countries ‹in the same Queen's› [replacing *Q.Eliz* deleted] reign; & of [followed by *my father* deleted] Sir Richard's owne ‹dispatches› during his 19 yeare Residence in ‹the Court of› France whither he was sent by Charles the First & continued by his successor.

On Sir Richard Browne (d. 1604), Elizabethan M.P. and officeholder, see Evelyn, *Diary*, vol. iv, p. 303n.

[44] Replacing *to* deleted.

[45] Followed by *on severall other subjects ‹& extravagances›* in Evelyn's copy.

[46] Evelyn's copy has *dayes of his Eclipse* instead of *& almost despaire*.

[47] Here, Evelyn's copy has the following, significantly different, passage:

> This Revolution [followed by *obliging me to be* [?] *& an* [?] deleted], & my Father in Laws attendance at Court (being eldest Clerk of the Counsel) obliging me to be almost perpetually at London, The Intercourse of ‹formal› Letters [followed by *&* deleted] frequent Visits, & constant Meeting at Gresham College ‹succeeding› was very seldome necessary.

[48] After *Seraphic-Love*, Evelyn's copy has the following passage: *which is Long & full of Civility & so may passe for a Compliment, with the Rest, long-since mingld among* [replacing *with* deleted] *my other packetts*. Boyle's letter is now lost.

[49] *no* in Evelyn's copy.

[50] Followed by *which* and *who* deleted.

[51] Replacing *the Rope* [?] deleted.

[52] Evelyn's copy here adds *& for severall ‹Engins &› Inventions*.

[53] Here an insertion was begun but abandoned after one letter, which deleted: *u*?

[54] Followed by *were* deleted.

[55] Replacing *gave* deleted.

[56] Evelyn's copy adds at this point: *‹This was›* [replacing *'Tis the* deleted] *the foundation of the ‹vast› estate he since Injoyd.*

⁵⁷ Evelyn's copy here adds *concerning the Bills of Mortality*.
⁵⁸ Followed by *he was* deleted.
⁵⁹ Evelyn's copy here adds: *& some who yet sat at the helm*.
⁶⁰ Here Evelyn's copy adds: *infinitly Industrious, Nothing was to hard for him* (there is a line through *Nothing...for*).
⁶¹ Replacing *&* deleted. Three words later, *Inferences* is followed by *passe* deleted.
⁶² Replacing *quakers* deleted; *Quaker* preceding *Anabaptist* appears in the margin.
⁶³ Followed by *&* deleted, as is *Countenances*.
⁶⁴ Followed by *Vicess* [?] deleted.
⁶⁵ Preceded by *dextrously* in Evelyn's copy.
⁶⁶ Followed by *poore Rich &* in Evelyn's copy.
⁶⁷ Followed by *as* [?] deleted.
⁶⁸ Followed by *& other uncommon Arts* in Evelyn's copy.
⁶⁹ Followed by *or* deleted. Later in the sentence, *he* is altered from *they*.
⁷⁰ Followed by *R* deleted.
⁷¹ Obscured by blot in the MS.
⁷² Replacing *stay* deleted.
⁷³ Followed by *& ‹of› the kindnesse which allways continued between them* in Evelyn's copy.
⁷⁴ Replacing *and* deleted.
⁷⁵ At the bottom of fol. 57, Evelyn has written *verte*.
⁷⁶ Followed by *&* deleted.
⁷⁷ At this point, instead of the phrase about Urquhart, Evelyn's copy of the letter has the following passage:

> & I am able to bring my own Pedigree from [followed by *Ann* deleted] one Evelyn, Nephew to Androgius, who brought Julius Caesar into Britain the second time: will [replacing *would* deleted] you not smile at this? whilst *Onslow, Hatton & Evelyn* came ‹I suppose, much at the same time› [replacing an illegible deleted word] out of Shropshire ‹into Surry, and ajacent Counties› (from places still retaining their names), some [preceded by *some* deleted] time during the Barrons Warrs.

⁷⁸ Followed by *so* deleted.
⁷⁹ Followed in Evelyn's copy by the alternative wording: *perswading me to apply a ‹miraculous› Plaster of his*.
⁸⁰ Followed by *Dion Cass: all* deleted. Evelyn's copy here has simply *Herodotus, Thucydides & the rest of that Classe*. Earlier, the sentence is there slightly differently worded, describing Evelyn's grandson as being *master of a ‹handsome› Style*.
⁸¹ Evelyn's copy has the following passage at the very end: *The Master of Trinity* [i.e. Richard Bentley] *was often* [?] *at St James's without being so kind as to visite the Clinic*

DOCUMENT 6C

George Stanhope to Evelyn, 6 September 1703, enclosing transcript of Roger Boyle's funerary inscription

⁸² Evidently altered from something else: *of* [?].

DOCUMENT 7

Thomas Dent's letter to William Wotton, 20 May 1699

[1] Replacing *if* deleted.
[2] Brackets deleted before *he never* and after *failed*.
[3] Followed by *as* [?] deleted.
[4] Replacing *thy*, deleted.
[5] Replacing an illegible deletion, apparently *sons* written twice.
[6] Followed by *was* deleted.
[7] Followed by *his constant custom* deleted.
[8] Followed by *for* deleted.
[9] Altered from *prepared*; *to* is lost in the binding.
[10] Replacing *of* deleted.
[11] Followed by *Letters* deleted.
[12] Altered from a different, illegible word.
[13] Replacing *Omen* deleted.
[14] Replacing *th*, deleted.

DOCUMENT 8

James Kirkwood's letter to William Wotton, 22 June 1702

[1] Altered from *my* [?] deleted.
[2] Followed by *This* repeated when Kirkwood decided to begin a new paragraph.
[3] Altered from another word: *like* [?].
[4] Altered from *one*.
[5] Preceded by *who have* [?] deleted.
[6] Altered from *bin*.
[7] Altered from *Kirck*.
[8] Altered from *of*.
[9] Altered from *abought*.
[10] Altered from *whi* [?].
[11] Altered from *bin*. Two words later, *on* is in fact *own*.
[12] Altered from *more*.
[13] Followed by a faint *D* in the MS.

DOCUMENT 9

Chapter from William Wotton's life of Boyle

[1] Altered from *Instruments*.
[2] Replacing *Æquilibrium of Liquors* deleted.
[3] Altered from *ways*.
[4] ( ) added above line in MS after *meant*.
[5] Followed by *are* deleted.

[6] Followed by *have* deleted. Earlier in the sentence, an extra stroke has been deleted at the end of *meant*.

[7] Replacing *I* [?] deleted.Earlier in the sentence, *useless* is followed by a deleted diagonal stroke in the text.

[8] Followed by *II* [?] deleted.

[9] Followed by *The* deleted.

[10] Followed by *a* [?] deleted.

[11] Replacing *part* deleted.

[12] Replacing *piece* deleted.

[13] Replacing *fastned* deleted.

[14] Followed by *wel* deleted (at end of line).

[15] Followed by *upwards* deleted.

[16] Followed by *no* deleted.

[17] Followed by *the Cylinder will be so full of Air, that* deleted.

[18] Replacing *no* deleted.

[19] Followed by *Wooll* deleted.

[20] Followed by *is* deleted. Four words later, *To* followed by *that* deleted.

[21] Just above *Domm* there is a small half circle.

[22] Replacing *a great deal* deleted.

[23] Followed by *makes an* deleted.

[24] Followed by *in* [?] deleted.

[25] Replacing *before* deleted.

[26] Followed by a cross in the text.

[27] Altered from *Volumes*.

[28] *which* is accidentally repeated in the MS.

[29] Replacing *of which that made one* deleted.

[30] Replacing *took another Glass* deleted.

[31] The page is damaged here, but the word appears to be *Now*.

[32] Followed by *so its* deleted. Four words later, *cemented* followed by *to its* [?] deleted.

[33] Replacing *a* deleted.

[34] Followed by *as* deleted.

[35] Followed by *half* deleted.

[36] Followed by *XIII* deleted.

[37] Followed by *were let down into the Receiver,* deleted.

[38] Followed by *in* deleted.

[39] Followed by *exh* deleted.

[40] Replacing *were* deleted.

[41] Replacing *Air* deleted.

[42] Followed by *to appearance* deleted.

[43] Followed by *& the Air perhaps not so thorog* deleted.

[44] Followed by *w* deleted.

[45] Replacing *Vacuum* deleted. Four words later, *be* followed by *as* deleted.

[46] Followed by *in a Letter* deleted.

[47] Followed by *would* deleted.

[48] Replacing *left in* deleted.

[49] Followed by *w* deleted.

50 Followed by *Plate* deleted.

51 Altered from *extreame*.

52 Followed by *which being well exhausted,* deleted.

53 Followed by *R* [?] deleted.

54 Followed by *Muscle* deleted.

55 Replacing *might* deleted. After *probably, be* is deleted.

56 Replacing *might be* deleted.

57 Followed by *togeth* deleted.

58 Followed by *hermetically* deleted.

59 Replacing *stopt* deleted.

60 Followed by deleted comma.

61 Altered from *Fishes.*

62 Replacing *its* deleted.

63 Followed by *from* deleted.

64 *be* is accidentally repeated in the MS.

65 Followed by *st* deleted.

66 Replacing *many* deleted.

67 Followed by *that* deleted.

68 Followed by *Heat* deleted.

69 Replacing *the open* deleted.

70 Followed by *perfectly* deleted.

71 Replacing *open* deleted.

72 Altered from *xvi.*

73 Followed by a deleted comma.

74 Followed by a small illegible deletion: *Us* [?].

75 Followed by *be* deleted.

76 Followed by the start of a letter deleted with a diagonal stroke.

77 Followed by *without* deleted.

78 Followed by *stem* deleted.

79 Replacing *it* deleted.

80 Followed by *a* deleted.

81 Altered from *had.*

82 Replacing *Myrrhe* deleted.

83 There is a mark above *quick* in the MS: *ea* [?].

84 Followed by *a* deleted.

85 Replacing *choke* deleted.

86 Followed by *ab* deleted.

87 Replacing *This* deleted. After *common, Air* has been deleted.

88 Followed by *He* in thick ink deleted.

89 Followed by *distances* deleted.

90 Replacing *several* deleted.

91 Replacing *see* deleted.

92 A punctuation mark has been deleted with a diagonal stroke after *sides.*

93 Followed by *those* deleted.

94 Followed by *equal* [?] deleted.

95 Followed by *reciprocal* [?] deleted.

96 Followed by *s* [?] deleted.

[97] Replacing *was* deleted.

[98] Followed by *ris* [?] deleted.

[99] The right hand edge of this page has been torn in two places so that the ends of several lines are missing and some words have wholly or partially disappeared. These are shown in square brackets.

[100] Replacing *that* deleted.

[101] Replacing *is* deleted.

[102] Followed by *this* deleted.

[103] The bottom of this page has been torn. After *strongly [wi]thout any [visible]* has been deleted. At the beginning of the first two lines of fol. 32 insertion marks appear.

[104] Followed by *ol* [?] deleted.

[105] Followed by *Receivers* deleted.

[106] Followed by *made* deleted.

[107] Followed by *Ba* deleted.

[108] *122* apparently altered from *222*. Later in the sentence, there is what appears to be a symbol after *Atmosphere's*.

[109] Followed by *wi* deleted.

BIBLIOGRAPHY

Ames, William, *Conscience With the Power and Cases thereof. Divided into V Bookes* (London, 1639).

Anglesey, Arthur Annesley, Earl of, *Memoirs*, ed. Sir Peter Pett (London, 1693).

Appleby, A. B., 'Nutrition and Disease: the Case of London, 1550–1750', *Journal of Interdisciplinary History*, 6 (1975), pp. 1–22.

Ashcraft, Richard, 'Latitudinarianism and Toleration: Historical Myth versus Political History', *Philosophy, Science and Religion in England 1640–1700*, eds Richard Kroll, Richard Ashcraft and Perez Zagorin (Cambridge University Press, Cambridge: 1992), pp. 151–77.

Aubrey, John, *Brief Lives, Chiefly of Contemporaries*, ed. Andrew Clark, 2 vols (Clarendon Press: Oxford, 1898).

Bahlman, D. W. R., *The Moral Revolution of 1688* (Yale University Press: New Haven, 1957).

Baillet, Adrian, *La Vie de Monsieur Descartes*, 2 vols (Paris, 1691).

Bannister, Mark, *Privileged Mortals: the French Heroic Novel 1630–60* (Clarendon Press: Oxford, 1983).

Barlow, Thomas, *Several Miscellaneous and Weighty Cases of Conscience* (London, 1692).

Barlow, Thomas, *The Genuine Remains*, ed. Sir Peter Pett (London, 1693).

Bayle, Pierre, *A General Dictionary, Historical and Critical*, eds J. P. Bernard, Thomas Birch, John Lockman, etc., 10 vols (London, 1734–41).

Beddard, Robert, 'Vincent Alsop and the Emancipation of Restoration Dissent', *Journal of Ecclesiastical History*, 24 (1973), pp. 161–84.

Bentley, Richard, *Works*, ed. A. Dyce, 3 vols (London, 1836–8).

Bentley, Richard, *Correspondence*, 2 vols (London, 1842).

Biographia Britannica: or, the Lives of the Most eminent Persons Who have flourished in Great Britain and Ireland, From the earliest Ages, down to the present Times, eds William Oldys and Joseph Towers, 6 vols (London, 1747–66).

Birch, Thomas, *The History of the Royal Society of London*, 4 vols (London, 1756–7).

Boas, Marie: see Hall, M. B.

Boswell, James, *Life of Johnson*, ed. George Birkbeck Hill, rev. L. F. Powell, 6 vols (Clarendon Press: Oxford, 1934).

Bottrall, Margaret, *Every Man a Phoenix: Studies in Seventeeth-century Autobiography* (John Murray: London, 1958).

Boulton, Richard, *The Works of the Honourable Robert Boyle, Esq., Epitomiz'd*, 4 vols (London, 1699–1700).

Boulton, Richard, *The Theological Works of the Honourable Robert Boyle, Esq., Epitomiz'd*, 3 vols (London, 1715).

Boulton, Richard, 'The Life of the Honourable Robert Boyle, Esq. ' in *Theological Works* (above), vol. i, pp. 1–372.

Boyce, Benjamin, *The Theophrastan Character in England to 1642* (Harvard University Press: Cambridge, Mass., 1947).

Boyle, Roger, *Parthenissa, A Romance*, 6 vols (Waterford, 1651; London, 1654–69).

Bramston, Sir John, *Autobiography*, ed. T. W. Bramston (Camden Society: London, 1845).

Budgell, Eustace, *Memoirs of the Lives and Characters of the Illustrious Family of the Boyles*, 3rd edition (London, 1737).

Buehler, C. F., 'A Projected but Unpublished Edition of the "Life and Works" of Robert Boyle', *Chymia*, 4 (1953), pp. 79–83.

Bunyan, John, *Grace Abounding to the Chief of Sinners and The Pilgrim's Progress*, ed. Roger Sharrock (Oxford University Press: London, 1966).

Burnet, Gilbert, *The Memoires of the Lives and Actions of James and William Dukes of Hamilton and Castleherald &c.* (London, 1677).

Burnet, Gilbert, *A Sermon Preached at St Dunstans in the West at the Funeral of Mrs Anne Seile* (London, 1678).

Burnet, Gilbert, *A History of the Reformation of the Church of England*, 3 vols (London, 1679–1715).

Burnet, Gilbert, *Some Passages of the Life and Death Of the Right Honourable John Earl of Rochester* (London, 1680).

Burnet, Gilbert, *The Life and Death of Sir Matthew Hale, Kt.* (London, 1682).

Burnet, Gilbert, *A Sermon Preached at the Funeral of Mr James Houblon* (London, 1682).

Burnet, Gilbert, *The Life of William Bedell* (London, 1685).

Burnet, Gilbert, *Some Letters. Containing, An Account of what seemed most remarkable in Switzerland, Italy, &c.* (Rotterdam, 1686).

Burnet, Gilbert, *A Sermon Preached at the Funeral of the Right Honourable Anne, Lady-Dowager Brook* (London, 1691).

Burnet, Gilbert, *A Sermon Preached at the Funeral of the Most Reverend Father in God John By the Divine Providence Lord Archbishop of Canterbury* (London, 1694).

Burnet, Gilbert, *History of My Own Time*, ed. Osmund Airy, 2 vols (Oxford, 1897–1900).

Burtchaell, G. D., and Sadleir, T. U., *Alumni Dublinenses* (Alex Thom & Co.: Dublin, 1935).

Butler, Martin, *Theatre and Crisis 1632–42* (Cambridge University Press: Cambridge, 1984).

Campbell, John, ed. and trans. J. H. Cohausen, *Hermippus Redivivus* (London, 1744).

Canny, Nicholas, *The Upstart Earl: a Study of the Social and Mental World of Richard Boyle, First Earl of Cork, 1566–1643* (Cambridge University Press: Cambridge, 1982).

Carter, Harry, *A History of the Oxford University Press, Volume I: to the Year 1780* (Clarendon Press: Oxford, 1975).

Caspari, Fritz, *Humanism and the Social Order in Tudor England*, University of Chicago Press, 1954; reprinted as 'Classics in Education no. 34' (Teachers College Press: New York, 1968).

Cherry, Bridget, and Pevsner, Nikolaus, *London 2: South* (The Buildings of England) (Penguin Books: Harmondsworth, 1983).

Ciampoli, Giovanni, *Prose di Monsignor Giovanni Ciampoli Dedicate All'Eminentissimo e Reverendissimo Signor Cardinal Girolamo Colonna* (Rome, 1649).

Clericuzio, Antonio, 'Carneades and the Chemists: a Study of *The Sceptical Chymist* and its Impact on Seventeenth-century Chemistry', in Hunter, *Robert Boyle Reconsidered* (below), pp. 79–90.

Collinson, Patrick, '"A Magazine of Religious Patterns": an Erasmian Topic Transposed in English Protestantism', *Studies in Church History*, 14 (1977), pp. 223–49.

Collinson, Patrick, *The Religion of Protestants: the Church in English Society 1559–1625* (Clarendon Press: Oxford, 1982).

Cook, H. J., *The Decline of the Old Medical Regime in Stuart London* (Cornell University Press: Ithaca, 1986).

Cook, H. J., 'Sir John Colbatch and Augustan Medicine: Experimentalism, Character and Entrepreneurialism', *Annals of Science*, 47 (1990), pp. 475–505.

Coward, Barry, *Oliver Cromwell* (Longman: London, 1991).

Crocker, T. C. ed., *Autobiography of Mary Countess of Warwick* (Percy Society: London, 1848).

Cudworth, Ralph, *The True Intellectual System of the Universe*, ed. J. L. Mosheim (trans. J. Harrison), 3 vols (London, 1845).

Curtis, T. C., and Speck, W. A., 'The Societies for the Reformation of Manners: a Case Study in the Theory and Practice of Moral Reform', *Literature and History*, 3 (1976), pp. 45–64.

Delany, Paul, *British Autobiography in the Seventeenth Century* (Routledge & Kegan Paul: London, 1969).

Dews, Nathan, *The History of Deptford*, 2nd edition (London and Deptford, 1884).

Digby, Sir Kenelm, *Private Memoirs* (London, 1827).

Ebner, Dean, *Autobiography in Seventeenth-century England* (Mouton: The Hague, 1971).

Evelyn, John, *Memoirs Illustrative of the Life and Writings of John Evelyn Esq., F. R. S.*, ed. William Bray, 2 vols (London, 1818).

Evelyn, John, *Miscellaneous Writings*, ed. William Upcott (London, 1825).

Evelyn, John, *Memoirs*, ed. William Bray, 5 vols (London, 1827).

Evelyn, John, *Diary and Correspondence*, ed. William Bray [and John Forster], 4 vols (London, 1850–2).

Evelyn, John, *Diary*, ed. E. S. de Beer, 6 vols (Clarendon Press: Oxford, 1955).

Fahie, J. J., *Memorials of Galileo Galilei* (Leamington and London, for the author, 1929).

Feingold, Mordechai, 'Galileo in England: the First Phase', in *Novita Celesti e Crisi del Sapere*, ed. P. Galluzzi (Florence, 1984), pp. 411–20.

Fisch, Harold, 'Bishop Hall's Meditations', *Review of English Studies*, 25 (1949), pp. 210–21.

Fisch, Harold, *Jerusalem and Albion: the Hebraic Factor in Seventeenth-century Literature* (Routledge & Kegan Paul: London, 1964).

[Fitzgerald, Robert], *Salt-Water Sweetned* (London, 1683).

Fitzmaurice, Lord Edmond, *The Life of Sir William Petty 1632–87* (London, 1895).

Foster, Joseph, *Alumni Oxonienses 1500–1800*, 4 vols (Oxford, 1891–2).

Foxcroft, H. C., *A Supplement to Burnet's History of My Own Time* (Clarendon Press: Oxford, 1902).

Frank, R. G., *Harvey and the Oxford Physiologists: a Study of Scientific Ideas and Social Interaction* (University of California Press: Berkeley and Los Angeles, 1980).

Fulton, J. F. *A Bibliography of the Honourable Robert Boyle*, 2nd edition (Clarendon Press: Oxford, 1961).

Galilei, Galileo, *Dialogues concerning Two New Sciences*, trans. Henry Crew and Alfonso de Salvio, intr. Antonio Favaro (Dover Publications: New York, 1954).

Garrison, J. D., *Dryden and the Tradition of Panegyric* (University of California Press: Berkeley and Los Angeles, 1975).

Gascoigne, John, *Cambridge in the Age of the Enlightenment* (Cambridge University Press: Cambridge, 1989).

Gassendi, Pierre, *The Mirrour of True Nobility & Gentility. Being the Life of the Renowned Nicolaus Claudius Fabricius, Lord of Peiresk*, Eng. trans. William Rand (London, 1657).

Glanvill, Joseph, *Plus Ultra: or, the Progress and Advancement of Knowledge Since the Days of Aristotle* (London, 1668).

Goldie, Mark, 'Sir Peter Pett, Sceptical Toryism and the Science of Toleration in the 1680s', *Studies in Church History*, 21 (1984), pp. 247–73.

Goldie, Mark, 'Priestcraft and the Birth of Whiggism', in *Political Discourse in Early Modern England*, eds Nicholas Phillipson and Quentin Skinner (Cambridge University Press: Cambridge, 1993), pp. 209–31.

Golinski, Jan V., 'Peter Shaw: Chemistry and Communication in Augustan England', *Ambix*, 30 (1983), pp. 19–29.

Greene, Thomas, 'The Flexibility of the Self in Renaissance Literature', in *The Discipline of Criticism*, eds Peter Demetz et al. (Yale University Press: New Haven, 1968), pp. 241–64.

Greyerz, Kaspar von, *Vorsehungsglaube und Kosmologie: Studien zu Englischen Selbstzeugnissen des 17 Jahrhunderts*, Veröffentlichungen des Deutschen Historischen Institutes London, vol. 25 (Göttingen and Zurich, 1990).

Grosart, A. B., ed., *The Lismore Papers*, 1st Series, 5 vols (London, 1886).

Grosart, A. B., ed., *The Lismore Papers*, 2nd Series, 5 vols (London, 1887–8).

Grotius, Hugo, *Opera Omnia Theologica*, 3 vols (Amsterdam, 1679).

Guerlac, Henry, 'Sir Isaac and the Ingenious Mr Hauksbee', in *Melanges Alexandre Koyré*, 2 vols (Hermann: Paris, 1964), vol. i, pp. 228–53.

Hale, Thomas, *An Account of Several New Inventions and Improvements Now necessary for England [with] The New Invention of Mill'd Lead* (London, 1691).

Hall, A. R., 'William Wotton and the History of Science', *Archives Internationales d'Histoire des Sciences*, 5 (1948), pp. 1047–62.

[Hall], Marie Boas, *Robert Boyle and Seventeenth-Century Chemistry* (Cambridge University Press: Cambridge, 1958).

Hall, M. B., 'Henry Miles, F. R. S. (1698–1763) and Thomas Birch, F.R.S. (1705–66)', *Notes and Records of the Royal Society*, 18 (1963), 39–44.

Hall, M. B., *Robert Boyle on Natural Philosophy: An Essay with Selections from his Writings* (Indiana University Press: Bloomington, 1965).

Hall, M. B., *Promoting Experimental Learning: Experiment and the Royal Society 1660–1727* (Cambridge University Press: Cambridge, 1991).

Haller, William, *The Rise of Puritanism* (Columbia University Press: New York, 1938).

Hardison, O. B., *The Enduring Monument* (University of North Carolina Press: Chapel Hill, 1962).

Harwood, John T., ed., *The Early Essays and Ethics of Robert Boyle*

(Southern Illinois University Press: Carbondale and Edwardsville, 1991).

Harwood, John, 'Science Writing and Writing Science: Boyle and Rhetorical Theory', in Hunter, *Robert Boyle Reconsidered* (below), pp. 37–56.

Hauksbee, Francis, *Physico-Mechanical Experiments on Various Subjects* (London, 1709).

Hearne, Thomas, *Remarks and Collections*, ed. C. E. Doble et al., 11 vols (Oxford Historical Society: Oxford, 1885–1921).

Herbert of Cherbury, Edward, Lord, *Life*, ed. J. M. Shuttleworth (Oxford University Press: London, 1976).

Hexter, J. H., 'The Education of the Aristocracy in the Renaissance' in *Reappraisals in History* (Longmans: London, 1961), pp. 45–70.

Hunter, Michael, *Science and Society in Restoration England* (Cambridge University Press: Cambridge, 1981); reprint edition (Gregg Revivals: Aldershot, 1992).

Hunter, Michael, 'The Problem of "Atheism" in Early Modern England', *Transactions of the Royal Historical Society*, 5th series 35 (1985), pp. 135–57.

Hunter, Michael, 'Science and Heterodoxy: an Early Modern Problem Reconsidered', in *Reappraisals of the Scientific Revolution*, eds D. C. Lindberg and R. S. Westman (Cambridge University Press: Cambridge, 1990), pp. 437–60.

Hunter, Michael, 'Alchemy, Magic and Moralism in the Thought of Robert Boyle', *British Journal for the History of Science*, 23 (1990), pp. 387–410.

Hunter, Michael, *Letters and Papers of Robert Boyle: A Guide to the Manuscripts and Microfilm* (University Publications of America: Bethesda, Md., 1992).

Hunter, Michael, 'Casuistry in Action: Robert Boyle's Confessional Interviews with Gilbert Burnet and Edward Stillingfleet, 1691', *Journal of Ecclesiastical History*, 44 (1993), pp. 80–98.

Hunter, Michael, 'The Conscience of Robert Boyle: Functionalism, "Dysfunctionalism" and the Task of Historical Understanding', in *Renaissance and Revolution: Humanists, Scholars, Craftsmen and Natural Philosophers in Early Modern Europe*, eds J. V. Field and F. A. J. L. James (Cambridge University Press: Cambridge, 1993), pp. 147–59.

Hunter, Michael, ed., *Robert Boyle Reconsidered* (Cambridge University Press: Cambridge, 1994).

Hunter, Michael, *The Royal Society and its Fellows 1660–1700: the Morphology of an Early Scientific Institution*, new edition (British Society for the History of Science: Oxford, 1994).

Hunter, Michael, 'John Evelyn in the 1650s: a Virtuoso in Quest of a Role', in id., *Science and the Shape of Orthodoxy: Intellectual Change in Late Seventeenth-century Britain* (The Boydell Press: Woodbridge, forthcoming).

Hunter, Michael, 'How Boyle Became a Scientist', *History of Science*, 33 (1995), forthcoming.

Hunter, Michael, 'Robert Boyle and the Dilemma of Biography in the Age of the Scientific Revolution', in *Telling Lives in Science: Studies in Scientific Biography*, eds Michael Shortland and Richard Yeo (Cambridge University Press: Cambridge, forthcoming).

Jacob, J. R., *Robert Boyle and the English Revolution: a Study in Social and Intellectual Change* (Burt Franklin: New York, 1977).

Johnston, G. P., 'Notices of a Collection of MSS Relating to the Circulation of the Irish Bibles of 1685 and 1690 in the Highlands and the Association of the Rev. James Kirkwood Therewith', *Papers of the Edinburgh Bibliographical Society*, 6 (1906 for 1901–4), pp. 1–18.

Joy, Lynn S., *Gassendi the Atomist: Advocate of History in an Age of Science* (Cambridge University Press: Cambridge, 1987).

Keill, John, 'Epistola ad Cl[arissimum] virum *Gulielmum Cockburn*, Medicinae Doctorem. In qua Leges Attractionis aliaque Physices Principia traduntur', *Philosophical Transactions*, 26 (1708), pp. 97–110.

Kellaway, William, *The New England Company 1647–1776* (Longmans: London, 1961).

Kelso, Ruth, *The Doctrine of the English Gentleman in the Sixteenth Century*, University of Illinois Studies in Language and Literature, 14 (1929).

Kenyon, J. P., *Revolution Principles: the Politics of Party 1689–1720* (Cambridge University Press: Cambridge, 1977).

Keynes, Sir Geoffrey, *A Bibliography of Sir William Petty* (Clarendon Press: Oxford, 1971).

Korshin, Paul, 'The Development of Intellectual Biography in the Eighteenth Century', *Journal of English and Germanic Philology*, 73 (1974), pp. 513–23.

Lansdowne, Marquess of, ed., *The Petty Papers*, 2 vols (Constable: London, 1927).

Lansdowne, Marquess of, ed., *The Double Bottom or Twin-Hulled Ship of Sir William Petty* (Roxburghe Club: Oxford, 1931).

Larcom, T. A., ed., *The History of the Survey of Ireland Commonly Called the Down Survey* (Dublin, 1851).

Levine, Joseph M., *The Battle of the Books: History and Literature in the Augustan Age* (Cornell University Press: Ithaca, 1991).

MacLeod, Christine, *Inventing the Industrial Revolution: the English Patent System 1660–1800* (Cambridge University Press: Cambridge, 1988).

Maddison, R. E. W., 'Robert Boyle and Some of his Foreign Visitors

[Studies in the Life of Robert Boyle, F. R. S., Parts 1 and 4]', *Notes and Records of the Royal Society* 9 (1951), pp. 1–35, 11 (1954), pp. 38–53.

Maddison, R. E. W., 'Salt Water Freshened [Studies in the Life of Robert Boyle, F. R. S., Part 2]', *Notes and Records of the Royal Society*, 9 (1952), pp. 196–216.

Maddison, R. E. W., 'A Summary of Former Accounts of the Life and Work of Robert Boyle', *Annals of Science*, 13 (1957), pp. 90–108.

Maddison, R. E. W., 'Robert Boyle and the Irish Bible', *Bulletin of the John Rylands Library*, 41 (1958), pp. 81–101.

Maddison, R. E. W., 'The Portraiture of the Honourable Robert Boyle', *Annals of Science*, 15 (1959), pp. 141–214.

Maddison, R. E. W., 'The Grand Tour [Studies in the Life of Robert Boyle, F. R. S., Part 7]', *Notes and Records of the Royal Society* 20 (1965), pp. 51–77.

Maddison, R. E. W., 'Galileo and Boyle: a Contrast', *Saggi su Galileo Galilei*, ed. C. Maccagni, 2 vols (G. Barbèra: Florence, 1967), vol. ii, pp. 348–61.

Maddison, R. E. W., *The Life of the Honourable Robert Boyle, F. R. S.* (Taylor & Francis: London, 1969).

Manget, J. T., ed., *Bibliotheca Chemica Curiosa*, 2 vols (Geneva, 1702).

Marshall, John, 'The Ecclesiology of the Latitude-men 1660–1689: Stillingfleet, Tillotson and "Hobbism"', *Journal of Ecclesiastical History*, 36 (1985), pp. 407–27.

McKeon, Michael, *The Origins of the English Novel 1600–1740* (Radius: London, 1988).

Mendelson, Sara H., 'Stuart Women's Diaries and Occasional Memoirs', in Mary Prior, ed., *Women in English Society 1500–1800* (Methuen: London, 1985), pp. 181–210.

Mendelson, Sarah H., *The Mental World of Stuart Women* (Harvester Press: Brighton, 1987).

Middleton, W. E. K., *The History of the Barometer* (Johns Hopkins Press: Baltimore, 1964).

Middleton, W. E. K., *The Experimenters: a Study of the Accademia del Cimento* (Johns Hopkins Press: Baltimore, 1971).

Middleton, W. E. K., ed., *Lorenzo Magalotti at the Court of Charles II: his 'Relazione d'Inghilterra' of 1668* (Wilfrid Laurier University Press: Waterloo, 1980).

More, L. T., *The Life and Works of the Honourable Robert Boyle* (Oxford University Press: New York, 1944).

Moreri, Louis, *The Great Historical, Geographical and Poetical Dictionary*, 2 vols (London, 1694).

Moreri, Louis, *The Great Historical, Geographical, Genealogical and Poetical Dictionary*, ed. Jeremy Collier, 2 vols (London, 1701).

Nethercot, A. H., *Abraham Cowley: the Muse's Hannibal* (Oxford University Press: London, 1931).

Newton, Sir Isaac, *Opticks*, reprint of 1730 edition, new edition (Dover Publications: New York, 1979).

Nichols, John, *Literary Anecdotes of the Eighteenth Century*, 8 vols (London, 1812–14).

North, Roger, *The Lives of the Norths*, ed. Augustus Jessopp, 3 vols (London, 1890).

North, Roger, *General Preface and Life of Dr John North*, ed. Peter Millard (University of Toronto Press: Toronto, 1984).

Osborn, J. M., 'Thomas Birch and the *General Dictionary* (1734–41)', *Modern Philology*, 36 (1938), pp. 25–46.

Oster, Malcolm, ' "The Beame of Divinity": Animal Suffering in the Early Thought of Robert Boyle', *British Journal for the History of Science*, 22 (1989), pp. 151–79.

Oster, Malcolm, 'Biography, Culture and Science: the Formative Years of Robert Boyle', *History of Science*, 31 (1993), pp. 177–226.

Pascal, Blaise, *The Physical Treatises*, trans. I. H. B. and A. G. H. Spiers, intro. Frederick Barry (Columbia University Press: New York, 1937).

Paul, C. B., *Science and Immortality: the Éloges of the Paris Academy of Sciences (1699–1791)* (University of California Press: Berkeley and Los Angeles, 1980).

Pepys, Samuel, *Private Correspondence and Miscellaneous Papers 1679–1703*, ed. J. R. Tanner, 2 vols (G. Bell and Sons: London, 1926).

Pepys, Samuel, *Naval Minutes*, ed. J. R. Tanner, Navy Records Society, vol. 60 (1926).

Pepys, Samuel, *Letters and the Second Diary*, ed. R. G. Howarth (J. M. Dent & Sons: London, 1932).

[Pet]t, [Pete]r, *A Discourse concerning Liberty of Conscience* (London, 1661).

Pett, Sir Peter, *The Happy Future State of England* (London, 1688).

Piper, David, *The English Face*, revised edition, ed. Malcolm Rogers (National Portrait Gallery: London, 1992).

Plomer, H. R., *A Dictionary of Printers and Booksellers who were at work in England, Scotland and Wales 1668–1725* (Bibliographical Society: Oxford, 1922).

Potter, Lois, *Secret Rites and Secret Writing: Royalist Literature 1641–60* (Cambridge University Press: Cambridge, 1989).

Principe, Lawrence, 'Boyle's Alchemical Pursuits' in Hunter, *Robert Boyle Reconsidered* (above), pp. 91–105.

Principe, Lawrence, 'Style and Thought of the Early Boyle: Discovery of the 1648 Manuscript of *Seraphic Love*', *Isis*, 85 (1994), 247–60.

Prynne, William, *Canterburies Doome Or the First Part of a Compleat History of the Commitment, Charge, Tryall, Condemnation, Execution of William Laud Late Arch-Bishop of Canterbury* (London, 1646).

Redondi, Pietro, *Galileo Heretic*, trans. Raymond Rosenthal (Allen Lane: London, 1988).

Sanderson, Robert, *De Obligatione Conscientiae Praelectiones Decem* (London, 1660).

Sanderson, Robert, *Ten Lectures on the Obligation of Humane Conscience*, trans. Robert Codrington (London, 1660) [this work has a duplicate title-page on which it is entitled *Several Cases of Conscience Discussed in 10 Lectures in the Divinity School at Oxford*].

Selden, John, *The Reverse or Back-face of the English Janus*, trans. Redman Westcot (London, 1682).

Shapin, Steven, 'Pump and Circumstance: Robert Boyle's Literary Technology', *Social Studies of Science* 14 (1984), pp. 481–520.

Shapin, Steven, and Schaffer, Simon, *Leviathan and the Air-Pump: Hobbes, Boyle and the Experimental Life* (Princeton University Press: Princeton, 1985).

Shaw, Peter, *The Philosophical Works of the Honourable Robert Boyle, Esq.*, 3 vols (London, 1725).

Sibbald, Sir Robert, *Memoirs*, ed. F. P. Hutt (Oxford University Press: London, 1932).

The Spectator, ed. D. F. Bond, 5 vols (Clarendon Press: Oxford, 1965).

Sprat, Thomas, *The History of the Royal Society (1667)*, ed. J. I. Cope and H. W. Jones (Routledge & Kegan Paul: London, 1959).

Spurr, John, '"Latitudinarianism" and the Restoration Church', *Historical Journal*, 31 (1988), pp. 61–82.

Spurr, John, *The Restoration Church of England 1646–89* (Yale University Press: New Haven and London, 1991).

Stauffer, D. A., *English Biography before 1700* (Harvard University Press: Cambridge, Mass., 1930).

Stauffer, D. A., *The Art of Biography in Eighteenth-century England* (Princeton University Press: Princeton, 1941).

Stewart, Larry, *The Rise of Public Science: Rhetoric, Technology and Natural Philosophy in Newtonian Britain 1660–1750* (Cambridge University Press: Cambridge, 1992).

Stubbs, Mayling, 'John Beale, Philosophical Gardener of Herefordshire', *Annals of Science*, 39 (1982), pp. 463–89; 46 (1989), pp. 323–63.

Swift, Jonathan, 'A Meditation upon a Broomstick', *A Tale of a Tub and other Early Works*, ed. Herbert Davis (Blackwell: Oxford, 1939), pp. 239–40.

Taylor, E. G. R., *The Mathematical Practitioners of Tudor and Stuart England* (Cambridge University Press: Cambridge, 1954).

Todd, Margo, *Christian Humanism and the Puritan Social Order* (Cambridge University Press: Cambridge, 1987).

Tombes, John, *Anti-Paedobaptism, Being a Full Review of the Dispute concerning Infant-Baptism*, 3 parts (London, 1652–7).

Townsend, Dorothea, *The Life and Letters of the Great Earl of Cork* (Duckworth & Co.: London, 1904).

Trevor-Roper, Hugh, *Catholics, Anglicans and Puritans: Seventeenth-century Essays* (Secker and Warburg: London, 1987).

Trott, F. J., 'Prelude to Restoration: Laudians, Conformists and the Struggle for "Anglicanism" in the 1650s' (University of London PhD thesis, 1992).

Tyacke, Nicholas, 'Religious Controversy in Restoration England', in *The History of the University of Oxford: III: the Seventeenth Century*, ed. N. Tyacke (Oxford University Press: Oxford, forthcoming).

Urquhart, Sir Thomas, *The Jewel*, ed. R. D. S. Jack and R. J. Lyall (Scottish Academic Press: Edinburgh, 1983).

Vane, Sir Henry, *The Substance of What Sir Henry Vane Intended to have Spoken upon the Scaffold on Tower-Hill, at the time of his Execution, being the 14th of June 1662* (London, 1662).

Walker, Anthony, *Eureka, Eureka. The Virtuous Woman Found. Her Loss bewailed, and character exemplified, in a Sermon* (London, 1678).

Walton, Isaak, *The Lives of John Donne, Sir Henry Wotton, Richard Hooker, George Herbert & Robert Sanderson*, ed. George Saintsbury (Oxford University Press: London, 1927).

Watkins, Owen C., *The Puritan Experience* (Routledge & Kegan Paul: London, 1972).

Webster, Charles, 'New Light on the Invisible College: the Social Relations of English Science in the mid-Seventeenth Century', *Transactions of the Royal Historical Society* 5th series 24 (1974), pp. 19–42.

Webster, Charles, *The Great Instauration: Science, Medicine and Reform 1626–60* (Duckworth: London, 1975).

Westfall, R. S., 'Newton and the Hermetic Tradition', *Science, Medicine and Society in the Renaissance*, ed. A. G. Debus, 2 vols (Science History Publications: New York, 1972), vol. ii, pp. 183–98.

Weyland, John, *The Hon. Robert Boyle's "Occasionall Reflections". With a Preface, &c.* (London, 1808).

Wilson, George, *Religio Chemici. Essays* (London, 1862).

Wojcik, Jan W., 'The Theological Context of Boyle's *Things above Reason*' in Hunter, *Robert Boyle Reconsidered* (above), pp. 139–55.

Wood, Anthony, *Athenae Oxonienses*, ed. Philip Bliss, 4 vols (1813–20).

Wood, Anthony, *Fasti Oxonienses*, ed. Philip Bliss, 2 vols (London, 1815–20).

Wood, Anthony, *Life and Times*, ed. Andrew Clark, 5 vols (Oxford Historical Society: Oxford, 1891–1900).

Wotton, William, *Reflections upon Ancient and Modern Learning*, 2nd edition (London, 1697).

Wotton, William, *The History of Rome, from the Death of Antoninus Pius, to the Death of Severus Alexander* (London, 1701).

Wotton, William, *A Letter to Eusebia: Occasioned by Mr Toland's Letters to Serena* (London, 1704).

Wotton, William, *A Defence of the Reflections upon Ancient and Modern Learning* (London, 1705).

Yeo, Richard, 'Genius, Method, and Morality: Images of Newton in Britain, 1760–1860', *Science in Context*, 2 (1988), pp. 257–84.

INDEX

180

Index

Index

Davenant, Sir William, xxxv, 69
Davies, Dr, c
Davity, Pierre, 15
Dee, John, c
Denham, Sir John, c
Dent, Thomas, xiii, xl, lxviii, lxx, 103–6
Deptford, xliii, 85, 91, 97, 100–2
Descartes, René, lxix, 85, 88
Desmond, James Fitzgerald, Earl of, 101–2
Dickinson, Edmund, 92
Digby, Sir Kenelm, xxi, xxxvii, 96
Diodati, Jean, xxvii, cv, 26
Diogenes Laertius, xix, xxvii, 26
Doctors Commons, 77
Domesday Book, xxiii
Douch, William, 10
Dublin, 4, 96
Dugdale, Sir William, 98
Dungarvan, Lord: see Burlington, Richard Boyle, 1st Earl of
Dunton, John, 62
Dury, John, 72–3
Dutch Wars, xxviii, 32–3

East India Company, lxviii, 27, 49
Edward VI, 97
Elizabeth I, 97, 101–2, 162
Ent, Sir George, 79
Epicurus, 87
Erasmus, Desiderius, 67
Eton College, xv, lxxvi, lxxxvi, ci, 5–6, 10, 23, 26, 28
Evelyn family, xliii, 163
Evelyn, John, xxxii, xxxvi, xxxvii, xl, lv, lxv, lxxv, lxxvii, lxxviii, 78
  Acetaria, 93
  correspondence with Wotton, xii–iii, xlvii–li, lxvi
  Elysium Britannicum, 93
  health, 92, 98–9
  history of trades, 93
  Kalendarium Hortense, 93
  letter to Wotton of 23 March 1696, xxix, xxxvii–viii, xlviii, lxxviii, lxxix, 84–90
  letter to Wotton of 27 Jan. 1702, xlii, xlviii–ix, li, xc, xcii
  letter to Wotton of 12 Sept. 1703, xli, xliii–iv, 91–9
  Numismata, 36
  Publick Employment, xlviii
  relations with Boyle, xliii, 85, 93, 96–7
  Sylva, 93
Evelyn, John, II, 85, 96
Evelyn, John, III, 98, 99
Evelyn, Mary, xxxvii, xliii
Everard, John, lxxi

Faithorne, William, ix
Faldoe, Charles, 10
Falkland, Lucius Cary, 2nd Viscount, xxxiv, 67, 69
Fell, John, xxiv–v, xxvi, lxx, cv, 33–4, 71
Felton, John, 96
Fenton, Edward, 97, 100, 101–2
Fenton, Sir Geoffrey, xliii, 97
Fenton, Sir Maurice, xliii, 96
Ferrara, 19
Finch family, xxxvi
Fitzgerald, Robert, xxxiv, lxviii, xcvi, 82
Florence, 19, 21, 28
Foxe, John, xx
France, 13–14, 22, 30–1, 86
Frank, Thomas, 109

Galilei, Galileo, xxi, xliv, lii, 19–20, 112–3
Gassendi, Pierre, *Life of Peiresc*, xxxiv, xxxvii, xlviii, xlix, li, lx, 62, 85
  Lives of the Astronomers, li
Gauden, John, lxxi, 80
Geneva, xv, xxvii, ci, 14, 18, 26, 27, 104
Genoa, 21
Gentleman's Journal, xxxi
Glanvill, Joseph, xxii, lxxi, 80
Glorious Revolution, xxiv, lxx, 78
Gloucester, William, Duke of, xli
Goddard, Jonathan, 96
Gomberville, Marin le Roy, Sieur du Parc et de, lxxxiii
Gonson, Thomasine, 97
Goodall, Charles, lv
Goodwin, Thomas, 71
Goodwin, Timothy, xliv
Goring, Lettice, Lady, 8, 11
Graunt, John, 94
Greatorex, Ralph, 114
Green, Ann, 94
Greenwich, xxxvi, 97
Grenoble, 17
Grotius, Hugo, xxxiv, 34, 49, 62
Guericke, Otto von, xliv, lii, 111, 113–4, 119
Guernsey, Heneage Finch, Baron (later 1st Earl of Aylesford), 99
Guicciardini, Francesco, xx, xliii

Hale, Sir Matthew, xxxi–ii, 79
Hale, Thomas, xxxiii, lxviii, 59–60, 81
Hall, Joseph, vii, xvi
Hall, Marie Boas, lxiii, lxiv
Halley, Edmond, lxxx
Hamilton and Castleherald, James and William Hamilton, Dukes of, xxxi
Hare, Francis, 99
Harrison, John, xxiv, cv, 6, 24

184

Index

Saleurre, 18
salt-water, purification of, xxxiv, lxviii, 82, 87
Saluzzo, Marquisat of, 14
Sancroft, William, xxxii, 78
Sanderson, Robert, lxxi, lxxiii, 74–5, 79, 87
Savoy, Duke of, 14
Sayes Court, 93, 97
Scaliger, J.J., l, 90
Scarburgh, Sir Charles, 92
Schaffer, Simon, lxiii–iv
Schott, Gaspar, xc, 113
Scotland, 65, 69
   distribution of bibles in, xiii, xli–ii, 34, 49, 87, 107–8
Seaman, William, 34
Seile, Anne, xxix
Selden, John, xxxiv, 62
Seneca, Lucius Annaeus, xix, xxvii, 26
Shaen, Sir James, xcix, 78
Shannon, Francis Boyle, 1st Viscount, xl–i, ci, cv, 5, 7, 13, 23, 79, 103, 104, 105
Shannon Park, 103
Shapin, Steven, lxiii–iv, lxxv
Sharrock, Robert, 92, 96
Shaw, Peter, lvii–viii, lx, lxii, xciii, xciv
Shipwrights, Company of, 59
Sibbald, Sir Robert, xix
Sidney, Sir Philip, xxi
Siena, 20
Skeffington, Clotworthy, later 3rd Viscount Massereene, 78
Slare, Frederic, 78
Sloane, Sir Hans, xlix, l–li, lii, lvii, lxxx
Smith, John, ix
Smith, Thomas, lviii–ix, 69, 78, 82
Society for Promoting Christian Knowledge, xc, 109
Socinianism, 63
Southampton, Thomas Wriothesley, 4th Earl of, 33, 50
Southwell, Sir Robert, cv, 31, 49
Spalding, Mr, 110
Spanish Armada, defeat of the, 101–2
Spectator, The, lxxii
Speke, Lady, lxxxiv
Spencer, Martha, 96
spiritual autobiography, genre of, xx
Sprat, Thomas, 80, 86
Stafford, Sir Thomas, 13
Stalbridge, Dorset, xiii, xv, xl, xli, ci, 9, 10, 11, 12, 105
Stanhope, George, xliii, 91, 98, 100–2
Sternhold, Thomas, 72
Stillingfleet, Edward, lxx, lxxiii, 105–6
Stoicism, xvi, xix, xx; *see* also Boyle, Robert

Strafford, Thomas Wentworth, 1st Earl of, lxxiii, lxxxiii, 76
Suetonius, 83
Swift, Jonathan, lvii, xcii
Switzerland, 18

Tacitus, 83
Taylor, Jeremy, xxi, 73–4
Temple, Sir William, xxxvi
Tenison, Thomas, xxxviii, xxxix, xli, cv, 77, 105
Thanet, Elizabeth, Countess of, 85
Thucydides, 99
Tillotson, John, xxix, xxx, 78
Toland, John, xlv
Tombes, John, 74
Torricelli, Evangelista, lii, 113
Tunbridge Wells, 91, 100
Turkey Company, 34, 49

Upcott, William, lxxx
Urban VIII, Pope, 20–1
Urfé, Honoré d', 14
Urquhart, Sir Thomas, 98
Ussher, James, xxvii, xxx, lxxxvi, cv, 27, 47

vacuum, experiments with, li–iv, 111–48
Vane, Sir Henry, lxxii, 63–5
Venice, 19, 31
Verona, 19
Vespasian, Emperor, 83
Vicenza, 19

W., J., lxxv
Wales, xlv
Waller, Edmund, 69
Waller, Sir Hardress, 96
Wallis, John, 96
Walton, Isaac, xcii
Ward, John, lviii–lxi, xcv
Ward, Seth, 73, 92
Warr, John, xxxiii, xxxix, xl, lviii–ix, lxxxii, xcv, 69, 77, 78
Warwick, Charles Rich, 4th Earl of, 106
Warwick, Mary Rich, Countess of, xx, lxxvi, lxxxiii, cvi
Westfall, R.S., lxxv
Weyland, John, xciv
White, Robert, xxxiv, 62
Wilkins, John, xliii, cvi, 73, 92
William III, 60, 81
Williams, John, xxxviii–ix, xlviii, l
Wilson, George, xcvi, c
Wise, Henry, 98
Wood, Anthony, xxxi, lv, lvii
Wood, Robert, c
Worsley, Benjamin, xv

187